INDRA'S PEARLS
THE VISION OF FELIX KLEIN

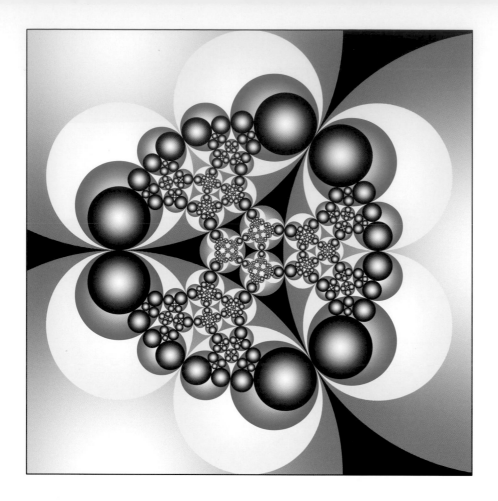

The ancient Buddhist dream of Indra's Net

In the heaven of the great god Indra is said to be a vast and shimmering net, finer than a spider's web, stretching to the outermost reaches of space. Strung at each intersection of its diaphanous threads is a reflecting pearl. Since the net is infinite in extent, the pearls are infinite in number. In the glistening surface of each pearl are reflected all the other pearls, even those in the furthest corners of the heavens. In each reflection, again are reflected all the infinitely many other pearls, so that by this process, reflections of reflections continue without end.

Cover picture: A mathematically generated picture foretold in the Buddhist myth of Indra's net? We sometimes call these *Klein Bubbles*. The smallest ones are *sehr klein*.

INDRA'S PEARLS

The Vision of Felix Klein

David Mumford, Caroline Series and David Wright

With cartoons by Larry Gonick

CAMBRIDGE
UNIVERSITY PRESS

CAMBRIDGE
UNIVERSITY PRESS

University Printing House, Cambridge cb2 8bs, United Kingdom

Cambridge University Press is part of the University of Cambridge.

It furthers the University's mission by disseminating knowledge in the pursuit of education, learning and research at the highest international levels of excellence.

www.cambridge.org
Information on this title: www.cambridge.org/9781107564749

© Cambridge University Press 2006

First published 2006
Reprinted 2008
First paperback edition with corrections 2015
Reprinted 2020

Printed in Singapore by Markono Print Media Pte Ltd

A catalogue record for this publication is available from the British Library

ISBN 978-0-521-35253-6 Hardback
ISBN 978-1-107-56474-9 Paperback

CONTENTS

PREFACE

What kind of a book is this?

This is a book about serious mathematics, but one which we hope will be enjoyed by as wide an audience as possible. It is the story of our computer aided explorations of a family of unusually symmetrical shapes, which arise when two spiral motions of a very special kind are allowed to interact. These shapes display intricate 'fractal' complexity on every scale from very large to very small. Their visualisation forms part of a century-old dream conceived by the great German geometer Felix Klein. Sometimes the interaction of the two spiral motions is quite regular and harmonious, sometimes it is total disorder and sometimes – and this is the most intriguing case – it has layer upon layer of structure teetering on the very brink of chaos.

As we progressed in our explorations, the pictures that our computer programs produced were so striking that we wanted to tell our tale in a manner which could be appreciated beyond the narrow confines of a small circle of specialists. You can get a foretaste of their variety by taking a look at the Road Map on the final page. Mathematicians often use the word 'beautiful' in talking about their proofs and ideas, but in this case our judgment has been confirmed by a number of unbiassed and definitely non-mathematical people. The visual beauty of the pictures is a veneer which covers a core of important and elegant mathematical ideas; it has been our aspiration to convey some of this inner aesthetics as well. There is no religion in our book but we were amazed at how our mathematical constructions echoed the ancient Buddhist metaphor of Indra's net, spontaneously creating reflections within reflections, worlds without end.

Most mathematics is accessible, as it were, only by crawling through a long tunnel in which you laboriously build up your vocabulary and skills as you abstract your understanding of the world. The mathematics behind our pictures, though, turned out not to need too much in the way of pre-liminaries. So long as you can handle high school algebra with confidence, we hope everything we say is understandable. Indeed given time and patience, you should be able to make programs to create new pictures for yourself. And if not, then browsing through the figures alone should give a sense of our journey. Our dream is that this book will reveal to our readers that mathematics is not alien and remote but just a very human exploration of the patterns of the world, one which thrives on play and surprise and beauty.

And how did we come to write it?

David M.'s story. This book has been over twenty years in the writing. The project began when Benoit Mandelbrot visited Harvard in 1979/80, in the midst of his explorations of complex iteration – the 'fractals' known as Julia sets – and the now famous 'Mandelbrot Set'. He had also looked at some nineteenth century figures produced by infinite repetitions of simple reflections in circles, a prototypical example of which had fascinated Felix Klein. David W. and I pooled our expertise and began to develop these ideas further in the Kleinian context. The computer rapidly began producing pictures like the ones you will find throughout the book.

What to do with the pictures? Two thoughts surfaced: the first was that they were unpublishable in the standard way. There were no theorems, only very suggestive pictures. They furnished convincing evidence for many conjectures and lures to further exploration, but theorems were the coin of the realm and the conventions of that day dictated that journals only publish theorems.[1] The second thought was equally daunting: here was a piece of real mathematics that we could explain to our non-mathematical friends. This dangerous temptation prevailed, but it turned out to be *much, much* more difficult than we imagined.

We persevered off and on for a decade. One thing held us back: whenever we got together, it was so much more fun to produce more figures than to write what Dave W. named in his computer `TheBook`. I have fond memories of traipsing through sub-zero degree gales to the bunker-like supercomputer in Minneapolis to push our calculations still further. The one loyal believer in the project was our ever-faithful and patient editor, David Tranah. However, things finally took off when Caroline was recruited a bit more than a decade ago. It took a while to learn how to write together, not to mention spanning the gulfs between our three warring operating systems. But our publisher, our families and our friends told us in the end that enough was enough.

You know that 'word problem' you hated the most in elementary school? The one about ditch diggers. Ben digs a ditch in 4 hours, Ned in 5 and Ted in 6. How long do they take to dig it together? The textbook

[1] Since then, the pioneering team of Klaus and Alice Peters have started the journal *Experimental Mathematics*.

will tell you 1 hour, 37 minutes and 17 seconds. Baloney! We have uncovered incontrovertable evidence that the right answer is (4+5+6)=15 hours. This is a deep principle involving not merely mathematics but sociology, psychology, and economics. We have a remarkable proof of this but even Cambridge University Press's generous margin allowance is too small to contain it.

David W.'s story. This is a book of a thousand beginnings and for a long time apparently no end. For me, though, the first beginning was in 1979 when my friend and fellow grad student at Harvard Mike Stillman told me about a problem that his teacher David Mumford had described to him: Take two very simple transformations of the plane and apply all possible combinations of these transformations to a point in the plane. What does the resulting collection of points look like?

Of course, the thing was not just to think about the shapes but to actually draw them with the computer. Mike knew I was interested in discrete groups, and we shared a common interest in programming. Also, thanks to another friend and grad student Max Benson, I was alerted to a very nice C library for drawing on the classic Tektronix 4014 graphics terminal. The only missing ingredient was happily filled by a curious feature of a Harvard education: I had passed my qualifying exams, and then I had nothing else to do except write my doctoral thesis. I have a very distinct memory of feeling like I had a lot of time on my hands. As time has passed, I have been astonished to discover that that was the last time I felt that way.

Anyway, as a complete lark, I tagged along with David M. while he built a laboratory of computer programs to visualize Kleinian groups. It was a mathematical joy-ride. As it so happened, in the summer of 1980, there was a great opportunity to share the results of these computer explorations with the world at the historic Bowdoin College conference in which Thurston presented his revolutionary results in three-dimensional topology and hyperbolic geometry. We arranged for a Tektronix terminal to be set up in Maine, and together with an acoustically coupled modem at the blazing speed of 300 bits per second displayed several limit sets. The reaction to the limit curves wiggling their way across the screen was very positive, and several mathematicians there also undertook the construction of various computer programs to study different aspects of Kleinian groups.

That left us with the task of writing an explanation of our algorithms and computations. However, at that point it was certainly past time for me to complete my thesis. Around 1981, I had the very good fortune of chatting with a new grad student at Harvard by the name of Curt McMullen who had intimate knowledge of the computer systems at the Thomas J. Watson Research Center of IBM, thanks to summer positions there. After roping Curt in, and at the invitation and encouragement of Benoit Mandelbrot, Curt and David M. made a set of extremely high quality and beautiful black-and-white graphics of limit sets. I would like to express my gratitude for Curt's efforts of that time and his friendship over the years; he has had a deep influence on my own efforts on the project.

Unfortunately, as we moved on to new and separate institutions, with varying computing facilities, it was difficult to maintain the programs and energy to pursue this project. I would like to acknowledge the encouragement I received from many people including my friend Bill Goldman while we were at M.I.T., Peter Tatian and James Russell, who worked with me while they were undergraduates at M.I.T., Al Marden and the staff of the Geometry Center, Charles Matthews, who worked with me at Oklahoma State, and many other mathematicians in the Kleinian groups community. I would also like to thank Jim Cogdell and the Southwestern Bell Foundation for some financial support in the final stages. The serious and final beginning of this book took place when Caroline agreed to contribute her own substantial research work in this area and her expository gifts, and also step into the middle between the first and third authors to at least moderate their tendency to keep programming during our sporadic meetings to find the next cool picture. At last, we actually wrote some text.

We have witnessed a revolution in computing and graphics during the years of this project, and it has been difficult to keep pace. I would also like to thank the community of programmers around the world for creating such wonderful free software such as TEX, Gnu Emacs, X Windows and Linux, without which it would have been impossible to bring this project to its current end.

During the years of this project, the most momentous endings and beginnings of my life have happened, including the loss of my mother Elizabeth, my father William, and my grandmother and family's matriarch Elizabeth, as well as the birth of my daughters Julie and Alexandra. I offer my part in these pictures and text in the hope of new beginnings for those who share our enjoyment of the human mind's beautiful capacity to puzzle through things. Programming these ideas is both vexing and immensely fun. Every little twiddle brings something fascinating to think about. But for now I'll end.

Caroline's story. I first saw some of David M. and David W.'s pictures in the mid-80s, purloined by my colleague David Epstein on one of his periodic visits to the Geometry Center in Minneapolis. I was struck by how pretty they were – they reminded me of the kind of lace work called tatting, which in another lifetime I would have liked to make myself.

I presumed that everyone else understood all about the pictures, and didn't pay too much attention, until a little while later Linda Keen and I were looking round for a new project. I had spent many years working on Fuchsian groups (see Chapter 6), and was wanting something which would lead me in to the Kleinian realm where at that time it was all go, developing Thurston's wonderful new ideas about three-dimensional non-Euclidean geometry (see Chapter 12). By that time, I had somehow got hold of Dave W.'s preprint[1] which described the explorations reported in Chapter 9. I suggested to Linda that it might fit the bill.

[1] *The shape of the boundary of the Teichmüller space of once- punctured tori in Maskit's embedding,* Unpublished preprint.

The first year was one of frustration, staring at pictures like the ones in Chapter 9 without being able to get any real handle on what was going on. Then one morning one of us woke up with an idea. We tried a few hand calculations and it seemed promising, so we asked Dave W. to draw us a picture of what we called the 'real trace rays'. What came back was a rudimentary version of the last picture in this book – the one we have called 'The end of the rainbow'. For me it was more like 'The beginning of the rainbow', one of the defining moments of my mathematical life. Here we were, having made a total shot in the dark, having no idea what the rays could mean, but knowing they had absolutely no right to be arranged in such a nice way. It was obvious we had stumbled on something important, and from that moment, I was hooked.

For another year we struggled to fit the rays into the one mathematical straight-jacket we could think of, but it just didn't quite work. One day, I ran into Curt McMullen and mentioned to him what we were playing with. 'Real trace', he pondered, 'That's the convex hull boundary'.[2] And with that clue, we were off. What Curt had told us was that to understand the two dimensional pictures we had to look in three-dimensional non-Euclidean space, real Thurston stuff, as you might say. Finally we were able to verify at least most of the two Daves' conjectures theoretically.

[2] See Chapter 12 for a bit more explanation and some pictures.

When the 19th century mathematician Mary Somerville received a letter inviting her to make a translation, with commentary, of Laplace's great book *Mécanique Céleste*, she was so surprised she almost returned the letter thinking there must have been some mistake.[3] I suppose I wasn't quite so surprised to get a letter from David M. asking me to help them write about their pictures, but it wasn't quite an everyday occurrence either.

[3] Fortunately, her enlightened husband convinced her otherwise. The book became a best seller and was even published illegally in the US!

Although I may perhaps write another book, I am unlikely ever again to have the chance to work on one which will be so much trouble and so much fun.

And don't think this book is the end of the story. If you flick through you will see cartoons of a rather portly character gluing up pieces of rubber into things like doughnuts. In fact all our present tale revolves about 'one-holed doughnuts with a puncture'. For the last few years, I have been trying to understand what happens when the doughnuts acquire more holes. The main thing I can report is – it's a lot more complicated! But the same wonderful structures, yet more intricate and inviting, are out there waiting to be tamed.

I would like to thank the EPSRC for the generous support of a Senior Research Fellowship, which has recently allowed me to devote much time to both the mathematical and literary aspects of this challenging project.

Guide to the reader

This is a book which can be read on many levels. Like most mathematics books, it builds up in sequence, but the best way to read it may be skipping around, first skimming through to look at the pictures, then dipping in to the text to get the gist and finally a return to understand some of the details. We have tried to make the first part of each chapter relatively simple, giving the essence of the ideas and postponing the technicalities until later. The more technical parts of the discussion have been relegated to the Notes and can be skipped as desired. Material important for later reference is displayed in Boxes.

The first two chapters, on Euclidean symmetries and complex numbers respectively, contain material which may be partially familiar to many readers. We have aimed to present it in a form suited to our viewpoint, at the same time introducing as clearly as possible and with complementary graphics the mathematical terminology which will be used throughout the book. Chapter 3 introduces the basic double spiral maps, called Möbius symmetries, on which all of our later constructions rest. From then on, we build up ever more complicated ways in which a pair of Möbius maps can interact, generating more and more convoluted and intricate fractals, until in Chapters 10 and 11 we actually reach the frontiers of current research. The entire development is summarised in the Road Map on the final page.

Words which have a precise mathematical meaning are in **bold face** the first time they appear. We have not always spelled out the intricacies of the precise mathematical definition, but we have also tried not to say anything which is mathematically incorrect. We have used a small amount of our own terminology, but in so far as possible have stuck to standard usage. Non-professional readers will therefore have to forgive us such terms as quasifuchsian and modular group, while readers with a mathematical training should be able to follow what we mean.

The book is written as a guide to actually coding the algorithms which we have used to generate the figures. A vast set of further explorations is possible for those readers who invest the time to program. This is prime hacking country! Because we hope for a wide variety of readers with many different platforms at their disposal, we have sketched each step in 'pseudo-code', the universal programming pidgin.

Inevitably we have suppressed a good deal of relevant mathematics and anyone wishing to pursue these ideas seriously will doubtless sooner or later have to resort to more technical works. Actually there are no very accessible books about Kleinian group limit sets[1], but there are plenty of texts which discuss the basics of symmetry and complex numbers. Some complex analysis books touch on Möbius maps and there is more in modern books on two-dimensional hyperbolic geometry. In the later part of the book we have cited a rather random collection of recent research papers which have important bearing on our work. These are absolutely not meant to be exhaustive, but should serve to help professional readers find their way round the literature.

[1] For readers with mathematical training the best introduction may still be Lester Ford's 1929 *Automorphic Functions*, Chelsea reprint, 1951.

Finally our Projects need some comment. They can be ignored: we aren't going to grade them or supply answers! Rather, we intend them as 'explorations' to tempt you if you enjoy the material and want to take it further. Some are fairly straightforward extensions or elucidations of material in the text and some involve open ended questions for which there is no definite answer. A few are definitely research problems. Others again explain details which are needed for full understanding or verification of the more technical points in our story. We have to leave it to the reader to pick and choose which ones suit their taste and mathematical experience.

Acknowledgements

We thank especially our cartoonist Larry Gonick for his uncanny ability to translate a complicated three-dimensional manipulation into an immediately evident cartoon. For historical background we are indebted to the St. Andrews History of Maths web site, tempered with many erudite details and healthy doses of scholarly scepticism from our friends David Fowler and Paddy Patterson. (All remaining errors, are, of course, our own.) Klein's own book *Entwicklung der Mathematik im 19. Jahrhundert* has also been an important source. We have read the Hua-Yen Sutra in the translation *The Flower Ornament Scripture* by Thomas Cleary, Shambhala Publications, 1993, and quotations are reproduced here with thanks. We should like to thank the Mathematics Departments of Brown, Oklahoma State, Warwick, Harvard and Minnesota for their hospitality. We should like to thank the NSF through its grant to the Geometry Center and EPSRC from their Public Understanding of Science budget for financial support. Finally we should like to thank our publisher David Tranah of Cambridge University Press, without whose constant prodding and encouragement this book would almost certainly never have seen the light of day.

Acknowledgments for the paperback edition

In this paperback edition, we would like sincerely to thank the many people who have read the book and sent corrections and suggestions, especially Yoshiaki Araki, Michael Barnsley, Peter Cromwell, John Guenther, Roger House, George Jackson, Arny Katz, Marius Kempe, Yohei Komori, Bill Margolis, Jon V. Pepper, Stuart Price and Masaaki Wada.

We would like particularly to thank Isabel Seliger for locating bibliographic information regarding one of the commentaries in Chinese Huayan Buddhist literature that explicates the metaphor of Indra's pearls.[1] The text elaborates the imagery of the reflecting pearls in ways that are strikingly close to contemporary mathematics. We are especially struck by this sentence in Tanabe's translation: "Within the boundaries of a single jewel are contained the unbounded repetition and profusion of the images of all the jewels." Could there be a better summary of the mathematics you will find below?

[1] *Calming and Contemplation in the Five Teachings of Huayan* (華嚴五教止觀 , Huayan wujiao zhiguan), in Taishō shinshū daizōkyō (Buddhist canon newly compiled under the Taishō reign era [1912–26]), eds. Takakusu Junjirō and Watanabe Kaigyoku (Tokyo: Taishō issai-kyō kankō-kai, 1914–22), vol. 45, no. 1867, pp. 513a28–513b21. The Chinese source text can be accessed online through the Chinese Buddhist Electronic Text Association (CBETA) website at www.cbeta.org/result /normal/T45/1867_001.htm. An English translation by George Tanabe appears in *Sources of Chinese Tradition: From Earliest Times to 1600*, vol. 1, ed. W. T. De Bary (New York: Columbia University Press, 1999), p. 473.

INTRODUCTION

I have discovered things so wonderful that I was astounded ... Out of nothing I have created a strange new world.

János Bolyai

With these words the young Hungarian mathematical prodigy János Bolyai, reputedly the best swordsman and dancer in the Austrian Imperial Army, wrote home about his discovery of non-Euclidean geometry in 1823. Bolyai's discovery indeed marked a turning point in history, and as the century progressed mathematics finally freed itself from the lingering sense that it must describe only the patterns in the 'real' world. Some of the doors which these discoveries flung open led directly to new worlds whose full exploration has only become possible with the advent of high speed computing in the last twenty years.

Paralleling the industrial revolution, mathematics grew explosively in the nineteenth century. As yet, there was no real separation between pure and applied mathematics. One of the main themes was the discovery and exploration of the many special functions (sines, cosines, Bessel functions and so on) with which one could describe physical phenomena like waves, heat and electricity. Not only were these functions useful, but viewed as more abstract entities they took on a life of their own, displaying patterns whose study intrigued many people. Much of this had to do with understanding what happened when ordinary 'real' numbers were replaced by 'complex' ones, to be described in Chapter 2.

A second major theme was the study of symmetry. From Mayan friezes to Celtic knotwork, repeating figures making symmetrical patterns are as ancient as civilization itself. The Taj Mahal reflects in its pool, floors are tiled with hexagons. Symmetry abounds in nature: butterfly wings make perfect reflections and we describe the tile pattern as a honeycomb. The ancients already understood the geometry of symmetry: Euclid tells us how to recognise by measurement when two triangles are congruent or 'the same' and the Alhambra displays many mathematically different ways of covering a wall with repeating tiles.

XV

The nineteenth century saw huge extensions of the idea of symmetry and congruence, drawing analogies between the familiar Euclidean world and others like Bolyai's new non-Euclidean universe.[1] Around the middle of the century, the German mathematician and astronomer August Möbius had the idea that things did not have to be the same shape to be identified: they could be compared as long as there was a definite *verwandschaft* or 'relationship' between every part of one figure and every part of the other. One particular new relationship studied by Möbius was inspired by cartography: figures could be considered 'the same' if they only differed by the kind of distortions you have to make to project figures from the round earth to the flat plane. As Möbius pointed out (and as we shall study in Chapter 3), these special relationships, now called Möbius maps, could be manipulated using simple arithmetic with complex numbers. His constructs made beautifully visible the geometry of the complex plane.[2]

[1] Non-Euclidean geometry was actually discovered independently and at more or less the same time by Gauss, Bolyai and Lobachevsky, see Chapter 12.

Towards the end of the century, Felix Klein, one of the great mathematicians of his age and the hero of our book, presented in a famous lecture at Erlangen University a unified conception of geometry which incorporated both Bolyai's brave new world and Möbius' relationships into a wider conception of symmetry than had ever been formulated before. Further work showed that his symmetries could be used to understand many of the special functions which had proved so powerful in unravelling the physical properties of the world (see Chapter 12 for an example). He was led to the discovery of symmetrical patterns in which more and more distortions cause shrinking so rapid that an infinite number of tiles can be fitted into an enclosed finite area, clustering together as they shrink down to infinite depth.

[2] The complex plane is pictured in Figure 2.1

It was a remarkable synthesis, in which ideas from the most diverse areas of mathematics revealed startling connections. Moreover the work had other ramifications which were not to be understood for almost another century. Klein's books (written with his former student Robert Fricke) contain many beautiful illustrations, all laboriously calculated and drafted by hand. These pictures set the highest standard, occasionally still illustrating mathematical articles even today. However many of the objects they imagined were so intricate that Klein could only say:

> The question is ... what will be the position of the limiting points. There is no difficulty in answering these questions by purely logical reasoning; but the imagination seems to fail utterly when we try to form a mental image of the result.[3]

[3] *The mathematical character of space-intuition*, Klein, *Lectures on Mathematics*, 1894, Reprinted by AMS Chelsea, 2000.

The wider ramifications of Klein's ideas did not become apparent until two vital new and intimately linked developments occurred in the 1970's.

The first was the growing power and accessibility of high speed computers and computer graphics. The second was the dawning realization that chaotic phenomena, observed previously in isolated situations (such as theories of planetary motion and some electronic circuits), were ubiquitous, and moreover provided better models for many physical phenomena than the classical special functions. Now one of the hallmarks of chaotic phenomena is that structures which are seen in the large repeat themselves indefinitely on smaller and smaller scales. This is called self-similarity. Many schools of mathematics came together in working out this new vision but, arguably, the computer was the *sine qua non* of the advance, making possible as it did computations on a previously inconceivable scale. For those who knew Klein's theory, the possibility of using modern computer graphics to actually *see* his 'utterly unimaginable' tilings was irresistible.

Our frontispiece is a modern rendering of one of Klein's new symmetrical worlds. In another guise, it becomes the *The Glowing Limit* shown overleaf. Peering within the bubbles, you can see circles within circles, evoking an elusive sense of symmetry alongside the self-similarity characteristic of chaos. Without the right mathematical language, though, it is hard to put one's finger on exactly what this symmetry is. The sizes and positions of the circles in the two pictures are not the same: the precise *verwandschaft* between them results from the distortion allowed by a Möbius map.

Klein's tilings were now seen to have intimate connections with modern ideas about self-similar scaling behaviour, ideas which had their origin in statistical mechanics, phase transitions and the study of turbulence. There, the self-similarity involved random perturbations, but in Klein's work, one finds self-similarity obeying precise and simple laws.

Strangely, this *exact* self-similarity evokes another link, this time with the ancient metaphor of Indra's net which pervades the *Avatamsaka* or *Hua-yen Sutra*, called in English the *Flower Garland Scripture*, one of the most rich and elaborate texts of East Asian Buddhism. We are indirectly indebted to Michael Berry for making this connection: it was in one of his papers about chaos that we first found the reference from the Sutra to Indra's pearls. Just as in our frontispiece, the pearls in the net reflect each other, the reflections themselves containing not merely the other pearls but also the reflections of the other pearls. In fact the entire universe is to be found not only in each pearl, but also in each reflection in each pearl, and so *ad infinitum*.

As we investigated further, we found that Klein's entire mathematical set up of the same structures being repeated infinitely within each other at ever diminishing scales finds a remarkable parallel in the philosophy and imagery of the Sutra. As F. Cook says in his book *Hua-yen: The Jewel Net of Indra*:

> The Hua-yen school has been fond of this mirage, mentioned many times in its literature, because it symbolises a cosmos in which there is an infinitely repeated interrelationship among all the members of the cosmos. This relationship is said to be one of simultaneous *mutual identity* and *mutual intercausality*.

The Glowing Limit. This illustration follows the mantra of Indra's Pearls *ad infinitum* (at least in so far as a computer will allow). The glowing yellow lacework manifests entirely of its own accord out of our initial arrangement of just five touching red circles.

In the words of Sir Charles Eliot:

> In the same way each object in the world is not merely itself but involves every other object and in fact *is* everything else.

Making a statement equally faithful to both mathematics and religion, we can say that each part of our pictures contains within itself the essence of the whole.

Perhaps we have been carried away with this analogy in our picture *The Glowing Limit* in which the colours have been chosen so that the cluster points of the minutest tiles light up with a mysterious glow. Making manifest of the philosophy of the Sutra, zooming in to any depth (as you will be able to do given your own system to make the programs), you will see the same lace-like structure repeating at finer and finer levels, worlds within worlds within worlds. The glowing pattern is a 'fractal', called the 'limit set' of one of Klein's symmetrical iterative procedures. How to understand and draw such limit sets is what this book is all about.

Like many mathematicians, we have frequently felt frustration at the difficulty of conveying the excitement, challenge and creativity involved in what we do. We hope that this book may in some small way help to redress this balance. On whatever level you choose to read it, be it leafing through for the pictures, reading it casually, playing with the algebra, or erecting a computer laboratory of your own, we shall have succeeded if we have conveyed something of the beauty and fascination of exploring this God-given yet man-made universe which is mathematics.

CHAPTER 1

The language of symmetry

You boil it in sawdust, you salt it in glue
You condense it with locusts and tape
Still keeping one principal object in view–
To preserve its symmetrical shape.

The Hunting of the Snark, Lewis Carroll

Symmetry, to a mathematician, encompasses much more than it does in everyday usage. One of the pioneers of this grander view was the distinguished and influential German mathematician Felix Klein. In 1872, on the occasion of his appointment to a chair at the University of Erlangen at the remarkably early age of 23, Klein proposed to the mathematical world that it should radically extend its received view of symmetry, to encompass things which had never been thought of as symmetrical before. Our quotation from Lewis Carroll, alias Charles Dodgson, mathematician and Fellow of Christ Church College, Oxford, was written only four years later. Perhaps Dodgson had heard about Klein's ideas and had them in mind as he composed his nonsensical verse.

In his historic short article,[1] the young Klein synthesized over fifty years of mathematical development in a new and profoundly influential way. It is today difficult to fully appreciate the significance of what he said, because his lecture crystallized one of those paradigm shifts which, after they have happened, seem so obvious that it is hard to imagine how anyone could ever have thought otherwise.

In a nutshell, Klein proposed viewing geometry as 'the study of the properties of a space which are invariant under a given group of transformations'. To study geometry, he said, one needed not only objects (triangles, circles, icosahedra, or much wilder things like the fractal pictures in this book), but also movements. In the classical Euclidean regime which had been around for over two millennia, these movements had always been rigid motions: pick up a figure and place an identical copy down in a new place. Klein's radical idea was

[1] With the erudite title *Vergleichende Betrachtungen über neuere geometrische Forschungen*, or 'Comparative considerations concerning recent researches in geometry', now universally known as the *Erlanger Programm*.

1

Felix Christian Klein, 1849–1925

Felix Klein was born in Düsseldorf in what was then the Prussian empire in 1849. He studied mathematics and physics at the University of Bonn. He started out on his doctoral work aiming to be a physicist, but was drawn into geometry under the influence of his supervisor Plücker. Plücker died in 1868, the same year that Klein finished his doctorate, and Klein was the obvious person to complete his advisor's unfinished work. This brought him to the attention of Clebsch, one of the leading professors at Göttingen, who soon came to consider the young Klein likely to become the leading mathematician of his day.

Klein received his 'call' to the University of Erlangen in Bavaria in 1872 and did his most creative work in the next 10 years. In 1875 he moved to the Technische Hochschule at Munich where he found many excellent students and his great talent for teaching came into its own. In the same year he married Anne Hegel, granddaughter of the famous philosopher. In 1880 he moved to the highly stimulating mathematical environment at Leipzig. Here he developed the deep theory to which this book is devoted, but it was also here that his delicate health first collapsed under the strain of an intense rivalry in this work with the brilliant young French mathematician Poincaré. Klein spent the years 1883-1884 plagued by depression and never fully recovered his mathematical powers. His work relating to our subject was developed at great length in two treatises written jointly with Robert Fricke over the period 1890-1912.

In 1886 Klein moved to a chair in Göttingen where he remained until his retirement. Besides his mathematical ability, Klein had very considerable managerial and administrative talents, and it was his skill and energy which built up the famous mathematical school at Göttingen which flourished as the world's leading mathematical centre until it was dismantled by Hitler in the 1930's. Klein's influence spread far, partly via the many foreigners who studied with him, among them the Americans Frank Cole and William Osgood, the Italians Luigi Bianchi and Gregorio Ricci-Curbastro, and some pioneering women like Mary Newson and Grace Chisholm Young. Around the turn of the century, he began to take a lively interest in the teaching of mathematics, encouraging the introduction of calculus into the school curriculum. He retired due to ill health in 1913 and died in Göttingen at the age of 76.

that other movements, which might stretch or twist the objects quite drastically, could be thought of as geometrical movements too. In this way geometry could be taken to encompass a much wider variety of set-ups than those previously conceived. Geometers should study those features

Photograph courtesy of the Archive of the Technical University of Braunschweig. Thanks also to Hans Opolka.

Robert Fricke, 1861–1930

Robert Fricke was born in Helmstedt, Germany. He studied and lectured in Göttingen before graduating in 1885 from Leipzig with a thesis written under Klein. His collaboration with Klein began with the publication of the two volumes of their first opus *Vorlesungen über die Theorie der elliptischen Modulfunktionen* in 1890 and 1892. During this period Fricke taught in two gymnasia in Braunschweig, and, more interestingly, was tutor to two sons of the Prussian Prince Regent Albrecht. Fricke did his 'Habilitation' in Kiel, following which in 1892 he lectured in Göttingen as a 'Privatdozent'. In 1894 he was appointed as Dedekind's successor in the Carolo-Wilhemina University in Braunschweig, and his always friendly relationship with Klein was cemented when he married Klein's niece Eleonora Flender later the same year.

Fricke was highly respected both as a mathematician and personally, working closely with Klein to develop much of the theory of what are now called Kleinian groups, the topic of our book. He played a leading role in the University administration, being Rektor from 1904–6 and again from 1921–3. His activity extended to state educational affairs where his experience as a school teacher was valued and he held several official posts. These many duties account for the long delay between the appearance of the first and second volumes (in 1897 and 1912 respectively) of the second opus *Vorlesungen über die Theorie der automorphen Funktionen*, of which Fricke was really the author, although with much input from Klein. In the final volume, Fricke took the opportunity to use new developments like Cantor's set theory and Brouwer's theory of dimension to solve a number of problems which had been unresolved in the past. He remained in post in Braunschweig until his death.

of the objects which the movements left unchanged, as so delightfully suggested in Carroll's verse.

There are two sides to the circle of ideas Klein was playing with: the idea of similar or symmetrical objects and the idea of 'transformation' or 'movement'. He brought these together using the idea of a **group**, a concept originally developed 50 years earlier by another very young mathematician, Évariste Galois. In his short career from 1829-1832, Galois saw that the solutions of a polynomial could be understood by defining their 'symmetries'; thus for example $+\sqrt{2}$ and $-\sqrt{2}$ can be considered as symmetrical solutions of the equation $x^2 = 2$. These ideas were way beyond the comprehension of any of his contemporaries and his work narrowly missed being completely lost.[1] Klein realised that, rather than trying to catalogue all possible kinds of symmetrical patterns and display them like the medieval Islamic builders of the Alhambra, the group concept gave a very simple and yet immensely powerful mathematical machinery for describing symmetries of all possible types.

The group concept simply describes the rules which govern the *repetitive* aspect of symmetry. For example, if you are allowed to make a move once, then you can make the same move again, and again, and again. Such repetition may bring you back to exactly the position from which you started (as in reflecting in a mirror, when two reflections bring you back where you began) or it may lay out an ever expanding mosaic of objects, for example tiles, in a regular pattern over a larger and larger area, like a vast floor.

In short, the usual way to think of symmetry is in terms of design and proportion, a slightly elusive quality of being balanced and correct.[2] Mathematicians, since Klein, have had at their disposal a more precise version: symmetry is balance created by repetitions of many movements of the same kind, specifically, by all the movements in some particular group. These two ideas, introducing the group idea into geometry and widening the class of movements to be studied, formed the background to Klein's own work. These were the themes he brought together in his famous grand plan.

A large part of Klein's later work became bound up with exhibiting and studying one particular new kind of symmetry which we are going to be explaining in this book. Before we enter these new realms, though, let's spend some time getting acquainted with the familiar Euclidean symmetries from Klein's new point of view.

A taxonomy of symmetry

Our first picture, Figure 1.1, shows a satellite photograph of the agricultural state of Iowa. Beside it is an idealized version in which we have perfected the symmetry. The landscape stretches out as far as the eye can see, broken up into a regular pattern of mile-wide farms, each containing one farmhouse, one pond and one tree. This world is so symmetrical that travelling one mile either north, south, east or west, you will reach a new position from which your view is completely indistinguishable from that which you had before.

What interests the mathematician studying symmetry is not so much the details of the figure, whether each field contains one sheep or two cows, as the movements or **motions** you have to make to implement the repeats. A motion is an easy thing to perform on a computer: draw a

[1] Galois' story is one of the great romances of mathematics, of which more on p. 20.

[2] In Chambers dictionary we find: 'exact correspondence of parts on either side of a straight line or plane, or about a centre or axis: balance or due proportion: beauty of form: disposition of parts'.

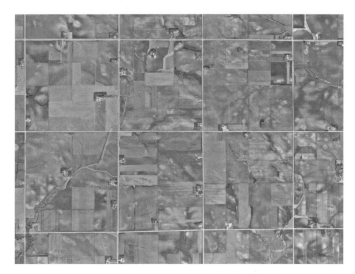

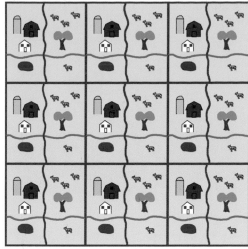

Figure 1.1. The real satellite photograph of Iowa on the left is not so very different from our idealized version on the right. Notice the brown 'north-south' and the green 'east-west' paths running across each farm.

[1] Railway lines, found in England, aren't usually so long and straight.

flower, and then copy your picture over to another place on the page. You can either think that you moved the flower, or, and this is the point of view that the mathematician prefers, that you picked up the whole page and put it down again so that the flower appears in its new place. In this way, the movement of the flower can be implemented by a definite motion of the whole plane. Figure 1.2 shows two flowers which have been transported around by **translation**. The flowers are quite different, but to the mathematician the underlying **translational symmetry** of the two pictures is exactly the same.

Symmetry is created by repeating or **iterating** the same motion a number of times. A pattern or object is **symmetrical** with respect to the motion if its individual points change position, but the pattern or object as a whole remains unchanged. The simplest symmetry is that of a shape which repeats itself infinitely often, moving the same fixed distance in the same fixed direction each time. A good example is a straight railroad track[1] across a flat prairie, extending forwards and backwards to the horizon as far as the eye can see, as in Figure 1.3. In your mind's eye, slide the entire track forward along its length just enough to move each tie or sleeper from its original position on the prairie to the position of the next. This is the motion of **translation**: the translation distance is the distance between ties (alias sleepers) and the direction is the direction of the track. After the motion, the track looks exactly as it did before, but, in fact, each tie has moved ahead and taken the place of the next. Notice that once again two things are involved here: an abstract movement, translation, and a physical object, the track. To say that an object has translational symmetry means that when it is physically translated to a new position, then although its parts are shifted, the view of the object as a whole is unchanged.

Figure 1.2. Two rows of flowers moved along by the same translations. The flowers are quite different but the symmetry of the two rows is the same.

Architects have made glorious use of translational symmetry. The visual effect of a repeating structure can be very graceful, as in the frieze also shown in Figure 1.3.

The starfish in Figure 1.4 is a good example of **rotational** symmetry. It has 12 arms, so if you rotate it by $360° \div 12 = 30°$ about its middle, each arm moves but the starfish looks just the same as it did before.

A third type of symmetry is **bilateral**, symmetry under **reflection**. This is the left-right symmetry of our bodies which you can see depicted in Leonardo's famous drawing in Figure 1.4. Imagine a vertical plane or mirror separating the left and right sides of a standing person and imagine moving every atom on the left side of the mirror to a point on the same horizontal level and at the same distance from the mirror on the right, and vice versa. Portraitists will object that the left side of the face expresses different facets of our personalities from the right and doctors will be confounded by the non-standard locations of colon and liver, but on the whole we can say that the body is unchanged. Most vehicles, like cars, boats, bicycles and planes, are nearly bilaterally symmetric, particularly on the exterior. Perhaps we subconsciously mould them on ourselves.

These three types of symmetries can often be seen in the same figure occurring in multiple ways. For instance, consider the wasps' nest shown in Figure 1.5. You should be able to spot each of the following types:

- translational symmetries in each of three different directions,
- rotational symmetry of 120° around points where 3 cells meet,
- rotational symmetry of 60° around the centre of each cell,
- reflectional symmetry in mirrors along edges where 2 cells meet,
- reflectional symmetry in mirrors through the midpoints of 2 opposite sides.

Figure 1.3. Two manifestations of translational symmetry: train tracks across the prairies and a frieze from the ancient Mexican city of Oaxaca.

Figure 1.3. Two manifestations of translational symmetry: train tracks across the prairies and a frieze from the ancient Mexican city of Oaxaca.

You can doubtless find other, more complicated symmetries, but all of them can be made from combinations of the ones above. Exactly how this is done mathematically we shall investigate below.

Before leaving the basic symmetry types, we would like to introduce Dr. Stickler, a long time associate of the authors, who will become a trusted guide in the pages which follow. You can see his photo in the top left frame of Figure 1.6, while the three other frames show him being moved around by translational, reflectional and rotational symmetries of the plane. As we extend our ideas about what we mean by

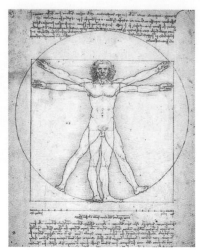

Figure 1.4. Left: A starfish displays rotational symmetry. This one has 12 arms, so it is symmetrical under rotation by $360° \div 12 = 30°$. Right: Leonardo's famous pen and ink study of the proportions of the male figure, showing nearly perfect bilateral symmetry.

Figure 1.5. A wasps' nest is a wonderful example of an object exhibiting multiple symmetries of all the three kinds.

symmetry, Dr. Stickler's rather staid progress here will be replaced by new kinds of motions which move him around in ever more exotic ways.

We have seen that there are two ways of thinking about symmetry. On the one hand, we can point to symmetrical objects, saying they are

Figure 1.6. In the top left frame you see a still photo of Dr. Stickler, while the other three frames show him being moved around by translations, reflections and rotations respectively.

examples of translation, rotational or reflectional symmetry as the case may be. However a second, deeper way of thinking about symmetry is to abstract from the picture and study the motions it embodies by themselves. The first way of thinking is more compelling because it gives you something tangible and visible, but the second is more fundamental because the abstraction contains no irrelevant detail, allowing us to focus on the underlying pattern itself.

Transformations of the plane

We have been talking in the last section about symmetries as 'movements' or 'motions' of the plane. Mathematicians, who even more than lawyers like to be extremely exact about their language, commonly speak in terms of a rather broader concept **transformation** or its common synonyms **mapping** or **map**. As in most of mathematics, these ordinary words are being used in specialised and very precise ways.

We had better explain carefully what is meant, because without using them it would be virtually impossible to write this book. In its widest sense, a **transformation** of the plane means simply a rule which assigns to each point P in the plane a new point Q. The rule might be: 'the new point is 3 inches to the left of the old one', or 'the new position is obtained by rotating 90° with respect to the centre point O'. The new point Q we get to is called the **image**[1] of the starting point P. Although a tremendous number of rules can be thought up, we shall only be thinking about rules whose effect can be undone. For example, the effect of the translation 'move 3 inches to the left', can be undone by the rule 'move 3 inches to the right', see Figure 1.7.

[1]Mathematicians have a predilection for giving special names to *everything* they talk about. They don't do this just to sound imposing: think of it like a surgeon laying out her instruments and checking everyone knows their correct names, to make sure she gets the right thing when she calls to the nurse.

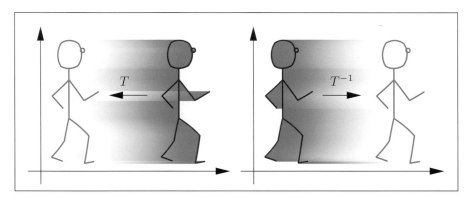

Figure 1.7. Dr. Stickler being transported by the transformation T which moves him 3 inches to the left. In the second frame you can see him transported by the inverse transformation T^{-1}, which moves him back 3 inches to the right.

A more concrete way to think of a transformation is to consider it as a procedure for physically moving the points of the plane to new locations. During the motion, the relative positions of points may get hugely distorted, but whatever figures or objects are in the plane get carried along and move to their new positions at the same time. You can see the effect of such a distortion on the classical bust of Paolina shown on the left in Figure 1.8. Even though her shape in the right hand picture has been radically altered, each of her features is still clearly identifiable. To create the right hand picture, we used a definite rule which told us exactly how to move each point in the left hand frame to a new point on the right. While leaving all her features intact, this transformation has fundamentally altered her shape.

To talk about this subject sensibly we need some notation to avoid the huge mouthfuls we found ourselves using above. We usually use letters like S and T to represent transformations and, following Euclid, use letters like P and Q to represent points in the plane. Like most mathematicians, we shall write $T(P)$ for the image of P under T; that is, $T(P)$ means 'the new point obtained by applying the rule T to the

Figure 1.8. The effect of a rather highly distorting transformation on the bust of Paolina. On the top left you see her graceful Grecian figure, but on the top right she has been transformed by a map which has created a rather ghastly caricature. The grids underneath show our rule. Following what happens in each square, you can trace the effect of the transformation we used on each different point in the plane. Roughly speaking, the rule is to rotate the region inside a certain circle clockwise through 45°, then to smooth things off by rotating the points in successively larger concentric rings through smaller and smaller angles until we reach an outer circle outside which every point remains fixed.

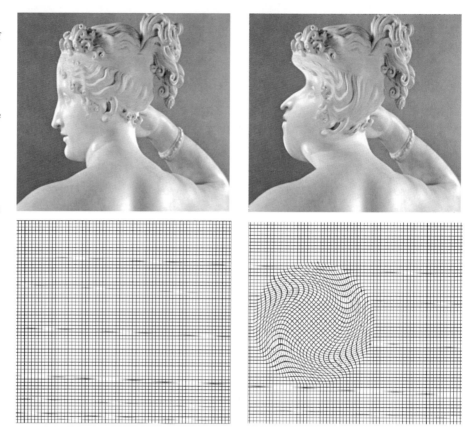

point P'. So, for instance, the transformation rule might be:

$T(P)$ is the point 3 inches to the left of P.

Often we want to think of this in an active way, so we say 'T **maps** P to $T(P)$'.

The rule which undoes the effect of T is known as the **inverse** of the map T. This is written T^{-1}, so the inverse of the rule:

$T(P)$ is the point 3 inches to the left of P

is

$T^{-1}(P)$ is the point 3 inches to the right of P.

Very often we want to study the effect of the rule or transformation T on a whole configuration, perhaps a circle C or a pair of parallel lines L and M. The notation $T(C)$ is convenient shorthand for 'the collection of all points $T(P)$ such that P is in C'. In our active language, we call $T(C)$ the image of C under the map T.

We have also been a bit vague about what we mean by 'the plane'. Anyone who has studied Euclid (or read the wonderful old book *Flatland* by A. Square[1]) probably has some mental image of this idealized two-dimensional world, but it is sometimes easier to describe things in formulas rather than often ambiguous words. As you no doubt learned in high school, standard Cartesian coordinates describe points in 'the plane' in terms of two numbers x, y by setting up:

[1] *Flatland, A Romance of Many Dimensions*, by Edwin A. Abbott, 1884, available in many modern editions.

- an origin, which can be any convenient point,
- two perpendicular lines through the origin, called the (horizontal) x-axis and the (vertical) y-axis.

Any point P in the plane is represented by two numbers x and y which measure its distance from the two axes, usually written as a bracketed pair (x, y). The distance x to the vertical or y-axis is taken to be positive if the point is on the right and negative if it is on the left; the distance y to the horizontal or x-axis is taken to be positive if the point is above the axis, negative if it is below.

As Descartes discovered, using algebra to do geometry can be a very powerful tool. Instead of describing a transformation by saying 'Move 3 inches to the left', we can say 'Move a point with coordinates (x, y) to the point with coordinates $(x - 3, y)$'. Decreasing x by 3 moves the point 3 units (in this case 3 inches) to the left, while not changing y keeps the point on the same horizontal line. Thus the transformation T defined by the rule:

$T(P)$ `is the point 3 inches to the left of` P

is expressed much more succinctly by the formula:

$$T(x, y) = (x - 3, y).$$

Thinking in terms of our dynamic language, we sometimes write

$$T : (x, y) \mapsto (x - 3, y),$$

read 'T maps (x, y) to $(x - 3, y)$'.

The symmetries of translation, rotation and reflection are all examples of transformations of the plane. For example, a translation can always be expressed by a formula

$$T(x, y) = (x + a, y + b).$$

This moves every point (x, y) in the plane a units right (or left if a is negative), and b units up (or down if b is negative). Reflection in the y-axis is the rule

$$T(x, y) = (-x, y).$$

This keeps each point on the same horizontal level, moving a point at

Figure 1.9. Cartesian coordinates can be used to locate any point in the plane. In this picture, you can see how the corner located at the point $(2, 1)$ gets rotated by the $90°$ rotation $T(x, y) = (-y, x)$ to its new position at $(-1, 2)$.

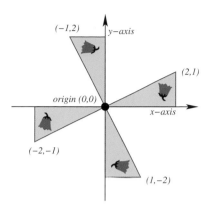

Box 1: Formulas for the three basic symmetries

Here are the general formulas for our three basic types of symmetries.

(1) Translation T which moves a point a units right and b units up:
$$T(x, y) = (x + a, y + b).$$

(2) Anticlockwise rotation R about the origin through an angle θ:
$$R(x, y) = (x \cos \theta - y \sin \theta, x \sin \theta + y \cos \theta).$$

(3) Reflection S across the vertical y-axis:
$$S(x, y) = (-x, y).$$

---

distance x units from the y-axis to a new point an equal and opposite distance on the other side.

The formula for rotation is a bit more complicated. You can see how to work out the formula for anticlockwise rotation by $90°$ about O in Figure 1.9. The rule is $T(x, y) = (-y, x)$.

The actual Cartesian formulas for the three basic symmetries are shown in Box 1. You will need to use some of these formulas if you want to make the computer programs we suggest, for example in Project 1.4 at the end of the chapter. If you haven't seen them before, we invite you to work through their derivation in Project 1.2.

Armed with our new language, we can give a precise meaning to what we meant in the last section by 'motion of the plane'. We meant a transformation with the property that it does not alter either the relative positions or sizes of any of the objects, so that in particular the distance between any two points P and Q is always the same as the distance between their image points $T(P)$ and $T(Q)$. It also means that any angle formed by three points $\angle PQR$ is equal to the angle formed by the three image points $\angle T(P)T(Q)T(R)$. Any motion of the plane with these properties is either translation, rotation, reflection or reflection in a line followed by translation in the direction of that line. The fourth type, called **glide reflection**, can be seen in the frieze in Figure 1.3.[1]

Figure 1.10.

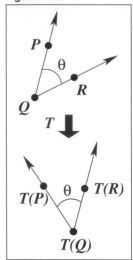

[1] Actually the angle preserving property follows from the distance preserving one. To see this, remember that the angles in a triangle are determined by the lengths of its three sides.

Composition, or making many transformations from a few

Let's go back to the hexagonal wasps' nest pattern in Figure 1.5, more mundanely familiar as a pattern of tiles on an institutional bathroom floor. We have seen that symmetries are really special kinds of motions or transformations of a pattern or figure which, while moving individual points, leave the pattern as a whole unchanged. In the case of the tiled bathroom floor, they are transformations with the very special property that they move the whole floor, putting it down on top of itself in such a way that all the tiles match up. Examining the floor after the event, you would see no sign that any change had taken place. In other words, a symmetry of the floor is a transformation T with the special property that T moves each tile U exactly into the position of some other tile V, in symbols $T(U) = V$.[2] This means that each point in U is transported by T to its new position in the image tile V. We want to include the possibility that $U = V$, meaning that as a whole the tile U stays in the same place, although it may be rotated or reflected by T so as to move some or all of its points. This will happen, for example, if we rotate by $60°$ about the centre of one of the tiles. We say for short that 'the transformation T **preserves** or **leaves invariant** the pattern of the tiles'. We have given precise mathematical meaning to the statement that T is a **symmetry** of the tiled floor.

[2] We are here ignoring any problems caused by the bathroom wall, and assuming the same tiled pattern extends in all directions as far as we wish.

Here's a new thing we can do with transformations. We are going to start with two transformations S and T of the plane, and make a new rule which describes a third. The new transformation will be called their **composition**, written symbolically just like a product: ST. The rule which defines ST is, quite simply,

first do T, and second do S.

In other words, take a point P. Use the rule given by T to get the point $T(P)$ to which T carries P. Now apply the rule given by S to the image point $T(P)$. We get to the point to which S carries $T(P)$, that is $S(T(P))$. This will be the rule which describes the transformation ST. In symbols:

$$ST(P) = S(T(P)).$$

Notice carefully the order here which is very important. Although you read the symbols from left to right, the rule ST means:

`first do T and then do S.`

If you wanted to do the operations in the other order:

`first do S and then do T`

you would have to write it symbolically as TS.

Composition of transformations is like a recipe: in a recipe each instruction calls for specific transformations to be carried out on the set of ingredients; the whole recipe is nothing but the composition of all these transformations. As in a recipe, the order matters. It is a very different matter to whip egg whites and then stir in cream, than to stir cream into egg whites and then try to whip up the lot. When the order really doesn't matter, we say that the transformations **commute**.

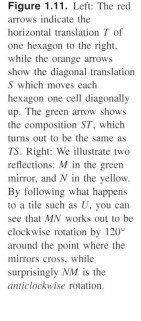

Figure 1.11. Left: The red arrows indicate the horizontal translation T of one hexagon to the right, while the orange arrows show the diagonal translation S which moves each hexagon one cell diagonally up. The green arrow shows the composition ST, which turns out to be the same as TS. Right: We illustrate two reflections: M in the green mirror, and N in the yellow. By following what happens to a tile such as U, you can see that MN works out to be clockwise rotation by 120° around the point where the mirrors cross, while surprisingly NM is the *anticlockwise* rotation.

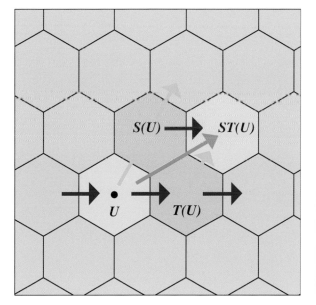

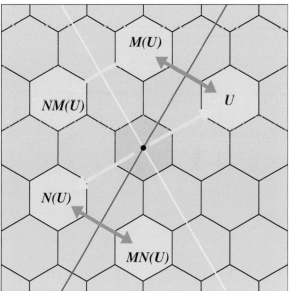

Suppose we start with transformations S and T which are symmetries of the hexagonal bathroom tiles. If U is a tile then, because T is a symmetry, each point in U is carried by T to a point in another tile V. Since S is a symmetry, points in V are carried by S to the points of a third tile W. Putting this together, we see that ST carries the points of U to the points of W. We conclude that ST is a symmetry: it also preserves the pattern of tiles.

Let's check this out in a few simple cases. The left part of Figure 1.11 shows an example where both S and T are translations. In the figure, the red arrows show the effect of the horizontal translation T which moves each hexagon onto its neighbour on the right, while the orange arrows show the effect of the diagonal translation S which moves each hexagon one cell diagonally up and a half-hexagon across. To work out the effect of the composition ST, we picked out the green tile U and marked in the tile $T(U)$ in pink. Then we found the tile $S(T(U))$, one row up from $T(U)$ and shifted a half-width to the right. You can check that carrying out the same procedure starting from any other tile produces an exactly similar effect. In other words, the composition ST is the translation shown by the green arrow, which takes U to the tile $S(T(U))$. If instead you worked with the composition TS, which moves U first to $S(U)$ and then to $T(S(U))$, you would come out with exactly the same result. In this case, we conclude that $ST = TS$.

Figure 1.12. In the left frame red arrows indicate the translation T, while the orange arrows now show the rotation R. Tracking the fate of the green tile U first under TR and then under RT, we find that this time the resulting tiles $T(R(U))$ and $R(T(U))$ are not in the same place, leading us to the conclusion that TR is not the same as RT. You can use the right frame to check that RT is rotation by 60° about the centre of a different tile V!

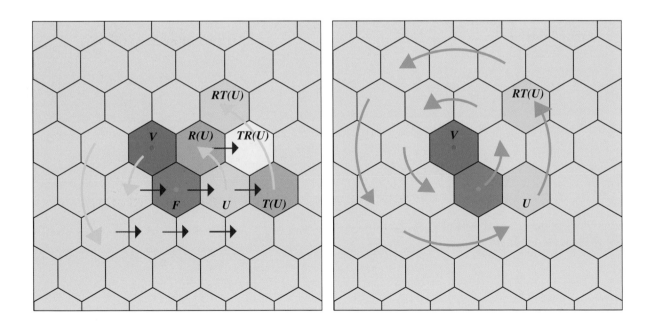

This looks easy, but sometimes the result of a composition is rather startling. In the Figure 1.12, T is the same translation as before. We have also introduced R, anticlockwise rotation by 60° about the central tile F. Remarkably, in this case RT works out to be a rotation of 60° too, but about the centre of a different tile V. In the figure, we track the fate of tile U, first under T and then under R. You can also see in the left frame that tile V moves one cell to the right by T and then back to where it started by rotation R. Try convincing yourself we are not fooling you by tracking the fate of a few more tiles under the composed map RT.

It is an interesting puzzle to work out general rules for describing the composition ST of any two symmetries S and T, but we don't have space to do this here. We suggest some more examples in Project 1.1.

Groups, or making mosaics without doing any work

In this book we shall be making lots of pictures of tilings and symmetries, so we need a good system for listing (and eventually calculating) all the transformations needed to put copies of the original tile down in their correct positions on the plane.

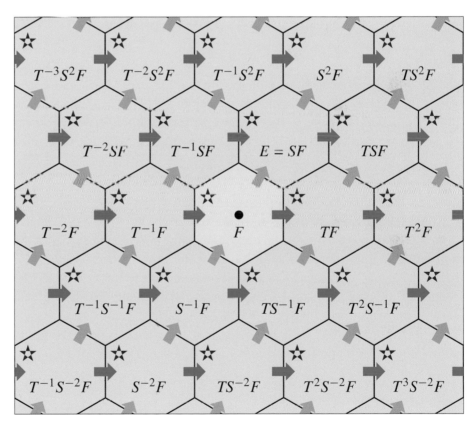

Figure 1.13. The decorated hexagonal tiling showing the label of each tile. The red stars are the **orbit** of the original red star in the initial tile F under the action of the group generated by S and T.

Suppose that you want to draw, or better write a program to draw, the tiling shown in Figure 1.13. Notice that all the tiles have a red star painted on one edge. The original hexagonal tiles each have a rotational symmetry of order six, and in addition can be reflected onto themselves in a variety of different ways. The painted tiles, on the other hand, have no symmetries at all. (Notice that the red star doesn't exactly point towards the nearby corner of the hexagon. If it did, we would have a reflectional symmetry.) This makes our problem a bit easier because it means that having decided on which floor space it should go into, the painted tile can be correctly fitted in just one way. We start by laying down our carefully prepared first tile which you can see in the middle of the figure labelled F for first. We want to systematically copy it over to other tiles using symmetries of the plane.

Focus first on the horizontal row of tiles containing our first tile F. This row surely has translational symmetry, so we shall have to do a bit of geometry to find out the distance between the centres of adjacent tiles on this row. In our scale drawing, we will assume the distance from the centre of a hexagon to an edge side is 1, so that the distance between one centre and the next is 2. Let T denote translation by 2 units to the right. The other tiles to the right along this row are just the translates

$$T(F), \ T^2(F), \ T^3(F), \ \ldots$$

and so on, where, to save a bit of writing, we use exponent notation and write, for instance, $T^3(F)$ for the composition $T(T(T(F)))$. If we want to move to the left we have to use the inverse translation T^{-1} which moves everything 2 units to the left. The tiles to the left of F are

$$\ldots, \ T^{-3}(F), \ T^{-2}(F), \ T^{-1}(F).$$

You can follow the labels on this row in Figure 1.13.

Now suppose we want to lay down the tiles in the next row up. We have picked out one of the tiles in this row by calling it E. We can reach E by translating F, this time in a diagonal direction. This is because the next row of hexagons is translated by distance one relative to the first. Since the centres of the tiles are distance 2 apart, you can calculate that the formula for this translation will be $S(x, y) = (x + 1, y + \sqrt{3})$.

Setting up S required some work, but look, now we can use T to get to all the other tiles along the row containing E. From right to left they are

$$\ldots, \ T^{-2}(E), \ T^{-1}(E), \ E, \ T(E), \ \ldots$$

rewritten on the figure as, for example, $T^{-2}S(F)$ and $TS(F)$.

We seem to have a system going. To get to the third row up you only have to repeat what you have just done. Apply S to E and then apply powers of T and you find the third row with entries like $T^{-2}S^2(F)$ and $T^{-1}S^2(F)$. If you want to get the row under F, you just have to apply the inverse transformation S^{-1} to F. This row has entries of the type $T^{-2}S^{-1}(F)$ and $T^{-1}S^{-1}(F)$. And so on. In fact, to lay out the complete array of tiles *we only have to use the two transformations S and T*, together with their inverses S^{-1} and T^{-1}. Writing in our exponent notation, we can get to any tile in the whole array by means of compositions like $T^{-5}S^3$ or T^lS^k, where k and l can be any positive or negative integers, including 0.[1]

This process of building up symmetries worked so well because, as we have seen already, if S and T are symmetries of some figure, then so is their composition ST, and so are their inverses S^{-1} and T^{-1}. Collections of transformations like this came up so often in the nineteenth century that eventually mathematicians realised it would be useful to give them a name. In formal mathematical language, a **group** of transformations is a collection with the two properties:

(1)   if S and T are in the collection, then so is ST;
(2)   if S is in the collection, then so is S^{-1}.

Usually the whole collection or group of transformations like this is symbolised by a single letter, often as not the letter G. The symmetries of Figure 1.13 are a good example: combine any two, and you get a third. We saw that every symmetry of this figure can be expressed as a composition of the basic transformations S, T and their inverses, so we say that the transformations S, T **generate** the group G. As our book progresses, the symmetries we work with will get more complicated, but the basic philosophy will always be the same. We shall start with a small number of generating transformations, and we shall compose or multiply them together in exactly this way.

Another example of a group comes from our Iowa picture with its identical square fields. This time, the basic symmetries are E, translation of the whole picture one mile east and N, translation by one mile due north. It is easy to see that any other symmetry of the picture can be expressed as a composition of E, N and their inverses $W = E^{-1}$ and $S = N^{-1}$. The set of all points you can get to starting at one point, for example what we might call the red 'home' farmhouse in the centre, and applying all possible compositions of E, N, E^{-1} and N^{-1} is called the **orbit** of the home farmhouse under the group. This picture of the orbit has made the group 'visible' in a beautiful geometrical way.

The symmetry group in this example is generated by the transformations E and N. Each farmhouse is labelled by exactly one symmetry, the one you have to use to reach it from the 'home' farmhouse. The square farms can be labelled by the same symmetries. Because they

Évariste Galois, 1811–1832
The beginnings of Group Theory

The idea of a group is credited to Évariste Galois who was, like us, engaged in studying symmetries, but symmetries of a rather different kind. His symmetries were ones which swapped around the roots of polynomial equations. Among other things, his discoveries *proved the impossibility* of solving problems which had puzzled the best mathematical minds for two millennia: you cannot trisect angles, nor 'duplicate' the unit cube (construct $2^{1/3}$), using ruler and compass alone.

Starting while still at high school, Galois developed his ideas entirely on his own during his late teens, but his life, troubled by political turmoil and his father's suicide, came to an abrupt and tragic end following a duel at the age of 20. The night before the fatal duel, Galois scribbled a famous letter to his friend Auguste Chevalier in which he outlined his discoveries and hoped that 'some men will find it profitable to sort out this mess'. The letter was published but Galois' manuscripts were lost or poorly understood, lying dormant for some years. The first person to realise their importance was Liouville, who finally arranged to have them published in 1846. From this time the idea of a group slowly gathered momentum, being taken up by Cauchy and then by Cayley (who we shall be meeting again in Chapter 4) in the 1850's.

The first person to embark on a systematic study of Galois' work and develop the theory of finite groups was Camille Jordan (1838–1921). His work culminated in 1870 with the publication of his famous book *Traité des substitutions et des équations algébriques*. This book, whose publication was a major mathematical event, at last contained a systematic development of all Galois' insights along with a systematic account of group theory up to that time.

In that same year, two young men who had only recently completed their doctoral studies journeyed to Paris and met with Jordan. These were Felix Klein and Sophus Lie. Their different ideas about how to integrate the idea of a group with those of geometry and calculus were to have a profound and lasting influence on all subsequent mathematics.

fit together without overlaps and cover everything, we say they tile or **tessellate** the plane.

Until you get used to it, the formalistic language used to explain a group looks rather abstract and obscure. In fact the idea was used for many years before it was condensed in this abbreviated way. The power of the group concept rests in the fact that it condenses all the myriad ways in which you might wish to combine symmetries into two simple rules.

The algebra of symmetry

This section is not all that exciting – and can be skipped over until needed – but there is a certain algebra involved in composing transformations which will be indispensable later on. We learnt in the last section that the inverse of a transformation T is simply the rule which undoes the effect of the rule defining T: if T moves a point P to a point Q, then T^{-1} moves Q back to P. We can express this neatly in terms of composition by the equation $T^{-1}T(P) = P$. This brings us to a special transformation, the **identity transformation** I. This is the rule which doesn't move any point at all, in symbols, $I(P) = P$. It may sound like a silly transformation to study, but it must be there for completeness, in just the same way as it was very hard to do arithmetic before the marvellous invention of the number 0. The equation $T^{-1}T(P) = P$ can be expressed in another way by saying

$$T^{-1}T = I.$$

It follows automatically that $TT^{-1} = I$. To see this, take an x. When we applied T, the point x must have come from somewhere, so let $x = T(y)$. Then $T^{-1}(x) = T^{-1}T(y) = I(y) = y$ and so

$$TT^{-1}(x) = T(y) = x = I(x).$$

Like its relative the number 0, composing the identity with anything has no additional effect, in other words,

$$TI = IT = T.$$

This is just as useful as, though not much more exciting than, the equation $0 + 3 = 3 + 0 = 3$!

Remember to say that T is a symmetry means that it preserves the pattern of the tiles. If T carries tile U to tile V, then just as surely T^{-1} will carry tile V to tile U. In other words, if T is a symmetry, then so is its inverse transformation T^{-1}. Even more obviously, the identity transformation I is a symmetry: this is partly why we said that being a symmetry should include the possibility that tiles can stay fixed!

Here is a little puzzle: if T is a transformation, what transformation undoes the effect of the inverse transformation T^{-1}? If you think for a minute of a few examples, you will see that it is nothing other than the transformation T. For example, if T is the translation 'move 3 inches to the left', then T^{-1} is the translation 'move 3 inches to the right'. To undo the effect of moving 3 inches to the right, you obviously have to move back three inches to the left. Otherwise said, T is the inverse of the inverse transformation T^{-1}. We would write this in symbols by saying that $(T^{-1})^{-1} = T$, no more surprising than the calculation $-(-2) = 2$.

As another example, let us work out the transformation which undoes the effect of the composition ST. If you do first T then S, then to undo what you just did you have to undo first S and then T. In other words, you would expect the inverse transformation to ST to be $T^{-1}S^{-1}$. We can check this by working out:

$$(T^{-1}S^{-1})ST = T^{-1}(S^{-1}S)T = T^{-1}IT = T^{-1}T = I.$$

We have displayed these important formulas in Box 2.

Tiles into tyres

Readers may well have come across the mathematical word **topology**, often described as 'rubber sheet geometry'. Explanations often begin with pictures of how to make a 'doughnut' or tyre-shaped surface. Anything with this shape, even if twisted or bent – but not torn – is called a **torus**. We have illustrated the procedure in Figure 1.14 where you see Dr. Stickler (in the three-dimensional world, he has put on a bit of weight) making a giant tyre.

There is a rather more sophisticated way of looking at this gluing construction which may be explained by going back to the infinite Iowa-like plain with identical farms illustrated in Figure 1.1. Repeating Dr. Stickler's procedure, imagine cutting out one field along its boundary and gluing its four sides in pairs to make a torus shaped

---

Note 1.1: **Bracketing compositions**

Once we have settled on the order in which maps are to be applied, they can be bracketed in any way you wish. The rule which expresses this is the **associative law** which says that

$$(ST)U = S(TU).$$

This means that doing first U, followed by the composition ST, has exactly the same effect as doing first TU, and following it with S. A more complicated example is that

$$(ST)(UV) = S(TU)V.$$

Ordinary language is not always associative. For example:

```
((The officer saw) (a thief))
(with binoculars).
```

is not at all the same as

```
(The officer saw) ((a thief)
(with binoculars)).
```

Box 2: Groups, compositions and inverses

A **group** of transformations is a collection with the two properties:

(1)   if S and T are in the collection, then so is ST;
(2)   if S is in the collection, then so is S^{-1}.

The effect of the composition ST is:

$$\texttt{first do } T \texttt{, then do } S.$$

To undo the effect of ST, invert each transformation and reverse the order:

$$\texttt{first do } S^{-1} \texttt{, then do } T^{-1}.$$

In symbols:

$$(ST)^{-1} = T^{-1} S^{-1}.$$

The inverse of an inverse is the thing itself, in symbols:

$$(T^{-1})^{-1} = T.$$

---

space-station containing only one farmhouse. If you wander 'north' on the space-station following the brown north-south path, then, instead of arriving at your neighbour's farmhouse, you return back at your own approaching from the 'south'. If you wander 'east' on the green east-west path, you arrive back at your own farmhouse, coming in from the 'west'.

The fantasy may occur to you: how can you tell if you are really living on a space station made of one farm or in a flat Iowa in which all the farms are absolutely identical? Such questions delight philosophers: there's no way to tell the difference if all the farms are truly the same! This gives us another way of thinking about Dr. Stickler's construction: instead of taking one field and 'gluing' its edges, you play God and declare that all the farms don't just look the same, they *are* the same because there is really only one farm. In mathematical terms, we can agree that all points in the same orbit of the group generated by the east and north translations E and N should be considered 'the same'.

As our fox wanders over the fields of Iowa from farm to farm, he traces some meandering path. Gluing up, we can trace the same path on the torus but now, since all the farms are the same (or if you prefer since there is only one farm), when the fox wanders over the boundary between one farm to the next, his path on the torus re-enters the same

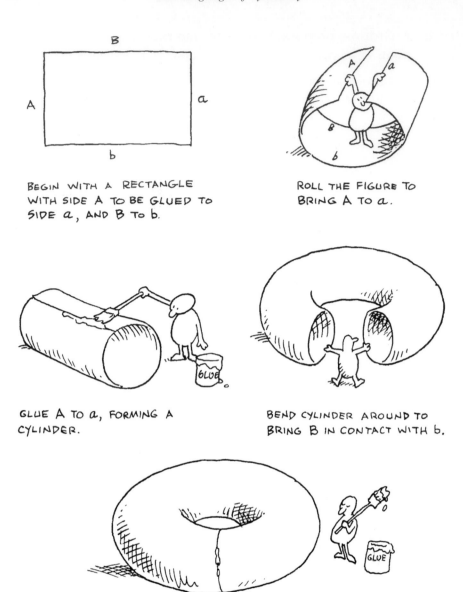

BEGIN WITH A RECTANGLE
WITH SIDE A TO BE GLUED TO
SIDE a, AND B TO b.

ROLL THE FIGURE TO
BRING A TO a.

GLUE A TO a, FORMING A
CYLINDER.

BEND CYLINDER AROUND TO
BRING B IN CONTACT WITH b.

GLUE B TO b, FORMING A DOUGHNUT-
SHAPED SURFACE OR TORUS.

Figure 1.14. Dr. Stickler creates a huge tyre from a square of rubber and a pot of glue. In the first frame is the large rubber sheet with which he begins. In the second and third frames he is rolling the sheet into a cylinder and pasting the left side to the right. In the fourth frame, he carefully bends the cylinder around so that its two ends come together, and in the last frame you see the final tyre. Notice that we have labelled the four sides of the rubber sheet a, b, A, B in such a way that the edge A is opposite the edge a and the edge B is opposite the edge b, so that a gets glued to A and b gets glued to B.

farm from the symmetrical place on the opposite edge. Any path which on the Iowa plain led from one farmhouse to another now leads from this single farmhouse back to itself. It is an interesting problem to work out just how many different ways there are of doing this, which you can learn more about if you skip ahead to Figure 9.8.

There is a wonderful portrayal of a three dimensional version of such an infinitely repeating universe in the story *The Library of Babel* by the great Argentinian writer Jorge Luis Borges. According to Borges, the Universe, otherwise known as the Library, is made up of an interminable distribution of identical hexagonal galleries. In each gallery, four of the six walls are covered by five long shelves, the remaining two sides each linking through narrow hallways to other galleries identical to the first and to all of the rest. Spiral stairways sink abysmally and soar upward to remote regions; in Borges' words: 'The Library is a sphere whose exact centre is any one of its hexagons and whose circumference is inaccessible.' While we have been unable to reproduce with mathematical exactitude just how these six sided galleries are supposed to link, his vision of infinite repetition in a universe of indistinguishable chambers is breathtaking and disquieting.

Later in the book, we shall see that more complicated groups of symmetries give us not only more complicated tiles, but also more complicated surfaces. Whenever we have a pattern of tiles invariant under the motions in some symmetry group, we can always try to use the group elements as a gluing recipe, which means you take a basic tile and stick two edges together whenever there is a transformation in the group which matches the points on one edge to the points on the other. Gluing up the sides in this way produces a surface, which we can think of as a model for a universe in which we, like the Doctor, are incapable of distinguishing between symmetrical points. In Project 1.3, you are invited to explore what happens when we try to glue up hexagonal tiles using the symmetries of the bathroom floor.

Conjugation, or how to keep things the same while changing your point of view

There is a one final link we need to make between geometrical pictures of tilings and the algebraic formalism of groups. Suppose we have two pictures, each made up of some set of tiles, which are related by a known transformation. That is, suppose we have a map S which carries a whole collection of tiles on one floor to another collection of tiles on another. We want to be able to take what we know about symmetries of the tiles on the first floor and convert it to information about the tiles on the second. In the Figure 1.15, on the left you can see a square tiling on the first floor, and on the right you can see the same tiling on the second floor, where it has been rotated by $45°$ relative to the first, so that the squares run diagonally instead of horizontally across the page. Thus S is rotation through $45°$. If we know the symmetries of the first picture, there must be a clever way to find the symmetries of the second without doing our calculations all over again from scratch.

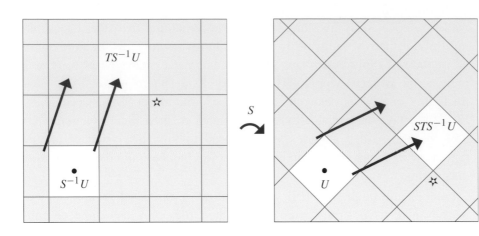

Figure 1.15. Two square tilings, the left one with the squares horizontal and vertical and the right with the picture rotated by 45 degrees.

Suppose T is a symmetry of the tiling on the left. We want to find the transformation which behaves just like T, but applied to the picture on the right. Let's call the new map $\hat{T}$. We may think of what we are doing as changing our viewpoint so that everything is carried to a new frame of reference by means of the transformation S. This means that every point P in the left hand picture is replaced by the transformed point $S(P)$ on the right. To find what the new map $\hat{T}$ does to a tile U, we first use S^{-1} to transport

---

Note 1.2: **Checking the conjugation formula**

In order to check the formula $\hat{T} = STS^{-1}$ in our example in Figure 1.15 we first need to find a formula for the conjugating map S. Fortunately we already wrote down formulas for the basic symmetries in Box 1 on p. 13. We need to apply the rotation formula with $\theta = -45°$. Using Pythagoras' theorem in a triangle with angles $90°, 45°, 45°$ we find $\cos 45° = 1/\sqrt{2} = \sin 45°$ so that $\cos(-45°) = 1/\sqrt{2}$ and $\sin(-45°) = -1\sqrt{2}$. Thus:

$$T(x, y) = (x + 1, y + 2),$$
$$S(x, y) = ((x + y)/\sqrt{2}, (-x + y)/\sqrt{2}),$$
$$S^{-1}(x, y) = ((x - y)/\sqrt{2}, (x + y)/\sqrt{2}).$$

Then we take a deep breath and work out

$$
\begin{aligned}
\hat{T}(x, y) &= STS^{-1}(x, y) \\
&= ST((x - y)/\sqrt{2}, (x + y)/\sqrt{2}) \\
&= S((x - y)/\sqrt{2} + 1, (x + y)/\sqrt{2} + 2) \\
&= (x + 3/\sqrt{2}, y + 1/\sqrt{2}).
\end{aligned}
$$

As predicted, it comes out that $\hat{T}$ is translation by the vector $(3/\sqrt{2}, 1/\sqrt{2})$, which is exactly the image of the original translation vector $(1, 2)$ under the conjugating map S!

the tile back to the original frame on the left, then apply the original map T, and finally use S to get us back to the new frame on the right. We have performed in order the maps S^{-1}, then T, then S, which means that we have discovered the formula

$$\hat{T} = STS^{-1}.$$

This will be very useful in many places in this book so, and although the formula looks complicated, it is well worth thinking through. The map $\hat{T}$ is called a **conjugate** of T. It behaves geometrically just like T, except that the original frame of reference has been moved by what we call the **conjugating map** S.

In Figure 1.15, T is translation in the direction $(1, 2)$ and S is clockwise rotation by $45°$ around the origin $(0, 0)$. It should be obvious from the figure that the conjugate transformation $\hat{T}$ is also a translation. The difference is that the direction of translation will have rotated clockwise by $45°$. In Note 1.2 we check this prediction against our formula $\hat{T} = STS^{-1}$.

Here are some general principles. Let S be any translation, rotation or reflection. If $\hat{T} = STS^{-1}$ then:

- for any translation T mapping 0 to P, then $\hat{T}$ is the translation which maps $S(0)$ to $S(P)$;

- if T is rotation around a point P, then $\hat{T}$ is the rotation around $S(P)$ through the same angle;

- if T is reflection in a mirror M, then $\hat{T}$ is reflection in the mirror $S(M)$.

Note 1.3: **Algebra of conjugation**

Another example of how conjugation works is the calculation

$$\hat{T}^{-1} = (STS^{-1})^{-1} = (S^{-1})^{-1}T^{-1}S^{-1} = ST^{-1}S^{-1}$$

which says that the inverse of a conjugate of T is just the conjugate of the inverse of T. You should be able to check that if T is a reflection then

$$\hat{T}^{-1} = ST^{-1}S^{-1} = STS^{-1} = \hat{T},$$

in other words, $\hat{T}$ has the same algebraic property as T.

Conjugation respects the composition of maps as well. In words, this means that the conjugate of a composition is the equal to the composition of the conjugates. The reader will probably agree at this point that it is easier to write a formula: what we have just said amounts in symbols to

$$S(TU)S^{-1} = (STS^{-1})(SUS^{-1}).$$

Let's get programming

Computer programmers who have been reading all this have probably been itching to get started drawing tiles and images with complicated symmetries. How is this done?

First of all, you will need to select a language and platform for handling your programs. That is largely a matter of personal taste and experience, and arguments about the choice can be very heated. We should admit – with a mixture of pride and embarrassment – that most of the complicated illustrations in this book were computed by programs written in the oldest and most conservative of all languages, namely FORTRAN. But we are not unaware of the evolution of programming in the last 50 years and we want to sketch a few of the issues to be considered in your selection. We'll outline four options.

Method I. A simple way to get started fast is to use one of the integrated packages which combine powerful mathematical commands with first class graphics. At this time, MATLAB, Maple and Mathematica are the three most developed products doing this. All of them include a wide array of basic two and three dimensional drawing commands, along with sophisticated enhancements like surface shading. Moreover, all of them have built-in libraries for doing complex number and matrix arithmetic[1] which we will be using throughout the book. These make it very easy to code the algorithms we describe concisely. The disadvantage is that programs written in these systems generally are 'interpreted', meaning that the English words in your code are only translated into actual machine-level instructions when the code is run, which in practice slows it down considerably. For many plots, this is not serious if you have a fast CPU, but for the very detailed fractal graphics, the time needed can expand to days. The other option, known as a 'compiled' language, translates your English code into machine-level instructions beforehand. This includes C, C++ and FORTRAN. Many compilers (especially FORTRAN) can optimize the machine-level translation very effectively, thereby obtaining extremely fast results. This can be crucial for the most elaborate plots. Another problem with these packages is that, in order to save you from having to declare all the variables you will use in advance, they do not optimize the use of memory very well and will rapidly fill up your available random access memory (RAM) unless you are careful.

[1] To be explained in the next two chapters.

Method II. Perhaps the next easiest method is to use one of the 'visual' program development packages which integrates a language such as Basic,

C, Pascal or Java with the graphical user interface of your computer. The advantage of this route is that the basic graphics operations will be integrated seamlessly into the programming language. They will be adapted to whatever platform you are using: PC, Mac or Unix/Linux/X-windows, so you need not learn the hair-splitting details of the window structure. However, be warned that you will probably have to master some aspect of object-oriented programming with which all recent graphical user interfaces (GUI's) are written. In these languages, data is elaborately structured and tagged as 'objects' which associate it with the particular procedures needed to use or modify it. For our applications, another draw-back is that the *only* traditional language to incorporate complex number arithmetic is FORTRAN. In all other languages, you must write your own package. Some are becoming commercially available but this is not a big market!

Method III. The traditional way of programming is to simply write code in a word processor, compile it, and link it with the graphics library on your operating system or graphics device. You first need to choose some language and get a compiler. High quality scientific graphics usually requires a separate graphics package (such as GL) which is 'linked' to an application program. This means that the application program may freely use all the commands defined in the graphics package. There is another layer of complexity in the choice of 'device' to which to output the graphics, which may be either a graphics screen or a printer. This approach has the great advantage that you know exactly what you are doing and can modify anything. The drawback is that it requires more study of the nuts and bolts of your computer system.

Method IV. Finally, there are some more exotic ways of proceeding. One is to produce the graphics directly in Adobe's beautiful page-description language PostScript, which has become something of a standard for high quality graphics. The language not only contains all the primitives for drawing but all the elements of a general programming language. You can either write concise programs which will execute entirely on your printer, or you can write your compiled code so that it outputs a text file of the PostScript commands which can be sent directly to any postscript printer. Finally, we believe that Excel enthusiasts can actually compute and plot our fractal shapes without leaving Excel – but we have not verified this!

Since we do not want to preclude readers from taking any of the above routes, we shall describe our code throughout the book using terms and notation which look like a pidgin programming language, but don't really refer to any actual one. This is usually called **pseudo-code**. For instance, names of variables will be given in `typescript`, just as if you were seeing them on a terminal. In computer languages, names of variables are often descriptive words, rather than a single letter; for example, we will use `gens` to describe the list of generators of a group. The use of square brackets to indicate an 'index' is a common convention, so a list which in printed mathematics looks like g_1, g_2, g_3 appears in our pseudo-code as `g[1]`, `g[2]`, `g[3]`. *(Watch out for the C language, though; lists or arrays begin with the index 0!)* When we need to divide our program into loops and other 'control' structures, we indicate this by common phrases like '`if ... then ...`,' '`do`,' '`while`' and '`for`'. With minor syntactic variations, such structures occur in just about all modern languages.

Making our plots. Almost all the plots in this book have a common form. They start with some 'seed' object, which might be simply a point or circle or hexagon, or which may be more elaborate, like a flower or Dr. Stickler, plotted in a basic reference position. Then there will be some set of symmetry transformations `g[1]`, `g[2]`, `g[3]` and so on, which we want to apply to the seed. Finally, we want to plot the symmetrical figure made up of all the images so obtained. The set of transformations needed to produce this 'ideal' symmetrical figure will almost always be infinite, either because it is conceived as on an infinitely large piece of paper (Iowa with infinitely many trees and fields in all directions) or because, in later chapters, some of our symmetries will shrink the seed, producing infinitely fine fractal detail. Thus the computer code will have a loop which will go on forever trying to generate an infinite number of symmetry transformations by composition (and inverse). Since in real life we have only finite screens and finite resolution, the unlimited loop must be stopped when the plotted seed either goes off the edge of the screen or is too small to see.

More formally, this means that our code will have the following structure:

- `STEP I. Initialize seed and generators of the group of symmetries.`
- `STEP II. Set up a loop for producing symmetries which could go on forever.`
- `STEP IIIa. At each step, create a new symmetry by composition with a pair of old ones.`
- `STEP IIIb. Add to the plot the shape obtained by applying the new symmetry to the seed.`

- STEP IV. Terminate the loop when all visible symmetric figures have been found.

Of course, it may be hard to be sure we have plotted exactly the visible symmetric figures and no more, but we will do our best. Also, in some versions, we will not plot the figure for every symmetry but only for the those 'near the boundary'. We shall come to this in Chapter 4.

How did we draw the symmetric figures in this chapter? Our seeds were either line segments or simple figures such as Dr. Stickler and the red star. Drawing them relied only on the simplest graphic commands: `line[x1,y1,x2,y2]`, `polyline[x,y]` and `fill[x,y]`. This is pseudo-code, but every graphics package has something similar: `line` draws a line segment from the point with coordinates (x_1, y_1) to (x_2, y_2), `polyline` draws a polygonal segments connecting in order a chain of points `(x[1], y[1])`, `(x[2],y[2])`, `(x[3],y[3])` and so on (where `x` and `y` are assumed to be arrays of the same length containing the coordinates), and finally `fill` closes the polygon and fills it with some previously chosen colour.

Drawing an image like our distortion of Paolina involves quite different commands. Typically, the image is represented in the computer by a rectangular array of 1-byte numbers called its pixels. To manipulate this picture to produce distortions like our one of Paolina, you will have to get access to this data. Since we aren't going to have many pictures like this in our book, we won't dwell on this here.

Drawing the hexagonal tiling. Let's give an example with some pseudo-code for drawing a hexagonal array like Figure 1.13. We will use 2 arrays `hexX`, `hexY` to hold the x- and y-coordinates of the vertices of the basic hexagon F. With the origin in the centre of the hexagon, the coordinates of the 6 vertices are:

$$\left(\frac{3}{2}, \frac{\sqrt{3}}{2}\right), \ (0, \sqrt{3}), \ \left(-\frac{3}{2}, \frac{\sqrt{3}}{2}\right), \ \left(-\frac{3}{2}, -\frac{\sqrt{3}}{2}\right), \ (0, -\sqrt{3}), \ \left(\frac{3}{2}, -\frac{\sqrt{3}}{2}\right).$$

Separating the $x-$ and $y-$coordinates, we thus set:
`hexX = [1.5, 0, -1.5, -1.5, 0, 1.5, 1.5]`
`hexY = sqrt(3)*[0.5, 1, 0.5, -0.5, -1, -0.5, 0.5]`.
In the pseudo-code in Box 3, we have declared arrays of numbers by adding square brackets to indicate the desired length. Thus, for example, `real hexX[7]` says we are declaring a list of 7 real numbers.[1] We have followed maths packages and object-oriented programming in allowing ourselves to multiply every element in an array by the same number, so that for example, `sqrt(3)*[0.5,1,...]` means the array `[sqrt(3)*0.5,sqrt(3),...]`.

[1] Theoretically a real number is given by an infinite decimal, but in the computer, `real` refers to a 'floating point number', an approximation decreed to be adequate by the IEEE standards committee.

Note that we repeat the starting point so that `polyline` closes up, that is, the command `polyline[hexX, hexY]` draws the whole hexagon.

The next thing we need is to enumerate the symmetries. On p. 18 we showed how compositions of the two basic translations S and T give the whole group of translations $S^k T^l$ which moves the hexagon so as to tile the whole plane. Translations in general are determined by two numbers a and b and the rule $(x,y) \mapsto (x+a, y+b)$. Thus we may describe translations by arrays of two real numbers. The two translations giving the hexagonal grid are $T(x,y) \mapsto (x+3, y)$ and $S(x,y) \mapsto (x+3/2, y+3\sqrt{3}/2)$. We use the arrays `genT = [3,0]` and `genS = [1.5,1.5*sqrt(3)]` to describe the translations in the pseudo-code. Then `k*genT + l*genS` will be the translation given by the composition $T^k S^l$ and will be denoted by `trans`.[1]

If we had an infinite sheet of paper, there would be an infinite number of symmetrical hexagons. The simplest way to make the plot finite is to specify termination conditions in advance. The program given here terminates the loop by only generating the translations $T^k S^l$ for k between `-kmax` and `+kmax` and l between `-lmax` and `+lmax`. It also sets `kmax`

[1] In our pseudo-code we have declared arrays of numbers by adding square brackets to indicate the desired length. Thus, for example, at the beginning of the code `trans[2]` says we are declaring a list of 2 numbers, while later on, `trans[2]` stands for the second entry in the array `trans`. This confusing convention is common practice in many computer languages.

Box 3: Drawing a hexagonal grid

```
real hexX[7], hexY[7];
real newhexX[7], newhexY[7];
real genT[2], genS[2], trans[2];
integer k, l, kmax, lmax;


hexX =[1.5,0,-1.5,-1.5,0,1.5,1.5];
hexY =sqrt(3)*[0.5,1,0.5,-0.5,-1,-0.5,0.5];
genT =[3,0];
genS =[1.5,1.5*sqrt(3)];
kmax =5;
lmax =5;


for k from -kmax to +kmax do {
for l from -lmax to +lmax do {
   trans = k*genT + l*genS;
   newhexX = hexX + trans[1];
   newhexY = hexY + trans[2];
   polyline[newhexX, newhexY];
} }
```

Declare all the variables we will need.

STEP I: Initialize the hexagon, generators and bounds.

STEP II+IV: Set up the loop and specify the termination conditions with kmax and lmax.

STEP IIIa: Compute a new translation.
STEP IIIb: Let it act on the figure.

Plot the new hexagon.
End loops over translations.

and `lmax` equal to 5. With this choice, the program plots 11 hexagons in each direction. (Keeping the choice of cutoffs in only the initialization step is good programming practice, so you can alter it by changing only one line.) A more sophisticated program might start from the seed hexagon and move left, right, up and down in some pattern, checking as it goes whether each new hexagon was outside the plotting area and using this to stop each part of the search when it went too far. More elaborate searches like this will be needed later on.

Projects

1.1: Composing maps using a figure

(1)   Run through the example on the left in Figure 1.12 again, this time doing first R and then T. Convince yourself that TR is another rotation, about the centre of which tile?

(2)   Still keeping the basic grid of hexagons, trace through the effects of two reflections N and M in the two mirrors in Figure 1.11. Check the claims in the caption about NM and MN.

1.2: The formula for rotation

We want to verify the formula for anticlockwise rotation R about the origin through an angle θ:

$$R(x, y) = (x \cos \theta - y \sin \theta, x \sin \theta + y \cos \theta).$$

As a simple check, substitute $\theta = 90°$ to recover the rule $(x, y) \mapsto (-y, x)$, see Figure 1.9. There are many ways to check the general formula. Here is one:

Step 1. Check that the formula correctly predicts what happens to $(1, 0)$. (Draw a diagram!)

Step 2. Do the same for the point $(0, 1)$.

Step 3. Prove that if $\mathbf{v}$ is a vector starting at the origin, then $R(t\mathbf{v}) = tR(\mathbf{v})$ for any real number t.

Step 4. Prove that if $\mathbf{v}$ and $\mathbf{w}$ are vectors starting at the origin, then

$$R(\mathbf{v} + \mathbf{w}) = R(\mathbf{v}) + R(\mathbf{w}).$$

Step 5. Put steps 1–4 together to conclude the formula.

In steps 1 and 2 we traced what happened to the two basic vectors $(1, 0)$ and $(0, 1)$. Step 3 says that the image of a point at distance t from the origin along $\mathbf{v}$ is moved by R to a point at the same distance t from the origin along the rotated vector $R(\mathbf{v})$, and step 4 says that the diagonal of a parallelogram with sides in directions $\mathbf{v}$ and $\mathbf{w}$ is rotated to the diagonal of a parallelogram with sides in directions $R(\mathbf{v})$ and $R(\mathbf{w})$. In these two steps we are showing that rotation R is a **linear** map, meaning that R maps straight lines to straight lines. Step 4 is explained pictorially in Figure 1.16.

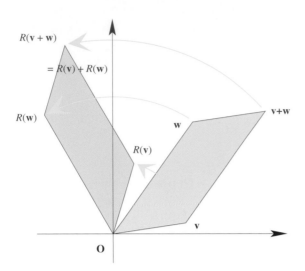

Figure 1.16. The pictorial explanation of why R is a linear map. It carries the vertices **v** and **w** of the blue parallelogram into the vertices $R(\mathbf{v})$ and $R(\mathbf{w})$ of the pink one. The point **v** + **w** is the corner of the blue parallelogram opposite the centre of rotation 0, and likewise $R(\mathbf{v}) + R(\mathbf{w})$ is the far corner of the pink one. On the other hand, since R carries the blue parallelogram to the pink one, the far pink corner must also equal $R(\mathbf{v} + \mathbf{w})$.

1.3: Gluing hexagonal tiles

Consider the hexagonal tiling in Figure 1.13. Cut out one hexagon F and imagine gluing up its three pairs of parallel edges. The map T glues the left and right edges, S glues the edges top-right and bottom-left, and $T^{-1}S$ the top-left and bottom-right. What surface does this create? You may actually be able to do this physically with some sufficiently elastic fabric. Stuff it to give shape to your exotic new toy, and send us a photo if you succeed!

1.4: Plotting rotations

This project assumes you have figured out how to implement some of our pseudo-code on your own system. You will have a program which will plot single images of your favourite shape under translation, rotation and reflection, and your program will be made so that you can vary the input data which tells the translation distance, the centre and angle of rotation, and the position and angle of the reflecting mirror. (You will need to use the formulas for the basic symmetries in Box 1.)

What happens when you try to rotate your figure through multiples of an angle which is not a whole submultiple of 360°? More concretely, plot all rotations by ka degrees where k goes from 0 to 20 (say) and a is *irrational*[1] – try $a = 100\sqrt{2}$. Keep increasing the upper limit for k and note how the plot fills in. There is a famous theorem which says you eventually fill in the entire circle.

[1] An **irrational** number is one which cannot be written as the ratio p/q of two whole numbers p and q, see Note 5.2 and p.310

1.5: Hexagonal tiles

We simplified the problem of hexagonal tiles in Figure 1.13 by adding a red star which 'broke the symmetry' of each tile. Suppose we want to lay down a hexagonal pattern with plain blue tiles. Now the pattern has rotational symmetry also, so that if we try to label some other tile U by the symmetry which carries F to U, then U will end up having several different names. We can resolve this problem by subdividing each tile into six identical triangles as in Figure 1.17. The symmetry corresponding to this subdivision is the $60°$ anticlockwise rotation R around the centre of F. We have labelled one of the six triangles B to indicate that it is our new basic tile, and coloured the images:

$$B, R(B), R^2(B), R^3(B), R^4(B), R^5(B).$$

Why is $R^6 = I$? Check from the picture that $R^{-1} = R^5$, $R^2 = R^{-4}$ and so on.

Figure 1.17. A tiling by hexagons subdivided into equilateral triangles. The map TS^2R carried the pink triangle labelled B to the pink triangle at upper right.

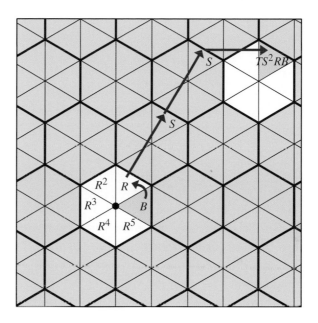

Convince yourself you can get to any other triangle in the picture by applying compositions of R, S and T. How can you reach the top pink triangle? Because there are no symmetries fixing any given triangle (except for reflections which we have disallowed because they turn tiles the wrong way up), each triangle is labelled by exactly one symmetry, the one you have to use to transport it from the basic tile B. A tile like this is called a **fundamental region** (or fundamental tile) for the action of the group generated by S, T and R.

1.6: A dodecagonal tiling

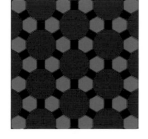

The tiling shown in the margin actually has exactly the same symmetries as the hexagonal tiling in Figure 1.17. See if you can identify a fundamental tile and then try writing a program to use it to tessellate as in the figure. You will have to wrap your plotting program in one or more loops so that you can draw multiple copies by repeatedly composing symmetries.

Many of Escher's prints contain other unusual tilings you can play with. For example, it is interesting to find a fundamental region for his beautiful 1941 etching *Shells and Starfish* which has been widely reproduced.

CHAPTER 2

A delightful fiction

*So progresses arithmetic subtlety,
the end of which, as is said, is as refined as it is useless.*

Ars Magna, Girolamo Cardano

The intricate fractal shapes we are aiming to draw are based on the algebra and geometry of complex numbers. Complex numbers are really not as complex as you might expect from their name, particularly if we think of them in terms of the underlying two dimensional geometry which they describe. Perhaps it would have been better to call them 'nature's numbers'. Behind complex numbers is a wonderful synthesis between two dimensional geometry and an elegant arithmetic in which every polynomial equation has a solution. When complex numbers were first dreamed of in the Renaissance, they were treated as an esoteric, almost mystical, concept. This aura of mystery persisted well into the twentieth century – the senior author's aunt Margaret Silcock (née Mumford), who studied mathematics at Girton College Cambridge in 1916, liked to describe them as a 'delightful fiction'. In fact we still use the term 'imaginary numbers' to this day. Modern scientists, however, take complex numbers for granted, FORTRAN makes them a predefined data type, and they are standard toolkit for any electronic engineer. Perhaps the most remarkable fact about complex numbers is that they are absolutely essential to modern physics. In the theory of quantum mechanics, not only can the universe exist probabilistically in two states at once, but the uncertain composite state is constructed by adding the two simple states together with complex coefficients, introducing a complex 'phase'. Our quotation above is from the sixteenth century Italian mathematician Girolamo Cardano who gave one of the first accounts of the square root of -1. He could not have been more wrong when he declared this exotic construction useless!

What is a complex number? Roughly speaking, it is any 'number' which involves the square root of a negative number, like $\sqrt{-2}$ or $-35.32 + 4 \times \sqrt{-10}$. At the time complex numbers were discovered, mathematicians were very interested in the problem of solving simple equations. Suppose you wanted a number x which had a particular property, for instance, the sum of the cube of x and twice x should equal 5. This is expressed by the equation $x^3 + 2x = 5$, which is called cubic because thinking geometrically it involves volumes, the highest power of x being 3. Already in Babylonian times, about 2000 BC, mathematicians had algorithms for solving their version[1] of **quadratic equations** like $ax^2 + bx + c = 0$. If this equation had any solution at all, it would be given by the high school formula $x = (-b \pm \sqrt{b^2 - 4ac})/2a$. Since the formula uses square roots, we run into a problem: suppose the formula asks us to take the square root of a negative number. The product of two positive numbers is positive, so the square of a positive number is positive. Inconveniently, the product of two negative numbers is also positive, hence no number has a negative square and negative numbers have no square roots.

The inconvenience that $(-1) \times (-1) = 1$ is not due to arbitrary choice. Algebraically, one can reason that since $1 + (-1) = 0$, multiplying by (-1) shows $(-1) + (-1) \times (-1) = 0$, hence $(-1) \times (-1) = 1$. You can think of it geometrically too: imagine yourself on a road looking straight ahead. You can locate a point in front of you by giving its positive distance from where you stand, while points behind you are specified by negative distances. Multiplying these distances by (-1) puts points which were in front of you behind and vice versa. Doing this twice brings a point back to where it started, in symbols, $(-1) \times (-1) = 1$. Thus *the squares of all ordinary numbers are positive (or zero)*. In fact, in the formula above, if $b^2 - 4ac < 0$, the quadratic equation has no solutions in ordinary numbers at all.

The story took a new twist when early Renaissance mathematicians found complicated algorithms for the solutions of cubic and quartic equations, that is equations involving terms like x^3 and x^4. Sometimes these algorithms involved taking square roots of negative numbers even when, in the end, the equation had an ordinary number solution! The pioneer was Girolamo Cardano who in 1545 published a systematic treatise *Ars Magna* on algebra as it was understood in his day.[2] Cardano felt that algebra should only deal with geometric quantities like length, area and volume, so he had trouble with negative numbers (no length is negative), fourth powers (areas are given by squares, volumes by cubes but it ends there because space is three dimensional) and, hardest of all, square roots of negative numbers, which he called 'some recondite third sort of thing'. But as you can see from his example in Note 2.1, the logic of the algebraic

[1] The Babylonians formulated quadratic equations as a system of equations like $x + y = a, xy = b$.

[2] The first solution of a cubic equation was by Scipione del Ferro (1465–1526). Priority for the full solution should go to Niccolò Tartaglia (1499–1557), who claimed Cardano published his work having previously promised secrecy. The question was hotly disputed by Tartaglia and Cardano's student Ferrario.

rules was just too strong, and he was unable to dismiss the idea of square roots for negative numbers, which really came down to doing a new kind of arithmetic in which you were allowed to take the square root of -1. The arithmetical rules for dealing with complex numbers, that is numbers involving $\sqrt{-1}$, were first dealt with systematically in Rafael Bombelli's *Algebra* published in 1572, in which $\sqrt{-1}$ is confusingly called the 'plus of minus'. The same ideas appear in the unpublished writings of Thomas Harriot (1560-1621) on algebra, in which he deals at length with roots called 'inexplicable and imaginary'. In effect, these mathematicians were saying: "Let's just suppose that there were a square root of -1 and see what we get." To their amazement, no obvious contradictions arose when they began to calculate using the ordinary rules of arithmetic. Felix Klein described the situation like this:[1]

> "Imaginary numbers made their own way into arithmetic calculations without the approval, and even against the desires of individual mathematicians, and obtained wider circulation only gradually and to the extent that they showed themselves useful."

But useful, in fact essential, they are.

Let's try to retrace what these early mathematicians might have done. Adopting the modern notation, we'll represent $\sqrt{-1}$ by the symbol i. (Some people, particularly electrical engineers, symbolize the same quantity by j.) The main thing you have to remember is that

[1] See his book *Elementary Mathematics from an Advanced Standpoint*, p. 56 of the translation by E. Hedrick and C. Noble, Dover, 1945.

Note 2.1: **Cardano's problem**

Why did Cardano become involved with complex numbers at all? He had a formula that almost always seemed to find the solutions of cubic equations. However in the case of the equation:

$$x^3 = 15x + 4,$$

which he knew had the solution $x = 4$, his formula gave him:

$$x = \sqrt[3]{2 + \sqrt{-121}} + \sqrt[3]{2 - \sqrt{-121}}.$$

It was Bombelli who figured out how this expression could possibly have value 4. He guessed that the two terms on the right hand side of this equation

must be complex numbers of the form $a + ib$ and $a - ib$. Since their sum is 4, we find $a = 2$. Thus he needed:

$$2 + ib = \sqrt[3]{2 + 11i}.$$

Cubing both sides of this equation, you can check that we must have $b = 1$, in other words, $(2 + i)^3 = 2 + 11i$. You can also check that $(2 - i)^3 = 2 - 11i$ so it makes sense to say that

$$\sqrt[3]{2 + 11i} + \sqrt[3]{2 - 11i} = (2 + i) - (2 - i) = 4.$$

So in Cardano's example, we have a problem with real numbers which has a real solution, but it is found by using complex arithmetic!

$i \times i = \sqrt{-1} \times \sqrt{-1} = -1$. Firstly, using numbers like $i/3$ and $\sqrt{2}i$, the arithmetic rule that $(ab) \times (ab) = a^2 \times b^2$ shows that $(i/3)^2 = -1/9$ and $(\sqrt{2}i)^2 = -2$. It is easy to see with this sort of calculation that all negative numbers now have square roots. Nowadays, the ordinary numbers we use like 2.16 and -3.14159 are called **real** and numbers like $3i$ or $-1.12i$ are called **purely imaginary**. If we add a real and an imaginary number we get a number like $-3.1 + 2.6i$ which is called **complex**. In fact **complex numbers** are just all the numbers expressible as $x + iy$ where x and y are real. To show that a complex number should be thought of as a single entity, we often represent $x + iy$ by a single symbol z. The (real) number x is called the **real part** of z, written $\operatorname{Re} z$, and y is called the **imaginary part**, written $\operatorname{Im} z$.

The miracle is that once we have allowed ourselves to take this bold step, we are done! Using the usual rules of arithmetic, we can mechanically add, subtract and multiply to our heart's content: no contradictions arise and no new kinds of 'numbers' will ever appear. Division is a bit harder, but there is a simple trick that makes it clear there is only one possible way it could work. Firstly, dividing $a + ib$ by $x + iy$ should produce another complex number $u + iv$ with the property that

$$(u + iv) \times (x + iy) = a + ib.$$

The question is how to determine u and v. The long-lost 'twin brother' of $x + iy$ arrives to save the day.[1] Multiply both sides by the complex number $x - iy$, called the **complex conjugate** of $x + iy$. By our rules above the righthand side is:

$$(a + ib)(x - iy) = ax + ibx - aiy - i^2by = (ax + by) + i(bx - ay).$$

On the lefthand side, the reunion of the brothers produces a joyful response:

$$(x + iy)(x - iy) = x^2 + iyx - xiy - i^2y^2 = x^2 + y^2,$$

an honest old-fashioned positive real number, given that x and y are not both zero. At last we have:

$$u + iv = \frac{ax + by}{x^2 + y^2} + i\,\frac{bx - ay}{x^2 + y^2}$$

which is another 'number' of the same kind. The complete rules for complex number arithmetic are displayed in Box 4.

[1] In 1572 Bombelli saw the two numbers $\pm i$ as a pair which he called the 'plus of minus' and the 'minus of minus'. So 4 *più di meno* 2 and 4 *men di meno* 2 meant $4 \pm 2i$.

Geometry of complex numbers

Most scientists and engineers are so used to complex numbers that it is hard nowadays to stand back and appreciate what a strange thing we have

---

Box 4: **Complex number arithmetic**

Complex numbers are manipulated according to the ordinary rules of arithmetic, always remembering that $i^2 = -1$. For example,

$$(3 + 4i) + (2 - 3i) = (3 + 2) + (4i - 3i) = 5 + i$$

and

$$(3 + 4i)(2 - 3i) = 6 - 9i + 8i - 12i^2 = 6 - i + 12 = 18 - i$$

and

$$\frac{1}{2 + i} = \frac{2 - i}{2^2 + 1^2} = \frac{2}{5} - \frac{i}{5}.$$

The formal rules are:

$$(a + ib) + (x + iy) = (a + x) + i(b + y)$$
$$(a + ib) - (x + iy) = (a - x) + i(b - y)$$
$$(a + ib) \times (x + iy) = (ax - by) + i(ay + bx)$$
$$1 \div (x + iy) = \frac{x}{x^2 + y^2} - i\,\frac{y}{x^2 + y^2}$$
$$(a + ib) \div (x + iy) = \frac{ax + by}{x^2 + y^2} + i\,\frac{bx - ay}{x^2 + y^2}.$$

---

done. Ordinary numbers have immediate connection to the world around us; they are used to count and measure every sort of thing. Adding, subtracting, multiplying and dividing all have simple interpretations in terms of the objects being counted and measured. When we pass to complex numbers, though, the arithmetic takes on a life of its own. Since -1 has no square root, we decided to create a new number game which supplies the missing piece. By adding in just this one new element $\sqrt{-1}$, we created a whole new world in which everything arithmetical, miraculously, works out just fine.

The world of complex numbers leads to elegant and profound connections between geometry, algebra and analysis. In fact, much of their importance stems from the fact that they give us a simple method of representing points in the plane. The germ of this idea is to be found in John Wallis' 1685 *Treatise on Algebra*, but was not fully and clearly developed until the work of the Norwegian surveyor Caspar Wessel (1745–1818)[1], and did not gain general acceptance until the work of Gauss (1777–1830)[2]. Let's go back to our description of points in front of us by positive numbers and behind as negative. The left frame of Figure 2.1 shows Dr. Stickler standing on an infinite flat plain facing

[1] A translation of Wessel's paper, with much interesting background about the history of complex numbers, is to be found in bicentennial commemorative volume *Caspar Wessel* by B. Branner and J. Lützen, published by the Royal Danish Academy of Sciences and Letters, C. A. Reitzel, 1999.

[2] See Gauss' *Collected Works*, Vol. 2, pp. 174–178. Although Gauss claimed that he already understood the complex plane in his first proof of the Fundamental Theorem of Algebra (see Note 2.3) in 1799, he did not actually publish anything on the subject until 1831.

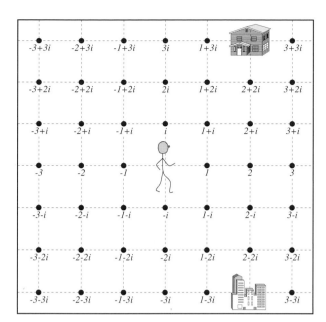

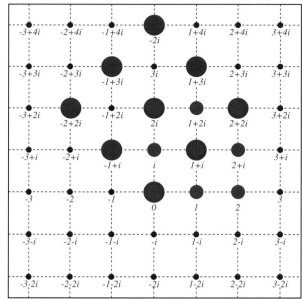

Figure 2.1. Left: Dr. Stickler stands on an infinite complex plain [*sic*], on which locations are marked by dots with complex numbers as addresses. Thus, his house is at $2 + 3i$, while his office is at $2 - 3i$. (You were expecting somewhat more austere accommodations?) Right: The nine red points are the products of the 9 blue points by $1 + i$. See how the grid is rotated and expanded by this operation.

[1] The complex plane is sometimes called the **Argand diagram**, after the rather obscure French mathematician Argand whose work on the geometry of complex numbers, originally published anonymously in 1806, also played a significant role.

along the horizontal x-axis, so that points directly ahead of him are marked out by positive numbers and points behind him with negative ones. Here's the idea: he can also describe points on the line stretching out at right angles to his left (the positive y-axis) by imaginary numbers $i, 2i, 3i, \ldots$; and points on the line to his right (the negative y-axis) by $-i, -2i, -3i, \ldots$. Even better, he can describe points ahead and to his left as complex numbers like $2 + 3i$, points ahead and to the right as $2 - 3i$, and so on. What this comes down to is this: the complex number $x + iy$ can be thought of as representing the point on the plane with coordinates (x, y). Dr. Stickler is standing at the origin $(0, 0)$. When we make this association, the coordinate plane is known as the **complex plane**.[1] The x-axis is now called the **real axis** because it represents those complex numbers $x + 0i$ which are actually ordinary real numbers, and the y-axis is called the **imaginary axis** because it represents all the numbers $0 + yi$ which are purely imaginary. In the complex plane, addition of complex numbers is the same as vector addition $(a, b) + (x, y) = (a + x, b + y)$. As you can see, this is just the same as complex addition $(a + ib) + (x + iy) = (a + x) + i(b + y)$.

Vector addition and subtraction is useful, but the beauty and power of complex numbers stems from the fact that the other algebraic operations of multiplication and division have an important geometrical interpretation as well. We could go on and just show you the formulas and figures, but we want you to read this book more interactively! Here's a little experiment.

If you haven't worked with complex numbers before, please try it to flex your mental muscles. In the complex plane, the nine complex numbers 0, 1, 2, i, $i+1$, $i+2$, $2i$, $2i+1$, $2i+2$ form a little 3×3 square with its bottom left corner at the origin. Now multiply all these nine numbers by $(1 + i)$ and plot your answers. This is easy, for instance $(1+i) \times (2i+1) = -1+3i$. The effect is startling: as you can see in the right frame of Figure 2.1, the resulting red dots again form a 3×3 square but compared to the original square of blue dots it is larger and turned on its side. In fact, the square has expanded by a factor of $\sqrt{2}$ and has rotated anti-clockwise through $45°$.

This simple experiment shows the general story: when complex numbers are thought of as points in the plane, multiplying by the complex numbers $x + iy$ has two effects: it stretches (or shrinks) everything by some factor r and rotates the plane around the origin through some angle θ. To understand this in more detail, we first need to look at the relationship between complex numbers and polar coordinates.

The **polar coordinates** of a point P in the plane are (r, θ), where r is the distance of P from the origin O, and θ is the angle the line joining P to O makes with the x-axis, measured anticlockwise. (Angles measured in the clockwise direction are counted as negative.) As you can see in Figure 2.2, if P has coordinates (x, y), then from Pythagoras' theorem it follows that $r = \sqrt{x^2 + y^2}$ and from trigonometry we get $\tan \theta = y/x$. Thus θ is given by the inverse of the tangent function applied to y/x, the **arctangent** of y/x.

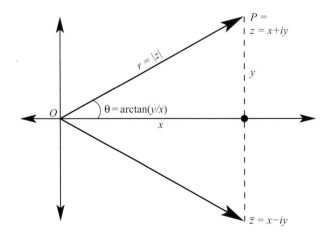

Figure 2.2. A complex number z can be described either via its real and imaginary parts x and y or via its modulus r and argument θ. The complex conjugate $\bar{z} = x - iy$ is obtained by reflecting in the real axis.

Now we can apply this to a complex number $z = x + iy$, thinking of z as a point $P = (x, y)$ in the plane. The distance r is called the **modulus** or **absolute value** of z, written $|z|$ (pronounced 'mod z') and the angle θ is called the **argument** of z, written $\arg z$. We should probably mention that, although angles are measured in degrees by ordinary people, mathematicians and scientists primarily use **radians**. The radian measure is based on the circumference being 2π times the radius of the circle, where π is the famous and mysterious number $3.14159\ldots$. A full circle of $360°$ measures 2π radians, an angle of $180°$ (or a half circle) is π radians, and $90°$ is $\pi/2$ radians, so one radian is about $57°$.

Suppose now that we are given the modulus $r = |z|$ and argument $\theta = \arg z$ of a complex number z. How can we recover the real and imaginary parts x and y so as to be able to write $z = x+iy$? Using trigonometry in the right-angled triangle in the previous figure, we see that $x = |z|\cos\theta$ and $y = |z|\sin\theta$, so that $z = |z|(\cos\theta + i\sin\theta)$. Thus we can go back and forth between writing a complex number using its real and imaginary parts, or using its modulus and argument. The form $z = |z|(\cos\theta + i\sin\theta)$ is called the **polar representation** of the z.

Using the polar representation, the geometry of complex multiplication looks easy: suppose z and w are two complex numbers. Then zw is that complex number whose modulus is the product of the moduli of z and w and whose argument is the sum of their arguments. In symbols:

$$|zw| = |z| \times |w|,$$
$$\arg(zw) = \arg z + \arg w.$$

These formulas can be proved using high school trigonometry as we have done in Note 2.2. They are deceptively simple: it is a remarkable fact that to multiply two complex numbers you just *multiply* the moduli and *add* the arguments. This idea is fundamental to manipulating complex numbers and, as we shall see in the next section, it imparts a geometrical significance to all the arithmetical operations we shall want to perform.

Using the polar form, we get a nicer expression for the complex number $1/z$. Suppose we write $z = |z|(\cos\theta + i\sin\theta)$ and we set $w = (1/|z|)(\cos(-\theta) + i\sin(-\theta))$. Then from our formula, zw has modulus $|z||w| = |z| \times 1/|z| = 1$ and argument $\arg z + \arg w = \theta + (-\theta) = 0$. We immediately conclude that $w = 1/z$, because our reasoning shows that w must be the number such that $z \times w = 1$.

Complex numbers with modulus one play a rather special role. If $z = x + iy$ has modulus 1, in symbols if $|z| = 1$, then by definition $x^2 + y^2 = 1^2 = 1$. This means that z lies on the circle of radius 1 with centre the origin, usually known as the **unit circle**. Using the polar representation, any such point z can be written in the form $z = \cos\theta + i\sin\theta$ with $\theta = \arg z$. The angle θ of course is the angle θ shown in Figure 2.2. For example,

$i = \cos 90° + i \sin 90°$. Again if $z = \frac{1+i}{\sqrt{2}}$ then $\left|\frac{1+i}{\sqrt{2}}\right| = \sqrt{1/2 + 1/2} = 1$ and $\frac{1+i}{\sqrt{2}} = \cos 45° + i \sin 45°$, which is the point on the unit circle making the angle 45° (or $\pi/4$ radians) with the x axis. From the multiplication formulas you can see that the effect of multiplying two points z and w on the unit circle is to produce another point zw on the unit circle whose argument is exactly the sum of the arguments of z and w!

When we wanted to divide a complex number by $x + iy$, its 'twin' or complex conjugate $x - iy$ came to the rescue. The complex conjugate also has a nice geometrical interpretation: you can see from Figure 2.2 that it is just the reflection of $x + iy$ in the real axis. From Dr. Stickler's perspective in Figure 2.1, complex conjugation interchanges points on his left with those on his right and vice versa. If we write $z = x + iy$, its conjugate is written $\bar{z}$. The operation of passing from z to $\bar{z}$ is called **complex conjugation**.[1] The complex conjugate relates to the modulus like this:

$$z\bar{z} = (x + iy)(x - iy) = x^2 + y^2 = |z|^2,$$

which explains the trick we used in the last section when we wanted to divide.

Complex conjugation has an equally nice interpretation in terms of the polar representation. Notice that z and its conjugate have the same modulus: $|z| = \sqrt{x^2 + y^2} = |\bar{z}|$. If $z = |z|(\cos\theta + i\sin\theta)$, then as we have seen, θ is defined by the equation $\tan\theta = y/x$. Replacing y by $-y$ we find that $\arg\bar{z} = \arctan(-y/x) = -\theta$. Thus $\bar{z} = |z|(\cos(-\theta) + i\sin(-\theta))$, confirming that $\bar{z}$ is the reflection of z in the horizontal x-axis. We have summarized these various handy formulas in Box 5.

You might ask if we can take the principle which led to complex numbers further. For example, can we find a complex number $z = x + iy$ such that $z^2 = (x + iy)^2 = i$? In other words, can we find a complex number which is the square root of the number i, or do we have to invent yet another sort of 'super-complex' number? It turns out we can always find square roots using the polar representation $z = |z|(\cos(\theta) + i\sin(\theta))$.

[1] Not to be confused with the other use of the word conjugation for changing maps which we discussed in the last chapter!

Note 2.2: **Multiplication in polar coordinates**

Suppose $z = r(\cos\theta + i\sin\theta)$ and $w = s(\cos\phi + i\sin\phi)$. Then using the trigonometric identities for the sine and cosine of the sum of 2 angles, we get:

$$\begin{aligned} zw &= rs(\cos\theta + i\sin\theta)(\cos\phi + i\sin\phi) \\ &= rs\,[(\cos\theta\cos\phi - \sin\theta\sin\phi) \\ &\quad + i\,(\cos\theta\sin\phi + \sin\theta\cos\phi)] \\ &= rs(\cos(\theta + \phi) + i\sin(\theta + \phi)). \end{aligned}$$

Box 5: Complex numbers and polar coordinates

Here are the main formulas used in the polar representation of complex numbers:

- The complex conjugate of $z = x + iy$ is $\bar{z} = x - iy$.
- The modulus or absolute value of $z = x + iy$ is the number $|z|$ found from the formula:
$$z\bar{z} = (x + iy)(x - iy) = x^2 + y^2 = |z|^2.$$
- The polar representation of the complex number $z = x + iy$ is
$$z = |z|(\cos\theta + i\sin\theta)$$
where $\theta = \arg z = \arctan y/x$.
- In polar form, the complex conjugate of $z = |z|(\cos\theta + i\sin\theta)$ is $\bar{z} = |z|(\cos(-\theta) + i\sin(-\theta))$.
- In polar form, complex numbers are multiplied by the rule:
$$|zw| = |z| \times |w|,$$
$$\arg(zw) = \arg(z) + \arg(w).$$
- If $z = |z|(\cos\theta + i\sin\theta)$ then $1/z = (1/|z|)(\cos(-\theta) + i\sin(-\theta))$.

From the formulas in Box 5 and Note 2.2:

$$\left(\pm\sqrt{|z|}\,(\cos(\theta/2) + i\sin(\theta/2))\right)^2 = |z|(\cos\theta + i\sin\theta) = z$$

and so the two complex numbers $\pm\sqrt{|z|}(\cos(\theta/2) + i\sin(\theta/2))$ are the two square roots of z. For example, suppose we want to find the two square roots of i. In polar form, $i = \cos\pi/2 + i\sin\pi/2$ and $|i| = 1^2 = 1$. Thus our formula gives

$$\sqrt{i} = \pm(\cos\pi/4 + i\sin\pi/4)$$

and as one can calculate using Pythagoras' theorem, $\cos\pi/4 = \sin\pi/4 = 1/\sqrt{2}$. Thus the square roots of i are $\pm(1 + i)/\sqrt{2}$.

Similar ideas allow you to find any root of any complex number, for example you might like to try finding the *three* cube roots of i. In fact, as explained in Note 2.3, it is a famous result that there is no need to introduce any 'super-complex' numbers to solve any polynomial equation created with any complex number coefficients at all.

Computing with complex numbers

From now on, almost every computation we make will use complex numbers. To do anything, therefore, we must set up a method for programming them. As is inevitable with computers, this is easy in some setups and harder in others. In the high level mathematics packages, in FORTRAN and in some releases of C++, complex number data types and complex arithmetic are built in. In most other environments, you will have to write your own library of complex number routines. An example for the language C can be found in the very useful book *Numerical Recipes in C*.[1]

[1] W. Press, S. Teukolsky, W. Vetterling and B. Flannery, *Numerical Recipes in C, The Art of Scientific Computing*, 2nd edition, Cambridge University Press, 1992.

Writing a complex arithmetic library works differently in traditional and object-oriented languages. In the traditional language C, for example, you first define a new data type `complex` (called a 'structure') so that a complex number `z` is an array of two real numbers. In some languages, the real and imaginary parts of a complex data structure `z` are obtained by functions such as `re(z)` and `im(z)` (or `real(z)` and `imag(z)`; many variations of the names occur). In *C*, the definition of the `complex` structure associates names to the real and imaginary parts, as in:

```
typedef struct double re, im; complex;
```

With this convention, the real and imaginary parts of a complex structure `z` are defined as double precision real numbers, and they are henceforth referenced as `z.re` and `z.im`.

Note 2.3: Solving polynomial equations

A **polynomial equation of degree** n is an equation of the form $ax^n + bx^{n-1} + cx^{n-2} + \cdots + fx + g = 0$, where n is a positive integer and $a, b, c, \ldots, f, g$ are any numbers, real or complex. For example, $x^3 + 2x - 5 = 0$ is a polynomial equation of degree 3, otherwise known as a **cubic equation**. A special case is the quadratic equation $ax^2 + bx + c = 0$. As we have seen, in complex arithmetic we can always find square roots even of negative or complex numbers. So the formula $x = (-b \pm \sqrt{b^2 - 4ac})/(2a)$ always yields a solution or **root** of this equation. The *Fundamental theorem of algebra* proved by d'Alembert in 1746 (with some gaps filled by Argand in 1814) says that a polynomial of degree n with complex coefficients always has a solution which is a complex number, and indeed that all possible solutions are either real or complex. For example the polynomial equation $x^5 + 4i x^4 - 9x^3 - 12i x^2 + 9x - 2 + 4i = 0$ has a complex root. (Polynomials like this will resurface in Chapter 9.) In other words, if you start in the complex numbers, you never have to go outside to find solutions to any polynomial equations: the complex numbers are, algebraically speaking, a closed system.

Then you write a library of the four basic arithmetic procedures cxadd(z,w), cxsub(z,w), cxmult(z,w) and cxdiv(z,w) which accept a pair of complex numbers as arguments and return the complex result. Thus a polynomial which would be written for real numbers as:

$$z*z/a + b*z - c$$

becomes for complex numbers:

cxadd(cxdiv(cxmult(z,z),a),cxsub(cxmult(b,z),c))).

Here is an example of a subroutine written with the C structure conventions:

```
complex cxmult(complex z,complex w){
    complex u;
    u.re = z.re*w.re - z.im*w.im;
    u.im = z.re*w.im + z.im*w.re;
    return( u );
}
```

The basic arithmetic procedures should be supplemented with many other necessary functions, such as abs(z), arg(z) which accept a complex number z and return the real numbers which are their absolute value and arguments. We'll also need a function cxmake(x,y) which accepts two real arguments x and y and returns the complex number z=x+iy with x and y as its real and imaginary parts.[1]

This can and has been done many times by mathematicians wishing to write C programs with complex numbers. Its main drawback is the way it makes formulas so ugly and uncheckable (almost as bad as LISP before it died a natural death). We are used to the so-called infix notation[2] in which arithmetic operators are written *between* variables and have familiar conventions about leaving out parentheses as much as possible, so that for example x*y + z is assumed to mean (x*y)+z and not x*(y+z). Some object oriented languages, on the other hand, such as C++, allow you to define complex numbers as a new *class* and to 'overload' the usual +,-,*,/ arithmetic operators so that they work on complex numbers in the same way that they work on real ones. Then you can write the usual expressions z*z/a + b*z -c with complex z,a,b,c in your code. Many distributions of C++ even include such a complex class.

To get started computing using complex numbers, you may like to try the first few projects at the end of the chapter.

[1] In the course of carrying out calculations, you may also find it useful to have routines for square roots, trigonometric functions (sines and cosines, and particularly *hyperbolic* sines and cosines), exponentials, logarithms, and power functions.

[2] By analogy with prefix and postfix.

Complex multiplication and mapping the plane

In Chapter 1 we discussed in detail how a transformation T of the plane can be viewed as a rule which assigns to each point P in the plane a new point Q. The operation of multiplication by a fixed complex number provides us with just such a rule. Fix a complex number a. Remembering that each point P can be represented by a complex number z, our rule can be written very simply:

$$T(z) = az.$$

The effect of such a T may be guessed at from the results of the experiment we carried out in Figure 2.1. Since maps like this are going to be very important to us, let's take as a first simple example the value $a = 2$. Applying T to the complex number $z = x + iy$ we find $T(z) = 2z = 2x + 2iy$, in other words the point with coordinates (x, y) maps to the point with coordinates $(2x, 2y)$. Now suppose that F is some shape in the plane. If we multiply the coordinates of all the points in F by 2, we obtain a similar figure $T(F)$ in which all the linear dimensions have been expanded by the factor 2. In fact T is a transformation of the plane which expands out from the origin O by a factor of 2 in all directions. Multiplying by $a = 1/2$ gives the inverse of T: the map $z \mapsto z/2$ pulls every point directly towards the origin, halving the distance from O and shrinking every object by half. You can see the effect of multiplication when $a = 1/2$ and $a = 3/2$ on the red fox in the top right quadrant of the left frame of Figure 2.3.

What would it mean to multiply by a negative number? If we try with -1, then the point $x + iy$ is transformed into the point $-x - iy$, otherwise said, (x, y) moves to $(-x, -y)$. This is the transformation which moves every point P across the origin O to the point Q on the opposite side at the same distance, in other words, reflection across the origin. We can get multiplication by -2.5 in two steps: first transform $x + iy$ into $2.5(x + iy)$, which causes expansion in all directions by a factor of 2.5, and then multiply by -1 which flips across the origin. This is what has happened to the large fox exiting the bottom left quadrant on the left. Notice that we could first shrink and then flip, or first flip and then shrink, getting the same answer either way. In mathematical language, multiplication is **commutative**, see p. 15. When we start combining different kinds of operations, for example translations and expansions, they may no longer commute.

So far, we have not done anything which required complex numbers; all we have said could have been done with vectors, and in more than two dimensions. The really interesting step is to multiply by a number which

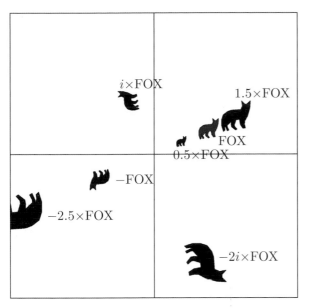

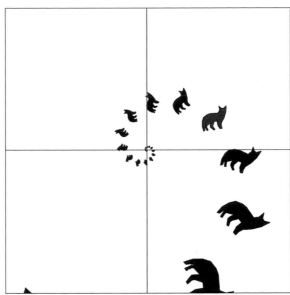

Figure 2.3. Left: The original red fox in the upper right quadrant is transported by the transformations $z \mapsto 1.5z, z/2, -z, -2.5z, iz$ and $-2iz$. Right: The fox is transported by the transformations $z \mapsto \ldots, a^{-2}z, a^{-1}z, az, a^1z, a^2z, \ldots$ and so on, where $a = 0.8 \times \left(\frac{\sqrt{3}}{2} + \frac{1}{2}i\right)$. Positive powers of a make him spiral into the origin; negative powers make him spiral away.

is truly complex. Just as we investigated the effect of multiplying by -2 in two stages, first a stretch and then a flip, so we can split up the question of understanding multiplication by a into two parts. Write a in its polar form $a = A(\cos\theta + i\sin\theta)$, where $A = |a|, \theta = \arg a$. Since A is real and positive, we already know that multiplication by A is expansion or contraction by this factor. How about multiplication by a complex number $\cos\theta + i\sin\theta$ of modulus one? From the formulas in Box 5, we see that for any z, the product $(\cos\theta + i\sin\theta)z$ has the same modulus as z, while its argument is $\theta + \arg z$. Thus $(\cos\theta + i\sin\theta)z$ is the point you get by rotating z about the origin anticlockwise through the angle θ. The combined operations of expansion (or contraction) and rotation cause spiralling in or out from O, as on the right in Figure 2.3.

This explains what happened in our experiment in Figure 2.1 when we multiplied the 3×3 grid of blue dots by $1 + i$. In polar coordinates, $1 + i = \sqrt{2}(\cos 45° + i\sin 45°)$, so the modulus of $1 + i$ is $\sqrt{2}$ and its argument is $45°$, confirming our conclusion that the grid turned through $45°$ and expanded by $\sqrt{2}$.

In the right frame of Figure 2.3 the multiplication factor is $a = \frac{4}{5}(\cos 30° + i\sin 30°)$. Denoting the fox by F, we have drawn his orbit F, aF, a^2F, a^3F, a^4F and so on. These image foxes make angles $0, 30°, 60°, 90°, 120°$ with the real axis and are at distances $1, 0.8, 0.64, 0.512, 0.4096$ from the origin. The orbit is spiralling into the origin, and the angle $\theta = 30°$ determines the rate of turning. If θ were negative, the spiral would turn backwards. The factor A, in this case 4/5, is the rate of contraction. If A were greater than one, the foxes would spiral away. This is exactly what happens if we *divide* instead of multiply by a. Dividing by powers of a is the same as multiplying by powers of $1/a$; thus $T^{-1} : z \mapsto z/a$ is the inverse of T. Applying T^{-1} causes the fox to spiral backwards, away from the origin, which you can also see in the figure.

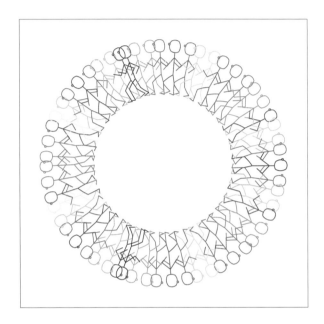

Figure 2.4. Dr. Stickler in the centrifuge. Dr. Stickler whirls around at angles equal to an irrational multiple of π, in this case $\pi \frac{\sqrt{5}-1}{2}$ radians. Each time he rotates by this angle, he changes colour. (Wouldn't you?) Eventually all the copies of Dr. Stickler would fill out a blurry ring. This was difficult for Dr. Stickler to endure, but after all it was in the interests of science.

Multiplication by powers of points on the unit circle gives an interesting exception. If $|a| = 1$, then there is no expansion or contraction, only rotation through an angle θ. All the points in the orbit of an initial point z_0 remain on the circle of radius $|z_0|$ and centre O, moving round and round with equally spaced jumps. There are two possibilities: either some orbit point eventually lands back on z_0, or not. In the first case, we start retracing our steps over the same path, as happened for example in Figure 1.6 (with $\theta = 72°$). In the second case, shown in Figure 2.4, orbit points (or rather Dr. Sticklers) keep piling up more and more thickly all the way round the circle.

Our pictures of foxes illustrate the **dynamics** of $z \mapsto az$. No matter what the value of the complex number a, the origin O is always a **fixed point**, because $T(O) = O$. If we start at some point other than O and **iterate**, we create an orbit which usually spirals into or out from the fixed point. The only exceptions are when either $|a| = 1$ and the orbits encircle O in concentric circular paths, or when a is real-valued, in which case the orbits head straight into or out from the origin with no spiralling. Whether they spiral or not, if the orbits head in towards O, (the case $|a| < 1$) we say that O is an **attracting fixed point** or **sink**, and if they head out away from O (when $|a| > 1$) we say that O is a **repelling fixed point** or **source**.

To gain more flexibility in constructing our pictures, we shall want to create maps which spiral into or out from fixed points other than O. To change the origin, we **conjugate** (not complex conjugate!), in the manner described in Chapter 1, by a translation $S : z \mapsto z+b$ which carries O to b. This means we need to study the conjugated map $STS^{-1}(z)$ which, as we easily calculate, is given by the formula

$$\hat{T}(z) = STS^{-1}(z) = ST(z - b) = S(a(z - b)) = a(z - b) + b.$$

It is easy to verify that, as expected, this maps fixes the point b. The dynamics of $\hat{T}$ is exactly the dynamics of T, translated by the amount b. Thus the orbit of any other point z under $\hat{T}$ spirals into or out from b, the direction, speed and twist of the spiral depending on a.

The map $\hat{T}$ has the form $z \mapsto Az + B$ where A and B are complex numbers. Such maps are sometimes called **affine**. In Project 2.4 we explore the effects of conjugating and composing affine maps in more detail. It turns out that, in the language of Chapter 1, the set of affine maps with $A \neq 0$ forms a **group**.

Complex exponentials

There is another perspective on the polar representation of complex numbers which is at the same time very beautiful and very useful. The fact that multiplying two quantities corresponds to adding their arguments may remind you of another very important bit of mathematics. Before the days of calculators, in order to multiply two large numbers one would take their logarithms and add. This is the principle of the slide rule: the method is based on the power rule

$$a^x a^y = a^{x+y}.$$

In other words, if you want to multiply powers you should add the exponents, as in $3^4 \times 3^2 = 3^6$. One of the most far reaching properties of complex numbers is that the rule relating multiplying numbers to adding arguments is in fact just a complex version of the same equation! The relation between complex numbers and exponentiation is expressed in an wonderful formula of Euler[1] which has good claim to be called one of the most important formulas in mathematics:

$$e^{i\theta} = \cos\theta + i\sin\theta.$$

[1] Leonhard Euler (1707–1783) was a master of calculations using complex numbers, which he brought to new heights. He did not, however, get as far as representing complex numbers geometrically in the plane.

Here $e = 2.71828 \cdots$ is the base of 'natural' or Naperian logarithms[1] and θ is measured in radians.

Whether this formula is considered to be a *definition* of exponentiation for purely imaginary values, or is a *theorem* based on requiring exponentiation to have certain basic properties, depends on one's point of view. The point is that it makes the basic power rule $a^x a^y = a^{x+y}$ consistent with the complex multiplication rule $\arg(zw) = \arg(z) + \arg(w)$, because according to our new formula, the equation

$$(\cos\theta + i\sin\theta)(\cos\phi + i\sin\phi) = \cos(\theta + \phi) + i\sin(\theta + \phi)$$

can be rewritten

$$e^{i\theta} \times e^{i\phi} = e^{i(\theta+\phi)}.$$

Let us try substituting some special values. For example, at 360° or 2π radians we get $e^{2i\pi} = \cos 2\pi + i\sin 2\pi = 1$, at 180° we get $e^{i\pi} = \cos\pi + i\sin\pi = -1$ and finally at 90°, the formula reads $e^{i\pi/2} = \cos\pi/2 + i\sin\pi/2 = i$. The formula $e^{i\pi} = -1$ has been described as one of the most beautiful in elementary mathematics, involving as it does all the fundamental quantities e, π, i, -1 in such a simple yet profound way.

[1] The number e can be defined as how much money you have after one year if you deposit one dollar in a savings account giving you 100% per annum interest *continuously compounded*. If they credited you with interest every month, you would have $2.613 \cdots = (1 + \frac{1}{12})^{12}$ dollars after 12 months; if they credited you with interest every day, you would have $2.714 \cdots = (1 + \frac{1}{365})^{365}$ dollars after 365 days; if they did so every hour, you get $2.71812 \cdots$ dollars. And so on. If the interest is 'continuous', you get e dollars!

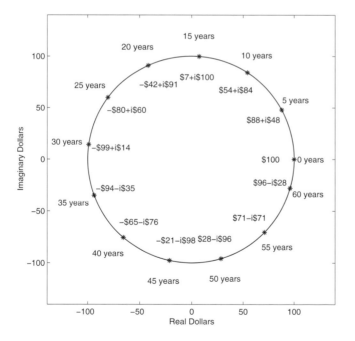

Figure 2.5. The history of Joe Bloggs' investment in the bank of imaginary interest. Each year, the point measuring his bank balance moves round by 0.1 radians or $18/\pi \doteq 5.73$ degrees. Because this number is irrational, no matter how long you go on, the starred point will never land back on the initial point marked 0 years.

There is an amusing and intuitive way to explain this family of esoteric looking formulas which goes like this. Suppose an imaginative and enterprising banker decides to offer an exciting new type of savings account – one which pays imaginary interest, at the rate of $10i\%$ each year. The public, fascinated by our book *Indra's Pearls*, wants to participate in this new financial offering. Joe Bloggs deposits 100 real dollars in such an account. After one year, he has earned 10 imaginary dollars in interest and his balance stands at $100 + 10i$ dollars. The next year he gets 10 more imaginary dollars and is thrilled: but to his chagrin, the imaginary balance of 10 imaginary dollars also earns interest of $0.1 i \times 10 i$ dollars, or -1 real dollars. So his balance after two years stands at $99 + 20i$ dollars. As the years go by, he keeps building up his pile of imaginary dollars but, as this gets bigger, he also sees the interest on this imaginary balance whittle away his real dollars at an ever increasing rate. In fact, if the bank used continuous compounding of interest rather than adding the interest once a year, then after 5 years Joe would have $88 + 48i$ dollars, having lost 12 real dollars in return for his 48 imaginary ones. Joe doesn't quite know what these imaginary dollars are good for, but maybe they aren't a bad deal in return for the 12 real ones he lost! Time passes and, after 10 years, he has $54 + 84i$ dollars and now his real money is bleeding away fast because of the interest on his imaginary balance. In fact, at 15 years, his balance is $7 + 99.5 i$ dollars and finally at 15 years, 8 months and 15 days he checks his balance, only to find he no real money at all, but 100 imaginary dollars. This length of time is in fact $10\pi/2$ years and what we have done is track his balance by Euler's formula. Explicitly, since continuous compounding is the same as using exponentiation, we have:

$$\text{(balance after } t \text{ years)} = \text{(initial deposit)} \times e^{\text{(interest rate)} \times t}$$
$$= 100 \times e^{0.1it}$$
$$= 100 \cos(0.1)t + 100i \sin(0.1)t.$$

Let's go on. More years elapse and now the interest on Joe's imaginary dollars puts him in real debt. *And* the interest on the real debt begins to take away his imaginary dollars. At 20 years, his balance stands at $-41 + 91i$ dollars, at 25 years $-80 + 60i$ dollars and at 30 years $-99 + 14i$ dollars. Finally at 10π years, which works out to be 31 years, 5 months, he finds himself 100 dollars in debt with no imaginary money. Not willing to give up, and finding the banker willing to extend him credit with only imaginary interest to pay, he perseveres and after about 47 years, finds that he has only imaginary debt now, and no real money either positive or negative. And now the interest on negative amounts of imaginary money is positive real money (because $(0.1)i \times (-100i) = +10$). So he finally begins to win back his real money.

On his deathbed, after 20π years, that is 62 years and 10 months, he has back his original deposit and has paid off his imaginary debt. He promptly withdraws this sum, sues his banker and vows never to have any truck with complex numbers again. His odyssey is traced in Figure 2.5. You may like to explore the difference between continuously and annually compounded interest further in Project 2.5.

Inversions and the Riemann sphere

Now that we have such a nice geometrical picture of the transformations of the plane formed by complex multiplication, perhaps we can do something similar with division. To be specific, let's try to understand the transformation T which takes a complex number z to its inverse $1/z$. If $z = r(\cos\theta + i\sin\theta)$, remember that the inverse in polar coordinates is:

$$\frac{1}{z} = \frac{1}{r}(\cos(-\theta) + i\sin(-\theta)).$$

Thus replacing z by $1/z$ results from two operations: replacing r by $1/r$ and θ by $-\theta$. We have seen that the second operation, replacing θ by $-\theta$, is complex conjugation or reflection in the real axis. The map of the plane which implements the first operation takes a point with polar coordinates (r, θ) to the point $(1/r, \theta)$. In terms of complex numbers, this is expressed by the formula $z \mapsto 1/\bar{z}$. This map is called **inversion in the unit circle**. The effect of this map is shown in Figure 2.6. It doesn't move points actually on the unit circle, but points inside the unit circle are moved to points outside and vice versa. The nearer a point is to the origin, the further away its image outside. It is natural to say that, under this map, the origin O itself is taken to 'infinity' and 'infinity' is taken back to O. The final result is that the map $T: z \mapsto 1/z$ is a composition: first invert in the unit circle and then reflect in the real axis. (Actually, in this particular case, the order you do the two operations doesn't matter. You could just as well compose them the other way round.)

Unlike the other maps we have studied thus far, you can see from Figure 2.6 that inverting distorts shapes in a fairly serious way. In Figure 2.3 the different foxes were different sizes but they all had the same shape. In our new picture, each fox is a significantly different shape from his twin on the other side of the unit circle. If you look closely, however, you will see at least one thing which does not change: the angles at the points of his ears. As we shall see shortly, this is a foretaste of a very important feature of all the maps we shall be studying from now on: although proportions in shapes may be hugely distorted, *angles remain the same*.

Playing with inversion suggests that it might perhaps be possible to introduce 'infinity' as a *bona fide* new 'complex number' which we could work with on the same basis as the other points in the plane. To get a geometric picture of how infinity fits in, stand a sphere on the usual infinite flat plane with its base the South Pole placed at the origin O as shown in Figure 2.7. Now imagine wrapping the plane up round the sphere, shrinking the distant areas more and more so they all fit, like wrapping pastry around an apple to make an apple dumpling. You will need to cap off your pastry shell, and we can do this by putting a single new point called '∞' or '**the point at infinity**' at the North Pole.

It turns out that we can extend almost all the ordinary rules of complex arithmetic quite consistently to include this new point. For example, for any complex number a, $a + \infty = \infty$ and $a \times \infty = \infty$. There are a few calculations which we do better to avoid, for example trying to work out

---

Note 2.4: **Inverting in a circle**

Inverting in a circle is a geometrical operation which can be defined for any circle in the plane. This figure shows the effect of inverting a red square in a black circle. Suppose that the centre of the circle is O and its radius is R. The rule which defines inversion is the following. Join the point P which you wish to map to O by a line, and suppose the distance from O to P is r. Then map P to the point P' on the same ray through O at distance R^2/r from O. The inversions of the four corners A, B, C, and D of the square are marked A', B', C' and D', respectively. Inversion bends the sides of the square into circular arcs. It is a flip which maps the inside of C to the outside and vice versa, and is often described as 'reflection' in the 'mirror' C. As with ordinary reflection, points on the mirror itself are left fixed. The centre O is mapped to 'infinity' and 'infinity' is mapped to O. We can get a nice formula for the inverse of a point z in a circle whose centre is the complex number a by conjugating by the map $z \mapsto T(z) = Rz + a$ which maps the unit circle onto the new circle. You first apply T^{-1}, then apply $z \mapsto 1/\bar{z}$ to invert in the

unit circle, then finally apply T to get the inversion in our circle. Performing this three-step manoeuvre produces the formula for inversion:

$$z \mapsto a + \frac{R^2}{\bar{z} - \bar{a}}.$$

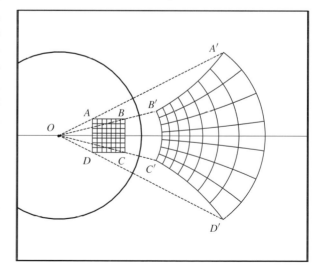

Figure 2.6. Our old friend the red fox and three of his variously coloured buddies are now inverted in the unit circle. The ear of our red friend – the fox *inside* the circle on the top right – nearly touches the circle, so stays nearly fixed. Note however how his proportions change: his image outside the circle is now cat-like. The silvery blue fox inside the circle on the bottom right is standing on the circle and his inversion looks like a reflection in a lake at his feet. On the left the true foxes are outside the unit circle and their inversions are inside. Note how the inversion of the grey fox on the bottom left is nearly all head because only his head is near the circle. On the other hand, the forepaws of the brown fox on the top left are the part closest to the circle so they become huge after inversion.

a value for $\infty - \infty$ is a dangerous thing. The complete list of rules is summarized in Box 6. This new picture of the complex numbers plus ∞ is called the **Riemann sphere**, after Bernhard Riemann (1826–1866) who introduced it in 1854. We have shifted from the flat earth perspective with a special point like Jerusalem at the centre, to a round earth in which all points are viewed on an equal basis. Riemann's picture is absolutely crucial to all the calculations, pictures and geometry we study from now on.

Note 2.5: **Insides and outsides**

Even a small child knows which is the inside and which is the outside of a circle on the plane. But which is the inside and which is the outside of a circle on the sphere? A circle on the sphere always divides the sphere into two parts. If the circle is small, one of these parts will be much smaller than the other and you might be pretty sure that when someone said 'inside' they meant the smaller part of the two. But what if the circle were the equator? Is the northern or the southern hemisphere the inside? Worse, when we start doing coordinate changes which allow us to treat ∞ like any other

point, then 'inside' and 'outside' may get reversed. For example, the effect of the map $J : z \mapsto 1/z$, as seen on the sphere, is to interchange the northern and southern hemispheres. Seen on the extended plane, J interchanges the inside of the circle of radius 1 centre 0, with its outside. The 'exterior' region $|z| > 1$ has just as much right to be called a disk as the 'interior' region $|z| < 1$. The upshot is, that once we start working on the Riemann sphere and treating ∞ like any other point, we must remember to be very clear what we mean if we refer to the inside or outside of a circle.

Box 6: **The arithmetic of infinity**

We can extend most of the rules of arithmetic to include ∞! Letting a denote an ordinary finite number, then we decree:

$$a + \infty = \infty$$
$$a - \infty = \infty$$
$$a \times \infty = \infty \quad \text{if } a \neq 0$$
$$\infty/a = \infty$$
$$a/\infty = 0$$
$$a/0 = \infty \quad \text{if } a \neq 0.$$

The results of the calculations $\infty \pm \infty$, $0 \times \infty$, $0/0$ and ∞/∞ are undefined and these operations are not allowed.

In order to move freely between our two pictures we need to spell out in detail exactly how they correspond, telling which complex numbers $x + iy$ get put precisely where on a concrete model of the Riemann sphere. In other words, we need a transformation or map which gets us between the sphere and the plane. This is exactly the problem of making an atlas, where you need a rule telling you how to take points on the round earth and represent them by points on a flat piece of paper. As you will discover by looking at any atlas, there are many 'projections' which may be used. In one way or another, all methods suffer defects of distortion: as any cook knows, it is really impossible to wrap a flat piece of pastry round a round apple without shrinking, stretching or folding in some way. A method preferred by mathematicians using complex numbers is called **stereographic projection**, used by Ptolemy as far back as 150 AD.[1] Under stereographic projection, relative distances and areas get badly distorted but as we shall see shortly, circles remain circles and angles do not change.

Stereographic projection works like this. Imagine yourself as a fly sitting on the surface of the earth (idealized as a sphere!), which as before we place on the complex plane with the South Pole resting at O. As in Figure 2.7, take a *very* long straight stick, pointing behind you, through the interior of the globe, towards the North Pole, and stretching out in front of you to where it hits the plane below. The point where it hits will be your image position on the plane. As you can see, the South Pole is mapped to itself, and horizontal latitude circles on the sphere are mapped to circles concentric with O on the plane. The further down in the southern

[1] The origins of stereographic projection seem to be lost in the mists of time. According to the great scholar Otto Neugebauer's *History of Ancient Mathematical Astronomy* there is plenty of evidence 'that the method of stereographic projection is not Ptolemy's invention but antedates him by at least two centuries'.

hemisphere the circle, the nearer its image circle is to O. There is one exception to this rule: the North Pole itself cannot be mapped to the plane in this way, in fact if we tried the stick would point out tangent to the sphere at the North Pole and only land on the plane 'at infinity'. This fits very nicely, because as we have seen the North Pole represents the point ∞ on the Riemann sphere. Using the method of the fly with a stick, any pattern on the sphere can be transformed into one on the plane and vice versa, for example, the bust of Paolina you see on the right. The actual equations which implement stereographic projection are explained in Note 2.6.

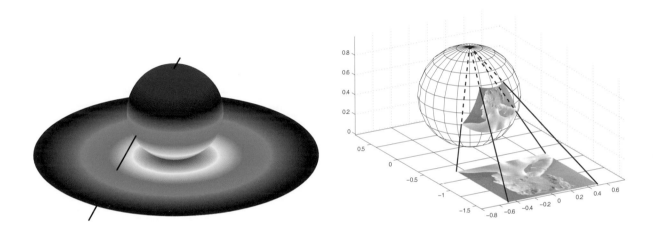

There is an elegant reason why stereographic projection between the complex plane and the sphere fits well with complex number geometry. One can see that the stereographic projection of a circle on the sphere must be an ellipse on the plane, because the set of all projection lines from the North Pole through points on the circle forms a cone. This cone is sliced by the complex plane in a conic section which is obviously an ellipse. It is not so obvious that the plane meets this cone in a circle. In other words, as was discovered by Thomas Harriot around 1590, *under stereographic projection circles on the globe map to circles on the complex plane*. Even better, *the angle in which two circles meet on the globe is exactly the same as the angle in which the image circles meet in the complex plane*. This will be tremendously important in everything we do from now on. We say that stereographic projection **preserves** angles and circles. You can find a nice proof in Hilbert and Cohn-Vossen's beautiful book *Geometry and the Imagination*.[1]

Figure 2.7. Two illustrations of stereographic projection. The colours in the first figure indicate how successively larger circles in the plane are wrapped higher and higher on the sphere. On the right, we see the classic bust of Paolina horizontally on the complex plane and 'lifted back' by stereographic projection onto the Riemann sphere.

[1] Chelsea, 1952.

Projects

2.1: Gaussian integers

[1] An integer is a whole number, positive or negative, like 3, −7, or even possibly 0.

Gaussian integers are those complex numbers for which the real and imaginary parts are integers, that is, numbers like $z = n + im$ where n and m are integers.[1] One of the most striking things about integers is that they factor uniquely into a product of prime numbers. A prime number is a positive integer which is not divisible by any smaller integers (other than 1). The primes form an infinite sequence which starts $2, 3, 5, 7, 11, 13, 17, \ldots$. Any positive integer is the product of a unique sequence of primes, thus $18 = 2 \cdot 3 \cdot 3$. Allowing multiplication by -1, we can also factor negative integers, for example $-84 = -2 \cdot 2 \cdot 3 \cdot 7$. The question is: what happens to prime factorization for Gaussian integers? We need to make some small modifications. Let's say that $n + im$ is 'positive' if $n > 0$ and $m \geq 0$, and say that it is prime if it is positive and not the product of two smaller Gaussian integers. (Here by definition, $p + iq$ is smaller than $n + im$ if $|p + iq| < |n + im|$.) As an example, try checking that $2 + i$ is prime. It is a theorem that with this new terminology, prime factorization still works! Any Gaussian integer is ± 1 or $\pm i$ times the product of a unique set of primes.

Note 2.6: **The equations for stereographic projection**

In three-dimensional coordinates, put the sphere of diameter 1 on the plane with the South Pole sitting at the origin $(0, 0, 0)$, the North Pole at $(0, 0, 1)$, and the complex number $z = x + iy$ at $(x, y, 0)$.

Start drawing a straight line from $(0, 0, 1)$ through a point (u, v, w) on the sphere and ending at the point $(x, y, 0)$ on the plane. Straight lines have the property that the rate of change of one coordinate with respect to another is constant. We can compute this rate of change for the first coordinate relative to the third coordinate in two ways:

$$\frac{u - 0}{w - 1} = \frac{x - 0}{0 - 1}.$$

The lefthand computation of this rate of change used the first and second points on our line, and the righthand used the first and third points. After the analogous computation for the second coordinate relative to the third, we have our formulas for stereographic projection from the sphere to the plane:

$$x = \frac{u}{1 - w} \quad \text{and} \quad y = \frac{v}{1 - w}.$$

To compute the inverse formulas for u, v, w in terms of x and y, we have to make use of the fact that (u, v, w) has distance $1/2$ from the centre of the sphere at $(0, 0, 1/2)$. The final formula turns out to be

$$(u, v, w) = \frac{(x, y, x^2 + y^2)}{x^2 + y^2 + 1}.$$

This formula may be used to prove circles on the sphere are mapped to circles in the complex plane in the following way. Circles on the sphere are the intersections of the sphere with planes given by an equation $au + bv + cw = d$. Substituting our formulas for u, v and w and clearing denominators results in the equation

$$ax + by = (d - c)(x^2 + y^2) + d$$

which is the equation of a circle in the complex plane.

This takes a bit of proving, but what is fascinating is to find the first dozen or so Gaussian primes and see how ordinary integers factor into Gaussian primes. It will turn out that many ordinary integers which are prime as integers are no longer prime as Gaussian integers. As a simple example, the prime number 5 now factors into the product of two primes $5 = (2+i)(2-i) = (-i)(2+i)(1+2i)$, and both $2+i$ and $1+2i$ qualify as Gaussian primes under our definition.

2.2: Computing complex square roots

We discussed the problem of finding the square root of a complex number $z = x+iy$ on p. 44. If you try following this through, you will find it involves having to compute $\theta = \arctan(y/x)$. To avoid this, we can proceed as follows. If $\sqrt{z} = u + iv$, we would like to find formulas for u and v which use only *real* square roots of positive real numbers. To do this, begin by expanding $z = x + iy = (u + iv)^2$ into the two real equations $x = u^2 - v^2, y = 2uv$. Now play around, eliminating v and getting a quadratic equation for u^2. You should get a unique solution with the extra restriction $u > 0$ (or $u = 0$ and $v \geq 0$). It should involve taking two square roots of positive real numbers.

Now write a program `complex cxsqrt(complex z)` which takes a complex number z and returns the above u by implementing these formulas.

2.3: The geometry of complex maps

We can write down maps of the complex plane to itself by using simple formulas like $T(z) = z^2$ and $T(z) = \sqrt{z}$. Visualizing how such a map behaves can be rather difficult because the input and output variables both have two real coordinates. The proper place to look at the 'graph' of such a map or function is in *four*-dimensional space, for which we still don't have terribly good glasses available. One way around this is to take a grid of horizontal and vertical lines in the z plane and then plot the images of these lines under the map T.

It is fun to carry this out for simple maps like those above. The patterns that appear are striking. Many computer mathematics systems have a built-in command for making this kind of plot; in Maple, for instance, it is called `conformal`. In the margin, we show a variation for the map $z \mapsto z + 1/z$ (which will play a crucial role in Chapter 3). The upper picture is the outside of the unit circle (shown in black) with colours corresponding to circles of increasing radius. The lower is the result of applying the map $z \mapsto z + 1/z$. The unit circle is squashed onto the horizontal line segment from -2 to 2, and the other circles are distorted into ellipses of varying eccentricity. The region inside the unit circle is also mapped onto the lower picture, because z and $1/z$ map to the same point.

2.4: Affine maps

Maps of the form $T : z \mapsto Az + B$ are sometimes called **affine**. In fact any affine map with $A \neq 0$ is conjugate (in the sense of Chapter 1) to the multiplication map $z \mapsto Az$, provided that $A \neq 1$. To see this, first show that T has exactly one fixed point Fix T by solving the equation $T(z) = z$. Now do a coordinate change using the translation $S(z) = z - \text{Fix } T$. The conjugated map we want is $\hat{T} = STS^{-1}$. Why does $\hat{T}$ have a fixed point at O? Check that $\hat{T}$ is the multiplication map $\hat{T} : z \mapsto Az$. Conclude that the action of T has the same dynamics as $\hat{T}$, but with the origin translated to $\frac{B}{(1-A)}$.

It is also interesting to try **composing** two affine maps. Does this create anything new? If $T : z \mapsto Az + B$ and $T' : z \mapsto A'z + B'$, try showing that TT' is another map of exactly the same kind. Then find the formula for the **inverse** map T^{-1} which undoes the effects of T and explain why the set of affine maps with $A \neq 0$ is a group.

2.5: Discrete and continuous spirals

The fox in the right frame of Figure 2.3 moved in discrete jumps. Can we modify this to make the fox move *smoothly* along a spiral path? Polar coordinates are the secret, because writing $a = |a|(\cos\theta + i\sin\theta)$ gives both the expansion factor $|a|$ and the angle of rotation θ. For example, we can halve each step by replacing $|a|$ by $\sqrt{|a|}$ and θ by $\theta/2$.

Choose any real number t and consider an expansion by the factor $|a|^t$ and a rotation by angle $t\theta$. Plot how the fox moves by these intermediate maps. You don't have to consider all possible values for t: taking just the fractions $t = k/10$ for k from -100 to 100 will give an interesting picture of 'continuous' motion.

Continuous spirals can be replaced by discrete ones too. In fact, after Joe Bloggs died, his son brought another suit against the bank, claiming that they never stated that the imaginary interest would be compounded continuously. He claimed the bank's advertisement implied that interest would be credited only once a year. This means that if his balance was z complex dollars at the beginning of a year, he would have $(1 + 0.1i)z$ complex dollars at the end. Why did his son sue over this apparently small detail? What would Joe have died with after 63 years had the interest been credited annually?

2.6: A tiling associated to inversion

Here is a nice way to visualize groups, tiling, inversions and stereographic projection all at the same time. This time, by using only 8 tiles, we shall be able cover the whole earth! Start on the complex plane with complex coordinate z and the transformations given by complex conjugation $z \mapsto \bar{z}$, inversion in the unit circle $z \mapsto 1/\bar{z}$ and reflection in the origin, $z \mapsto -z$. Composing these in all possible ways gives the list:

$$T_1(z) = z, \quad T_3(z) = -z, \quad T_5(z) = 1/z, \quad T_7(z) = -1/z$$
$$T_2(z) = \bar{z}, \quad T_4(z) = -\bar{z}, \quad T_6(z) = 1/\bar{z}, \quad T_8(z) = -1/\bar{z}.$$

Show that these 8 transformations form a group in which every transformation has order one or two.

Skip ahead to p. 88 and you will see a tessellation of the plane by 8 tiles exhibiting exactly this group of symmetries.[1] Each tile is moved to another by each of the 8 transformations – pick a tile and a T and see how this works. The tiles no longer all have the same shape, because we are using transformations which stretch and shrink in rather complicated ways. Transported onto the sphere using stereographic projection, however, they do all look the same.

[1] This tiling is closely related to a regular octahedron. You can make similar tilings related to the other 'Platonic solids': the tetrahedron, the cube, the dodecahedron and the icosahedron. Klein made a famous study of the icosahedron in which he showed how functions invariant under its symmetry group could be used to solve quintic, that is degree 5, equations.

CHAPTER 3

Double spirals and Möbius maps

"First accumulate a mass of Facts: and then *construct a Theory.* That, *I believe, is the true Scientific Method."*
I sat up, rubbed my eyes, and began to accumulate Facts.

Sylvie and Bruno, Lewis Carroll

We now come to a key ingredient of our fractal constructions: the maps we use to make them. As we have seen, one of Klein's fundamental ideas was that any group of transformations can be used to create symmetry. Classically, we think about symmetry in terms of the Euclidean motions of translation, rotation and reflection. But symmetry can also be created from maps which distort, stretch and twist. In this chapter, we shall learn about a beautiful class of maps called Möbius maps, which stretch and twist in just the right controlled way. These are the maps which generate all our fractal pictures, and under which, as in Klein's vision, all our pictures are symmetric.

So what exactly are Möbius maps? The fox picture Figure 2.3 illustrated how maps like $T(z) = az + b$ represent a spiralling expansion or contraction from a fixed point or source. On the other hand, Figure 2.6 showed how the map $T(z) = 1/z$ turns the unit circle inside out and distorts the fox's shape in quite startling ways.[1] Klein taught us that the logic of symmetry demands that when you have two maps, you should always try to compose them. When you try doing this for these various fox maps, you always get a map with equation

$$T(z) = \frac{az + b}{cz + d}.$$

[1] Figure 2.6 actually shows the effect of the map $z \mapsto 1/\bar{z}$. To see the effect of $T(z) = 1/z$ we have additionally to reflect the inverted foxes in the real axis, see the discussion on p. 54.

These are the Möbius maps, and they do for the Riemann sphere what the affine maps $T(z) = az + b$ do for the complex plane. This chapter is devoted to familiarizing ourselves with how they work.

From single to double spirals

We start our study of Möbius maps with a geometrical description of some special examples. Figure 2.3 showed a fox being moved about by powers of the map $T(z) = 0.8(\cos 30° + i \sin 30°)z$. In this section, we will work instead with $T(z) = (1 + 0.4i)z$. This expands by a factor $\sqrt{1^2 + 0.4^2}$, or about 1.077, and rotates through an angle whose tangent is 0.4 that is, about 22°. As we saw, this pushes points out along paths which look like the arms of a spiralling galaxy, moving ever further away from the fixed point at 0.

Instead of a succession of shots of the fox, we might draw a continuous red spiral through the points in one orbit, for example the points

$$\ldots, T^{-2}(1), T^{-1}(1), 1, T(1), T^2(1), \ldots.$$

Such a picture is shown in Figure 3.1, where you can also see the spiral 'lifted' by stereographic projection back to the Riemann sphere. Notice how the green and yellow lines in the polar coordinate grid in the plane are transferred to the green and yellow latitude and longitude lines on the sphere. Imagine T sliding points outwards from the origin along the red spiral arms.

Figure 3.1. A spiral map lifted to the sphere. The spiralling action of the transformation $T(z) = (1 + 0.4i)z$ is shown simultaneously on the complex plane and the Riemann sphere. The green and yellow lines form the coordinate grids: latitude and longitude on the sphere and polar coordinates in the plane. Note the appearance of a second fixed point at ∞ when the red spiral orbit is stereographically 'lifted' back onto the sphere. To generate this spiral, see the discussion in Project 3.1.

On the sphere, the eye immediately picks out an important feature that is not so apparent on the flat view, namely that T not only has a source at the South Pole where the red spiral starts, but also a sink at the North Pole where the spiral ends. (Think of the spiral flow of water draining out of a plug hole.) The North Pole corresponds to the added point ∞ on the plane. In just the same way that 0 is a source, because going backwards along the orbit the points $z, T^{-1}(z), T^{-2}(z), \ldots$ come ever closer to 0, so we are justified in thinking of ∞ as a sink for T. Starting from any complex number z (with the single exception of the source 0), the points $z, T(z), T^2(z), \ldots$ move ever outwards towards ∞. Everything is consistent with the rules for arithmetic with infinity on p. 57: ∞ is a fixed point because $T(\infty) = (1 + 0.4i)\infty = \infty$.

Our red flow paths are exactly what cartographers call **loxodromes**, composed from Greek words for 'running obliquely'. The name comes from navigation: a loxodrome on a sphere is a path which cuts every line of longitude at the same angle. They are the routes a ship would travel if it maintained a constant bearing. Figure 3.2 shows a ribbon of loxodromes swirling around the sphere (with apologies to the mathematical artist M.C. Escher who made a very similar woodcut).

Figure 3.2. Sphere spirals. The coloured path is a ribbon of loxodromic spirals on the sphere (Actually, this statement is slightly inaccurate. Can you see why?)

As we saw in the last chapter, introducing the operation $z \mapsto 1/z$ allows one to treat the point ∞ on the same footing as any other point in the complex plane. This means that we ought to be able to create new double spirals which spiral out from any point and spiral in to another. Let's try to replace the map T, which has a source at 0 and a sink at ∞, by a map $\hat{T}$ which has a source at -1 and a sink at 1. The dynamics of $\hat{T}$ will serve as a prototype for all the maps we iterate from now on. Rather than just guessing, we can find $\hat{T}$ by the method of conjugation from Chapter 1. Besides helping us with the present problem, knowing how to apply this method in our new context will add a powerful tool to our bag of computational tricks.

The first step is to find a change of coordinates which moves the fixed points 0 and ∞ into their new locations at -1 and 1. Following the rules for arithmetic in the extended complex plane, you can check that the transformation

$$R : z \mapsto \frac{z - 1}{z + 1}$$

moves 0 to -1, and ∞ to 1. The transformation R will play the same role as the conjugating map which we called S on p. 25.

The conjugated map $\hat{T} = RTR^{-1}$ should be a map with all the same features as T but moved to their new positions by R. In particular, it should move points in spirals emanating from the new source $R(0) = -1$ and disappearing into the sink $R(\infty) = +1$. You can actually check that the source -1 is a fixed point of $\hat{T}$ by the calculation $\hat{T}(-1) = RTR^{-1}(-1) = RT(0) = R(0) = -1$, and similarly for the sink $+1$.

To play with the new map $\hat{T}$ any further, we need to work out a formula. The slightly more general calculation done in Note 3.1 shows that

$$\hat{T}(z) = \frac{(1 - 5i)z + 1}{z + 1 - 5i}.$$

By plugging in numbers, you can use this formula to work out where any point is carried by various powers of $\hat{T}$.[1]

It is hard to make much sense of a list of numbers like this. Much better to draw a picture; better yet, let's take a walk! Figure 3.3 is a time-elapsed photo of our good friend Dr. Stickler taking a stroll under the influence of $\hat{T}$. He has been following the path outlined by the red and green spirals for a very long, indeed infinite time. Starting at the source at -1, the centre of the left hand spiral, he is doomed to continue walking forever, getting ever nearer $+1$, the right hand sink. While Dr. Stickler

[1] For instance, the orbit of 0 is:
$\ldots, -.156 - .631i,$
$-.086 - .397i,$
$-.038 - .192i, 0,$
$.038 + .192i, .086 + .397i,$
$.156 + .631i, \ldots.$

Note 3.1: **Conjugating the standard spiral** $T(z) = kz$.

In our shorthand for composing maps, RTR^{-1} means the map $RTR^{-1}(z) = R(T(R^{-1}(z)))$. We need a formula for R^{-1}. If $w = R(z)$, then $z = R^{-1}(w)$. Thus inverting comes down to solving $w = \frac{z-1}{z+1}$ for z in terms of w. A little bit of algebra gives the answer: $R^{-1}(z) = \frac{1+z}{1-z}$. As a check we can just substitute:

$$R^{-1}R(z) = R^{-1}\left(\frac{z-1}{z+1}\right) = \frac{1 + \frac{z-1}{z+1}}{1 - \frac{z-1}{z+1}} = z.$$

Now we can calculate:

$$RTR^{-1}(z) = RT\left(\frac{z+1}{-z+1}\right) = R\left(\frac{kz+k}{-z+1}\right)$$
$$= \frac{\frac{kz+k}{-z+1} - 1}{\frac{kz+k}{-z+1} + 1}$$
$$= \frac{z(1+k) + k - 1}{z(k-1) + 1 + k}.$$

In the special case $k = 1 + 0.4i$, this gives (using $(2+0.4i)/0.4i = 1 - 5i$):

$$RTR^{-1} = \frac{(2 + 0.4i)z + 0.4i}{0.4iz + 2 + 0.4i} = \frac{(1 - 5i)z + 1}{z + 1 - 5i}.$$

may not notice any difference, as outside observers we see that he grows and shrinks like Alice in Wonderland. Even worse, his proportions change radically: as he moves round the source on the left, his head blows up alarmingly, while as he rounds the sink on the right, his legs extend while his head shrinks to pin-head size. Our new map is more complicated than most of the rather simple transformations we have studied thus far: it is **non-linear**, which means that it distorts shapes in complicated ways. Similar non-linear effects will be a feature of most of the symmetries we shall be studying from now on.

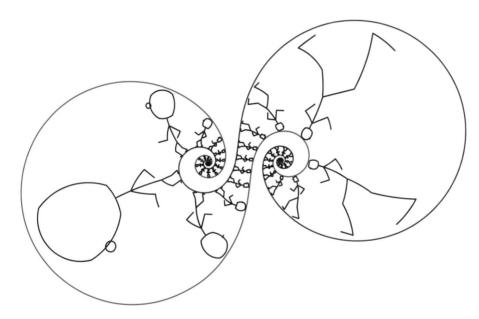

Figure 3.3. Dr. Stickler takes a spiral walk. The iterations of the doubly spiralling transformation $z \mapsto \hat{T}(z)$ applied to Dr. Stickler.

Once again, we can gain more understanding of this double spiral motion by viewing it on the Riemann sphere. In Figure 3.4, we used stereographic projection (strictly speaking, the inverse of stereographic projection) to lift the action of $\hat{T}$ from the plane to the sphere. The colour coding is as before, but the sphere is oriented in a new way, with the yellow longitude lines meeting at the ends of a horizontal axis running through the sphere from left-front to right-back.

The rotated green and yellow latitude and longitude circles in Figure 3.4 project to new green and yellow circles in the plane. (Remember that stereographic projection maps circles to circles, so any circle on the sphere projects into a circle on the plane.) The family of yellow circles are just the circles in the plane through the fixed points -1 and 1. The green circles all encircle one or other of these two points.[1] If you look carefully you will see that -1 and 1 are *not* the centres of the green

[1] The projection of the vertical green circle is an intermediate case. It projects to the y-axis and divides the family of green circles into two halves.

Figure 3.4. A double spiral on the plane and the sphere. The double spiral action of $\hat{T}$ on the complex plane and lifted to the Riemann sphere is shown in red, and the green latitude and yellow longitude lines are oriented in a new way adapted to the fixed points of $\hat{T}$.

circles. The two families of circles cut at right angles, because the same is true of the longitude and latitude circles on the sphere.

Comparison of Figures 3.1 and 3.4 reveals a striking fact: viewed on the Riemann sphere, the two spiral motions are identical up to a 90° rotation. In the language of conjugation, this rotation conjugates the 'lift' of the transformation T which produced the spirals on the sphere in Figure 3.1 into the 'lift' of the transformation $\hat{T}$ which produced the double spirals on the sphere Figure 3.4. Visually, this is obvious from the two figures and it can also be verified using algebra.

In summary, we have produced an interesting new type of map $\hat{T}$ by stereographically projecting the double spiral motion generated by a transformation of the type $T(z) = kz$ back to the sphere, rotating the sphere, and then projecting back to the plane. This geometrical description enabled us to understand $\hat{T}$ much more easily than if we had just plugged numbers into an algebraic formula. You can find some further exploration of maps like these in Project 3.1. In the next section, we shall look at what happens if we replace T by a translation $z \mapsto z + a$.

From translations to scallop shells

In Chapter 2, the simple operation of complex addition led us to study translations of the plane. What does this operation look like if we project back onto the Riemann sphere? In Figure 3.5 you see a family of blue parallel lines in the direction of the translation $T : z \mapsto z - 1.5$, together with their lifts by stereographic projection to the sphere. Each blue line lifts to a circle which goes through the North Pole. Since the blue lines are all parallel in the plane, on the sphere they make a scallop shell family of circles all tangent at their common meeting point N.

There is another way of visualizing the translation T. We have coloured

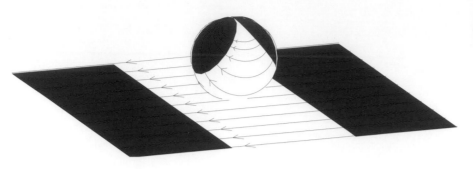

Figure 3.5. A translation lifted to the sphere. The blue lines are the trajectories of translations in the negative *x*-direction in the plane and the blue circles are the same trajectories on the sphere after lifting by stereographic projection. Translation carries the right red region to the white strip, which in turn maps to the left red region.

the two half planes bounded by $x = -0.75$ and $x = 0.75$ red, leaving a strip of width equal to the translation length 1.5 in between. As we translate, the lefthand red part and the white part, which you can think of as together forming the 'outside' of the red half plane on the right, move forwards in the direction of the arrows and cover the lefthand red half plane exactly. If we run the translation backwards, the 'outside' of the lefthand red half plane translates exactly onto the red part on the right.

What does this look like on the sphere? Stereographic projection carries the red half planes to disks tangent at the North Pole, the white strip between them becoming a crescent shape sometimes called a 'lune'. Seen on the sphere, translation moves everything outside the righthand red disk forward in the direction of the arrows and shrinks it in size so that its image is exactly the red disk on the left, while going back in time the inverse translation takes the outside of the left red disk to the inside of the one on the right. Further forward and backward images of these half planes would make a second scallop shell pattern of circles on the sphere, all tangent at the North Pole and all at right angles to the first family of blue circles.

We get yet another view of translation by conjugating by the map $J(z) = 1/z$. This is shown in Figure 3.6. Since J interchanges 0 with ∞, it maps the blue lines, which we can think of as a family of circles tangent at ∞, to a family of circles all tangent at the origin. Another way to think of this is that we rotate the sphere interchanging the North and South Poles, and then stereographically project back onto the plane. Once again, we get two families of tangent circles meeting at the same point, each forming a scallop shell pattern, with each circle in one family at right angles to each circle in the other. The conjugated map $\hat{T} = JTJ^{-1}$, whose formula we can work out as $\hat{T}(z) = \frac{z}{1.5z+1}$, has trajectories which move along the blue circles. The red disks map into the tangent disks shown in the figure. Thus the effect of $\hat{T}$ is to map the whole region outside one red disk so

Figure 3.6. A conjugate of a translation. This illustrates the conjugation of a translation by the map $z \mapsto 1/z$. The origin is fixed and the blue circles are trajectories of the map. The colours show the successive powers of the map: the small bottom green region is expanded to the yellow region, then to the red, then to the whole outer white region (including ∞). This is then contracted first to the top red region, then to the top yellow region then to the green. This figure also results from stereographic projection of the map in Figure 3.5 after turning that sphere upside down.

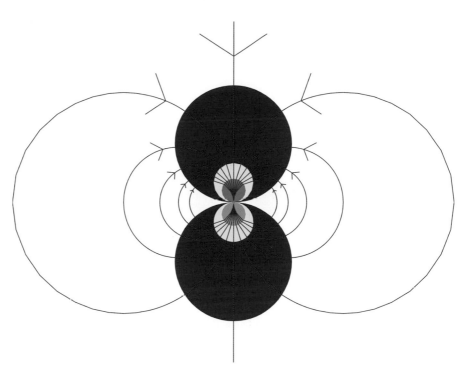

that it exactly covers the other. Running the map backwards reverses this process, mapping the outside of the second red disk to the inside of the first.

Iteration of the original map T on the plane pushes the left hand red half plane forward inside itself. Thus repeating the map $\hat{T}$ pushes the red disk forward to another tangent disk, shown in yellow, which is nested inside. Successive iterations of $\hat{T}$ push the disk further and further inside itself, which means that under forward iteration all orbits eventually get nearer and nearer to 0. Iterations of $\hat{T}^{-1}$ push the disks backwards so they again nest down to 0, this time from the opposite side.

The algebra of linear fractional transformations

The reader who has been checking the calculations in the last sections may rapidly begin to wish for a proper machine. Such a machine becomes an essential tool when we start making computer pictures which involve iterating many different maps many times. In the remainder of this chapter we shall develop a systematic theory of all maps of the form

$$T : z \mapsto \frac{az + b}{cz + d}$$

where the coefficients a, b, c, d are arbitrary complex numbers subject to the one condition that $ad - bc \neq 0$. These are called either **linear fractional transformations** (because they are fractions of linear expressions) or **Möbius transformations**, after the nineteenth century German mathematician August Möbius, who first studied such transformations systematically, showing that they are the most general transformations of the extended complex plane which map circles to circles. (As usual, lines are included here as a special kind of circle.) All the special examples studied thus far are of this type.

There is a mechanical procedure for composing linear fractional transformations using the algebra of 2×2 matrices. Understanding this fully involves a hefty dose of algebra, and it's possible to understand the rest of the book qualitatively without working it through. However for the reader who actually wants to implement our programs or follow the technical derivations later, this section and the next more geometric one will be among the most important reference points of the book.

First of all, let's see that T makes sense for any z which is either a complex number or ∞. We can work out the value for $z = \infty$ as follows:

$$T(\infty) = \frac{a\infty + b}{c\infty + d} = \frac{a + b/\infty}{c + d/\infty} = \frac{a + 0}{c + 0} = \frac{a}{c}.$$

The other thing to worry about is what happens when the denominator is 0, that is when $z = -d/c$:

$$T(-d/c) = \frac{-ad/c + b}{-cd/c + d} = \frac{-ad + bc}{-cd + dc} = \frac{-ad + bc}{0} = \infty.$$

This works because the assumption that $ad - bc \neq 0$ rules out any difficulties with the illegal value $0/0$. We shall see shortly that exactly the same assumption ensures that T has an inverse. The quantity $ad - bc$ is called the **determinant** of T.

How do linear fractional maps compose? Remembering our convention that TT' means first do T' and then do T, we just calculate out: if

$$T(z) = \frac{az + b}{cz + d} \quad \text{and} \quad T'(z) = \frac{a'z + b'}{c'z + d'}$$

August Ferdinand Möbius, 1790–1868

August Möbius conducted the Pleissenburg observatory in Leipzig from 1816 until his death in 1868. Late in his career, he became simultaneously professor of mathematics in the University. In his student days in 1813-4 in Göttingen, he attended some of Gauss's lectures on astronomy.

Möbius developed the theory of his transformations in a paper *Die Theorie der Kreisverwandschaft in rein geometrischer Darstellung* (The theory of circle relationships in a purely geometrical setting), published in 1855. In this work he placed great emphasis to what he called a *'relationship' (Verwandschaft)* between objects. Extending the idea of Euclidean congruence and similarity between figures, he considered two figures to be 'related' if you could get from one to the other by the kind of transformation made up of successive translations, magnifications and inversions that we are considering here. This allowed him to develop all kinds of properties of figures invariant under such transformations, foreshadowing Klein's *Erlanger Programm* discussed in Chapter 1. In fact Möbius' ideas about 'relationships' were precursors of the later concept of 'group'.

Möbius is perhaps more famous for his work on projective geometry. His ideas about 'relationships' had been introduced much earlier in his book *Der barycentrische Calcül* published in Leipzig in 1827, in which he used what are now known as *barycentric coordinates* to represent not only ordinary points of the plane but also ones at infinite distance, enabling him to state many beautiful properties of maps which preserve straightness and parallelism. He also did much to establish the basic theory of topology or rubber sheet geometry, in which 'relationship' is extended to have the even wider meaning of shapes which are obtained one from another by any kind of continuous deformation, stretching or shrinking which does not actually involve making and mending rips or tears. Möbius classified topologically all possible two-dimensional surfaces. You will likely have come across the strange surface called a Möbius band. He also worked in number theory, giving his name to the Möbius function and Möbius inversion formula which still play an important role.

then

$$T(T'(z)) = \frac{a \dfrac{a'z + b'}{c'z + d'} + b}{c \dfrac{a'z + b'}{c'z + d'} + d}$$

$$= \frac{(aa' + bc')z + (ab' + bd')}{(ca' + dc')z + (cb' + dd')}.$$

This shows that the composition TT' is again a linear fractional transformation.

The rule for composition looks rather messy. However it is really quite simple if we introduce the basic algebraic idea of representing T with the **matrix**

$$\begin{pmatrix} a & b \\ c & d \end{pmatrix}$$

formed from the coefficients of T. 'Matrix' is just a fancy term for a rectangular array of numbers, in this case a 2×2 square.

The algebra of 2×2 matrices is summarized in Box 7. The rule for multiplication is shown using arrows: you choose any row of the first matrix and any column of the second and, moving across, take products and add. The coefficients in our formula for the composition of two linear fractional transformations are exactly the same as those given in Box 7 for the matrix product! *Composing two linear fractional transformations is the same as multiplying the corresponding 2×2 matrices.*

Notice that if we multiply all the coefficients a, b, c, d in a linear fractional transformation by the same complex number t, the map T doesn't change. This useful fact just uses the identity

$$\frac{taz + tb}{tcz + td} = \frac{az + b}{cz + d}.$$

Let's try inverting the map $T : z \mapsto (az + b)/(cz + d)$. To do this, we set $T(z) = w$ and solve the equation $w = (az + b)/(cz + d)$ for z in terms of w. Multiplying up and rearranging we find that $z(cw - a) = -dw + b$ so that $T^{-1}(w) = (dw - b)/(-cw + a)$. So T^{-1} has the matrix $\begin{pmatrix} d & -b \\ -c & a \end{pmatrix}$. This is not quite the inverse *matrix* because

$$\begin{pmatrix} a & b \\ c & d \end{pmatrix} \begin{pmatrix} d & -b \\ -c & a \end{pmatrix} = (ad - bc) \begin{pmatrix} 1 & 0 \\ 0 & 1 \end{pmatrix}.$$

The factor $D = ad - bc$ multiplies all entries inside the matrix on the right. The factor is the determinant of the matrix for T. The matrix inverse of $\begin{pmatrix} a & b \\ c & d \end{pmatrix}$ is $\begin{pmatrix} d/D & -b/D \\ -c/D & a/D \end{pmatrix}$. Because multiplying the matrix coefficients by a complex number doesn't change the transformation, the matrix inverse gives the same *map T^{-1}*.

In Chapter 1 we introduced the idea of a group of transformations of the plane. Remember that a collection of transformations is called a group provided that:

(1) if S and T are in the collection then so is ST;
(2) if S is in the collection, then so is S^{-1}.

Box 7: **Matrix algebra and Möbius maps**

Matrices are usually introduced as maps which act on vectors by the formula

$$\begin{pmatrix} a & b \\ c & d \end{pmatrix} \begin{pmatrix} x \\ y \end{pmatrix} = \begin{pmatrix} ax + by \\ cx + dy \end{pmatrix}.$$

Such maps compose using matrix multiplication:

$$\begin{pmatrix} a & b \\ \xrightarrow{} \\ c & d \end{pmatrix} \begin{pmatrix} a' & \big| & b' \\ c' & \big\downarrow & d' \end{pmatrix} = \begin{pmatrix} aa' + bc' & ab' + bd' \\ ca' + dc' & cb' + dd' \end{pmatrix}.$$

The arrows have been put in only to indicate how you compute the product: each number in the matrix product is a 'dot product' of a *row* of $\begin{pmatrix} a & b \\ c & d \end{pmatrix}$ with a *column* of $\begin{pmatrix} a' & b' \\ c' & d' \end{pmatrix}$. The matrix $\begin{pmatrix} 1 & 0 \\ 0 & 1 \end{pmatrix}$ is known as the **identity matrix** and is often written *I*.

The determinant of the 2×2 matrix $M = \begin{pmatrix} a & b \\ c & d \end{pmatrix}$ is $D = ad - bc$. A matrix M has an **inverse** M^{-1} such that $MM^{-1} = M^{-1}M = I$ provided that $D \neq 0$, given by the formula

$$M^{-1} = \begin{pmatrix} d/D & -b/D \\ -c/D & a/D \end{pmatrix}.$$

A Möbius map is formed from the matrix M by setting

$$T(z) = \frac{az + b}{cz + d}.$$

One can compose and invert Möbius maps using the algebra of 2×2 matrices. Multiplying all coefficients of M by the same complex number does not affect the map. By dividing through by $\sqrt{D}$, one can always arrange that M has determinant 1.

We can replace 'the plane' in this definition by 'the extended complex plane' or Riemann sphere. So we have just shown that both of these two statements are true for the collection of all linear fractional transformations. In other words, we have just verified that *the set of linear fractional transformations forms a group*.

Here is a trick which very often makes inverting and other calculations easier. Because $\begin{pmatrix} d & -b \\ -c & d \end{pmatrix}$ and $\begin{pmatrix} td & -tb \\ -tc & td \end{pmatrix}$ define the same Möbius transformation, it is often handy to choose the number t so as to make the determinant come out as 1. Start with a matrix $\begin{pmatrix} a & b \\ c & d \end{pmatrix}$ with determinant $D = ad - bc$. Change the coefficients by dividing by $\sqrt{D}$, replacing a by $a/\sqrt{D}$ and so on. We get the same map and the new determinant $(ad - bc)/D$ simplifies to 1. Thus if $T = \begin{pmatrix} a & b \\ c & d \end{pmatrix}$ and $ad - bc = 1$, then $T^{-1} = \begin{pmatrix} d & -b \\ -c & a \end{pmatrix}$. From the formulas in Box 7, one can calculate that the product of matrices with determinant 1 also has determinant 1.[1] Very often, we shall just assume the matrix coefficients have been chosen at the outset with $ad - bc = 1$.

[1] This is done by checking that the determinant of the product of two matrices is the product of their determinants.

What we have just said shows that the subset of 2×2 complex matrices with determinant 1 itself forms a group, usually denoted in mathematical texts by the rather mysterious notation $SL(2, \mathbb{C})$. Decoding, L stands for linear, the 2 indicates their size and $\mathbb{C}$ shows that the entries are complex numbers. The 'S' stands for 'special' and indicates that $ad - bc = 1$.

Some linear fractional transformations are already familiar friends. First of all, taking $a = d = 1$ and $b = c = 0$, we get the map $T : z \mapsto (1 \cdot z + 0)/(0 \cdot z + 1)$, which is a complicated way of writing the identity map $z \mapsto z$. Secondly, it includes the affine maps $z \mapsto az + b$ coming from the arithmetic of complex numbers, which we studied in Chapter 2. These can be expressed in the form $T : z \mapsto (az + b)/(0 \cdot z + 1)$, given by a 2×2 matrix with $c = 0$. Since $T(\infty) = a/c$, maps with $c = 0$ are exactly those such that $T(\infty) = \infty$. In other words, our old friends $z \mapsto az + b$ are those Möbius maps for which ∞ is a fixed point. Important subclasses are the **pure translations** $z \mapsto z + b$ and the **pure scalings** $z \mapsto az$ (for which we shall often replace the letter a by the letter k). The coefficient matrix for translation $z \mapsto z + b$ is just $\begin{pmatrix} 1 & b \\ 0 & 1 \end{pmatrix}$, which is already normalized with determinant 1. To get determinant 1 for the scaling map $z \mapsto kz$, we have to write the matrix in the rather surprising form $\begin{pmatrix} \sqrt{k} & 0 \\ 0 & 1/\sqrt{k} \end{pmatrix}$.

Linear fractional transformations also include the map $J : z \mapsto 1/z$ which we have already found useful for interchanging the North and South Poles. The obvious way to write the coefficient matrix is $\begin{pmatrix} 0 & 1 \\ 1 & 0 \end{pmatrix}$, but watch out! This matrix has determinant -1, so a correct normalized form would be either $\begin{pmatrix} 0 & i \\ i & 0 \end{pmatrix}$ or $\begin{pmatrix} 0 & -i \\ -i & 0 \end{pmatrix}$. Details like this will reappear in our algebra time and again.

How many Möbius maps are there? This is a pretty vague question but one answer is this: there are exactly enough of them to allow you to move any three points to any other three. This means that if P, Q, R is a triple of distinct points in the extended complex plane and if P', Q', R' is another, then there is one and only one Möbius map T such that $T(P) = P'$, $T(Q) = Q'$ and $T(R) = R'$. We run through how to verify this in Project 3.2.

All the programs we shall be using from now on have variables which are Möbius maps. They will use two basic operations: composing two

---

Box 8: **mobius_on_point**

This routine inputs a matrix $\begin{pmatrix} a & b \\ c & d \end{pmatrix}$ and a complex number z and outputs the image point $T(z) = \frac{az+b}{cz+d}$.

The first version assumes the programming language accommodates arithmetic of complex numbers in 'infix' notation.

```
struct matrix {complex a,b,c,d};
```
Define a 'matrix' datatype corresponding to $\begin{pmatrix} a & b \\ c & d \end{pmatrix}$.

```
complex mobius_on_point(matrix T, complex z){
    return (T.a * z + T.b)/(T.c * z + T.d);
}
```
Calculate $T(z) = \frac{az+b}{cz+d}$.

The second version supposes you are using a language like C, and cannot do complex arithmetic in 'infix' style. In this version, we show how to treat the case $z = \infty$ by assuming that complex numbers are either pairs of real numbers or have a special value `inf`. Note `a==b` means test for whether `a` and `b` are equal and return true or false (whereas `a-b` replaces the present value of `a` by that of `b`).

```
complex mobius_on_point(matrix T, complex z){
    if z == inf, then
        if T.c != 0 then
            return( cxdiv(T.a,T.c) )
        else return( inf )
    else {
        num = cxadd(cxmult(T.a,z),T.b);
        den = cxadd(cxmult(T.c,z),T.d);
        if den == 0, return inf;
        else return cxdiv(num,den);
    }
}
```
If the given point z is ∞.

If $c \neq 0$, return a/c,

otherwise, return ∞.

If $cz + d = 0$, return ∞.

Möbius maps and applying a Möbius map to a point. As we have already discussed, how this is implemented depends entirely on your platform. If you use a package like MatLab or Maple, you don't have to worry because not only complex numbers, but also matrix algebra is built in. If you use FORTRAN, complex arithmetic is provided but you will need to write subroutines for matrix operations. If you use C or C++, you should probably define structures or classes which implement 2×2 complex matrices and their arithmetic. To give the flavour, we give C-like pseudo-code for a routine `mobius_on_point` in Box 8.

Geometry and dynamics

To work with linear fractional transformations, we need not only to manipulate them algebraically, but also to understand their geometrical and dynamical effects. In fact we shall be constantly passing back and forth from algebra to geometry, using each point of view to clarify and implement the other.

When we studied the map $z \mapsto (z - 1)/(z + 1)$ we needed to use the fact that it maps any line or circle to another line or circle. Since as we have seen a line is really nothing other than a circle through the point ∞ or the North Pole, from our expanded viewpoint lines and circles are really the same thing. To simplify, from now on when we refer to a circle we (almost always) include the possibility that it may actually be a straight line.

The beautiful fact is that *any* Möbius map T carries circles to circles. This means that if C is any circle or a line, then $T(C)$ will be either a circle or a line. Watch out! It is usually *not* true that if P is the centre of C, then $T(P)$ is the centre of $T(C)$. We saw an example in Figure 3.4, in which the map $\hat{T}$ maps the green circles one into another. If $\hat{T}$ also sent centres to centres, then all the circles in the green family would have to be concentric, which is definitely not the case.

Another closely related property of Möbius maps is that they preserve angles. We are used to measuring the angle between two lines, but we can equally well measure the angle between two circles by measuring the angle between the tangent lines where the two circles meet. Saying that *T* **preserves angles** means that if θ is the angle between two circles C and C', then θ is also the angle between the two image circles $T(C)$ and $T(C')$. Actually you can even be more precise than this: Möbius maps are **orientation preserving**, which means that if, say, you measure θ clockwise from C_1 and C_2, then θ (rather than $-\theta$) is also the *clockwise* angle from $T(C_1)$ to $T(C_2)$. You can find an explanation of why this happens in Note 3.2.

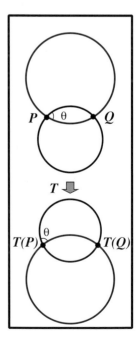

As we have seen, Möbius transformations usually alter the distance between points, and there is actually a relatively simple formula which tells us how. Suppose that as usual $T(z) = (az + b)/(cz + d)$. Then, for any two points z and w, we find with a bit of algebra that

$$|T(z) - T(w)| = \frac{|z - w|}{|cz + d||cw + d|}.$$

(Notice that $|z-w|$ is a convenient way of expressing the distance between the complex numbers z and w.) In words, the map T may either expand or contract distances, and it does so by multiplying them by the factor $1/|cz + d| \cdot |cw + d|$. It follows, for instance, that a very small circle around z is mapped to a very small circle around $T(z)$, the radius of the second circle being obtained from the first by dividing by the factor $|cz + d|^2$.

The next step in understanding the dynamics of the map $T : z \mapsto$

Note 3.2: **Why Möbius maps preserve circles and angles**

It is possible to prove the circle and angle preserving properties of a Möbius map by direct computation, writing down the equation of a circle and calculating its image directly. This gets complicated partly because circles may map to lines and vice versa. A neater method is to observe that if both T and S have each of these two properties, then so does their composition TS. Now we already know that the maps coming from complex arithmetic, translations and scalings, have these two properties. So also does the inverting map $J : z \mapsto 1/z$, as we saw in Chapter 2, when studying stereographic projection. To 'reduce' the general case to these basic examples, we only need to prove that $T : z \mapsto (az + b)/(cz + d)$ can be written as a composition of translations, scalings and the map J. A little bit of algebra reveals how to do it:

$$
\begin{aligned}
T(z) &= \frac{az + b}{cz + d} = \frac{c(az + b)}{c(cz + d)} \\
&= \frac{a(cz + d) + bc - ad}{c(cz + d)} \\
&= \frac{a}{c} - \frac{1}{c(cz + d)}.
\end{aligned}
$$

(The last part is simplified using $ad - bc = 1$.) In other words, this formula says that to get $T(z)$ from z, you first multiply by c, then add d, then invert, then multiply by $-1/c$, and finally translate again, this time by a/c. This sounds complicated, but we can immediately conclude that since adding, multiplying by a constant, and inverting all preserve circles and angles, so does T. Incidentally, this also shows that if we wanted to have a universe of transformations in which we can keep on composing transformations as often as we want and which contains translations, scalings and inversions, then this universe must be exactly the group of linear fractional transformations. The collection of mappings introduced by Möbius is not as arbitrary as you might think!

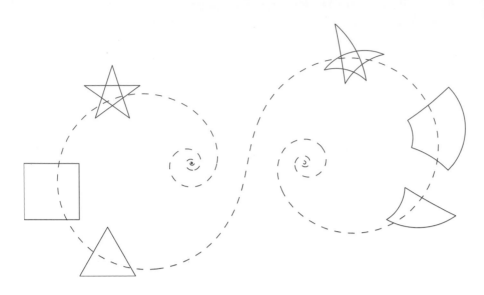

Figure 3.7. An equilateral triangle, a square and a five-pointed star are mapped to their images by a loxodromic transformation. Notice how the straight sides of the original polygons become circular arcs but that the angles remain 60°, 90° and 36°. The map used is the 16th power of the one studied at the beginning of the chapter.

$(az + b)/(cz + d)$ is to get a handle on its fixed points. In dynamics a fixed point of a transformation is one which does not move when you apply the map, so in our case it would have to satisfy the fixed point equation $T(z) = z$. To find the fixed points, therefore, we have to solve the equation $z = (az + b)/(cz + d)$. This is a quadratic equation $cz^2 + dz = az + b$, and thus usually has two solutions (just one in special cases when a square root is zero). Thus *every Möbius map has either one or two fixed points*. The actual formula for the fixed points is worked out in Note 3.3.

Note 3.3: **The fixed point formulas**

The fixed points of the transformation $T(z) = (az + b)/(cz + d)$ are the solutions to

$$z = \frac{az + b}{cz + d},$$

which simplifies to the quadratic equation $cz^2 + (d - a)z - b = 0$. The fabled quadratic formula favours us with the roots

$$z_\pm = \frac{a - d \pm \sqrt{(d - a)^2 + 4bc}}{2c}$$

where we use the $\pm$ subscript to match the fixed points with the choice of sign of the square root. Of course, this formula behaves rather poorly if c is 0. In this case, either $a \neq d$ and T has two fixed points ∞ and $b/(d - a)$, or $a = d$ and T is either the identity, or parabolic with a single fixed point ∞.

Expanding and rearranging $(d - a)^2 + 4bc$ yields $(a + d)^2 - 4ad + 4bc$. We are assuming that $ad - bc = 1$. The expression $a + d$ plays a crucial role in the geometry of T; it is called the trace of T, and denoted $\mathrm{Tr}\, T$. In terms of the trace,

$$z_\pm = \frac{a - d \pm \sqrt{(\mathrm{Tr}\, T)^2 - 4}}{2c}.$$

Often the best way to understand the dynamics of a transformation is to conjugate into a coordinate system in which its behaviour can be seen in a particularly simple form. This idea works particularly well for Möbius maps, and will lead us to a fundamental three-way classification of all Möbius transformations (other than the identity) into three types called **loxodromic**, **parabolic** and **elliptic**. We have already met examples of the first two: the transformation $\hat{T}$ which generated Figure 3.3 was loxodromic and the one which generated Figure 3.6 was parabolic. In fact by definition, all parabolic transformations are conjugate to translations $T(z) = z + a$, all elliptic transformations are conjugate to rotations $T(z) = kz$ with $|k| = 1$, and all loxodromic transformations are conjugate to scaling maps $T(z) = kz$ with $|k| > 1$. (Sometimes loxodromic transformations are called **hyperbolic** if k is real.) One can see why this is so by implementing a procedure which is really the reverse of how we arrived at our two initial examples: we conjugate by a map which takes the fixed points of T to ∞ and to 0. The various possible outcomes when you do this are explained in flow-chart form in Figure 3.8. Notice that only the parabolic maps have a single fixed point; all other kinds have two. The number k is called the **multiplier** of T.

Let's look at some pictures showing the three types. At the top of Figure 3.9, we show the effect of a loxodromic transformation T side by side with the conjugate scaling $z \mapsto kz$. The left frame shows the scaling map and you can see how it expands circles centred at the origin by a factor $|k|$ (about 2.5 in this example), spinning them round by an angle $\arg k$ as they go (here about 90°). The blue lines show the trajectories, which spiral away from the source at 0 and in towards the sink at ∞. To get the righthand picture of the original map T, we used a Möbius map which carried 0 and ∞ to the source and sink of T. Notice that the white ring in the lefthand picture is still really a ring on the right: remember that the point at ∞ should be added in. As we shall see in the next section, this white region is a tile for the group of powers T^n of T. In fact T and T^{-1} map the white region each onto one of the two red regions, T^2 and T^{-2} map the white region onto the two yellow regions, and so on.

The bottom of Figure 3.9 shows the two special cases in which the multiplier k has modulus 1 (so that T is elliptic) and when it is real (so that T is hyperbolic). The elliptic case is on the left. The particular transformation we have chosen satisfies $T^8 = I$, which comes from setting $k = \cos 45° + i \sin 45°$. The trajectories of the conjugated map $z \mapsto kz$ move along concentric circles centred on the fixed point 0, and to get the picture we conjugated by a map taking 0 and ∞ to the two fixed points of T. The

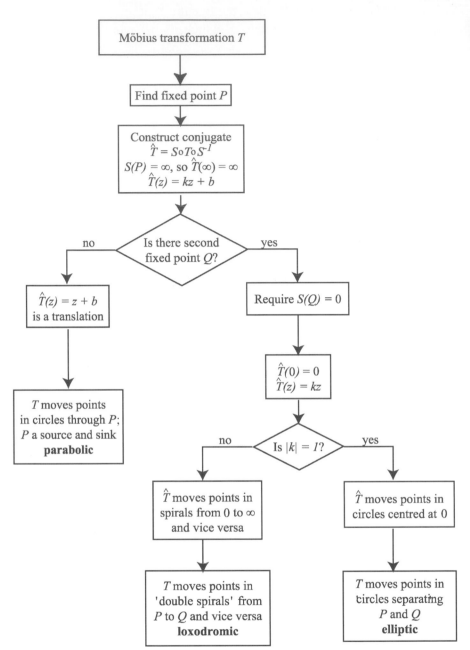

Figure 3.8. The algorithm for the three-way classification of Möbius transformations. The conjugating map S carries one fixed point of T to ∞. We use the fact that any transformation with a fixed point at ∞ is an affine map $z \mapsto kz + b$. If T has two fixed points, then S takes the second fixed point to 0.

points move around on the blue circles. The fixed points are called **neutral** since they are neither 'sources' nor 'sinks'. In the right frame you see the hyperbolic case in which k is real-valued. This time, points move straight along the blue lines with no spin at all.

It is easy to decide the type of T just by looking at its matrix, using a

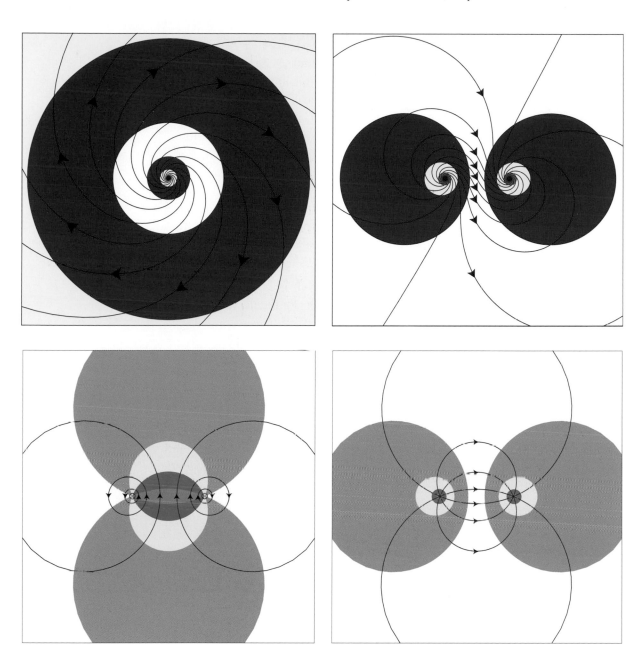

Figure 3.9. Möbius transformations. Above we have two loxodromic transformations while below are an elliptic and a hyperbolic transformation. See the text for discussion.

simple but magical number called its **trace**, which is defined very simply as Tr $T = a+d$. We already met Tr T in the fixed point formula in Note 3.3. The reason trace is so important is that the trace of a conjugate STS^{-1} is the same as the trace of T. In fact, it is simple to check that Tr $AB =$ Tr BA for any two matrices A and B. This shows:

$$\text{Tr } S(TS^{-1}) = \text{Tr}(TS^{-1})S = \text{Tr } T.$$

Box 9: **Classification of Möbius maps by trace**

Loxodromic maps have one source $\text{Fix}^- T$ and one sink $\text{Fix}^+ T$. Points spiral out from the source and spiral into the sink. They are characterized by the requirement that $\text{Tr}\, T$ is not between -2 and 2 and are conjugate to $T(z) = kz$, $|k| > 1$.

Hyperbolic maps are a special case of loxodromics in which points move not in spirals but in circles through $\text{Fix}^- T, \text{Fix}^+ T$. For them $\text{Tr}\, T$ is real and not between -2 and 2, and they are conjugate to $T(z) = kz$, k real, $k > 1$.

Elliptic maps have two neutral fixed points and points move around circles round the fixed points. For them $\text{Tr}\, T$ is real and strictly between -2 and 2, and they are conjugate to $T(z) = kz$, $|k| = 1$.

Parabolic maps have one fixed point $\text{Fix}\, T$ which is both source and sink. For them $\text{Tr}\, T = \pm 2$, and they are conjugate to translations $T(z) = z + a$ with multiplier $k = 1$.

The wonderful thing about traces is that you can use them to read off the type of a transformation mechanically. The result is described in Box 9 and proven in Note 3.4. In fact there is a nice formula relating the trace to the multiplier. You can find it in Note 3.5, where it is used to give a method for identifying which is the sink and which the source.

Tiles and tori for a loxodromic transformation

Back in Chapter 1, we used tiles to make geometry out of a group of transformations. The whole plane had to be covered by non-overlapping copies of one original tile, and every tile had to be obtained from the original tile by applying a symmetry in the group. Such a collection of tiles was said to tessellate the plane.

Of course, tiles laid out like this must meet along their edges. This means that associated to any edge of any tile is some symmetry in the group which maps the chosen edge to another edge of the same tile. We used maps like this to give a 'gluing recipe' for matching the edges of the original tile in pairs: points on two edges of the tile are glued if they are carried one to the other by a symmetry in the group. If we imagine the tile made of rubber, then by twisting, bending and stretching we can make the glued-up tile into a surface in three-dimensional space. This surface looks like the original tessellated plane except that there is now only one point for each orbit in the group, because any two points which are both images of one common point in the plane are now considered to be 'the same'. In Chapter 1, the plane was Iowa and the tiles were the individual farms. We cut out one farm, glued the two edges together in accordance with the symmetries in the group, and came up with a doughnut shaped surface called a torus containing only a single farm.

We can adapt this idea to the maps we are looking at in this chapter. Hand in hand with a single loxodromic transformation T goes the group consisting of all powers, positive or negative, of T. The tiling associated to this group fills up the whole Riemann sphere with the exception of the two fixed points of T. The basic 'tile' is a ring which, when glued up following the recipe from Chapter 1, makes a torus. This idea will lead us in the next chapter to a much more interesting 'swiss cheese tile' which glues up into a 'pretzel' or two-holed torus.

So let T be a loxodromic map, for example the one in the top right-hand frame in Figure 3.9. As we have already explained, viewed on the Riemann sphere, the white region is really a ring. This will be our basic tile F. (If you are uneasy about this, try looking in the lefthand frame first. The colour coding in the two frames is exactly the same.) As we have

---

Note 3.4: Calculating traces of different types of Möbius maps

Since traces are unchanged under conjugation, we can work out the classification by conjugating our transformation to any convenient position we please. Let's first deal with parabolic maps, those with a single fixed point. If we conjugate to move this fixed point to ∞, the new map must be of the form $z \mapsto az + b$ for some constants a and b. (This is because we must have $c = 0$, and we can then assume $d = 1$, see Note 3.3.) To avoid a second fixed point in addition to ∞, necessarily $a = 1$. The matrix is then $\pm \begin{pmatrix} 1 & b \\ 0 & 1 \end{pmatrix}$ so the trace is ± 2.

In all other cases, there are two distinct fixed points, so we conjugate to position these at 0 and ∞. It is not hard to work out that the new map must have a diagonal matrix $\begin{pmatrix} a & 0 \\ 0 & 1/a \end{pmatrix}$. The trace is $a + 1/a$, and the transformation is $z \mapsto a^2 z$, incidentally yielding a formula for the multiplier $k = a^2$ in terms of the trace $\mathrm{Tr}\, T$.

This new map is loxodromic provided that a^2, and hence also a, has modulus not equal to 1. With great presence of mind, we illustrated long ago in Project 2.3 that the mapping $a \mapsto a + 1/a$ carries the unit circle precisely onto the line segment from -2 to 2. In particular, the points $a = \pm 1$ are mapped to $a + 1/a = \pm 2$, and all the points *not* on the unit circle are mapped to points *not* on the line between -2 and 2. This translates precisely to the statement that our Möbius map is elliptic if its trace is real and strictly between -2 and 2; the identity (or parabolic) if the trace is ± 2; and loxodromic or hyperbolic in all other cases.

seen, $T(F)$ and $T^{-1}(F)$ are the two red rings surrounding the two fixed points of T. The tiles $T^2(F)$ and $T^{-2}(F)$ are yellow. Applying powers of T gives successive nested rings, so the picture is actually showing us a tiling for the group consisting of all powers of T. Each tile $T^n(F)$ is ring-shaped provided it is viewed on the Riemann sphere. In both pictures, all the tiles taken together fill up all the sphere with the exception only of the two fixed points of T.

The blue flow lines indicate how the two circular edges of the tile F have to be glued. It is easiest to see this in the left frame. Each point P of the inner circular boundary of the white ring is connected by a blue spiral to the point of the outer circular boundary. This point is exactly the image of P under T, in other words, the point $T(P)$. Thus our recipe tells us to glue the boundaries of the white tile together by matching points joined by the blue spiral. The result is seen in Figure 3.10: since we can allow the ring's rubbery nature to take out the twist caused by the spiralling in the blue flow lines, we get a torus.

As in Chapter 1, there is another way to think about this gluing. Imagine starting not from one ring but from the whole Riemann sphere minus the two fixed points of T. If we consider all the points

---

Note 3.5: **What's a sink and what's a source?**

Suppose the fixed points of the loxodromic map T are z_+ and z_- as described in Note 3.3. To distinguish between the source and the sink, we have to find the multiplier k. To do this, define the map

$$S(z) = \frac{z - z_-}{z - z_+}$$

which sends z_+ to ∞ and z_- to 0. Then STS^{-1} fixes both 0 and ∞, and consequently has the form $STS^{-1}(z) = kz$, where k is the multiplier of T. Replace z by $S(z)$ to obtain $k\, S(z) = STS^{-1}S(z) = S(T(z))$.

Now we do a sneaky thing to calculate k: we put in the ever-so-useful number ∞ for z. Then $S(\infty) = \frac{\infty - z_-}{\infty - z_+} = 1$ and $T(\infty) = \frac{a}{c}$ (let's give ourselves some relief by assuming c is not zero). This gives

$$k = \frac{\frac{a}{c} - z_-}{\frac{a}{c} - z_+}.$$

Plugging in our fixed point formulas and applying some elbow-grease reveals that

$$k = \frac{\frac{a+d+\sqrt{\mathrm{Tr}(T)^2 - 4}}{2c}}{\frac{a+d-\sqrt{\mathrm{Tr}(T)^2 - 4}}{2c}}$$

$$= \frac{\mathrm{Tr}\, T + \sqrt{\mathrm{Tr}(T)^2 - 4}}{\mathrm{Tr}\, T - \sqrt{\mathrm{Tr}(T)^2 - 4}}$$

$$= \left(\frac{\mathrm{Tr}\, T + \sqrt{\mathrm{Tr}(T)^2 - 4}}{2} \right)^2.$$

It's important to note that the square root here is exactly the same as (that is, has the same sign as) the square root in the fixed point formulas in Note 3.3. If $|k| > 1$, then z_+ is the attracting fixed point (or sink) which we have called Fix$^+$, and the repelling fixed point (or source) Fix$^-$ is z_-. If $|k| < 1$, the definitions of Fix$^+$ and Fix$^-$ are reversed.

Figure 3.10. This time Dr. Stickler is making his huge tyre out of a rubber ring. We have labelled the boundaries of the ring which are to be matched by the symbols A and a, but we might just as well have used T and T^{-1}. From the next chapter, we shall be using the $A - a$ notation systematically for the generating transformations of rather more elaborate groups.

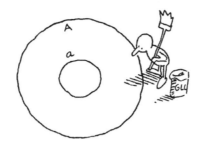

BEGIN WITH A FLAT RING OR ANNULUS, WHERE CIRCLE A IS TO BE GLUED TO a.

ROLL UP RING. REMEMBER, WE MAY STRETCH THE SURFACE AS MUCH AS WE LIKE.

ALMOST DONE NOW!

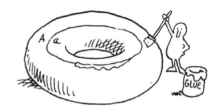

GLUE TO FORM A TORUS.

$\ldots, T^{-1}(z), z, T(z), T^2(z), T^3(z), \ldots$ in the orbit of one point z to be 'the same', then what you get is a torus. This is the same as the result of gluing, because no matter where you are on the sphere, some power of T always moves you back to some point in the original white ring. Because it has one hole, a torus is also called a **surface of genus one**.

Recipes for Möbius maps

Möbius maps are a very flexible class of transformations, and there will be many occasions when we need a Möbius map which does this, that or the other special thing. In this section we assemble some useful recipes Grandma taught us[1] for concocting exactly the map we need. As with any cookery book, you may not want to study all the details now, but keep the recipes handy for reference as you proceed.

[1] In Chapter 8, Grandma will excel herself by revealing her much more elaborate recipe for generating almost all the fractal images we shall ever want to draw.

Recipe I: Maps which carry the real axis to itself. Suppose you need a map which carries points on the real axis to other points on the real axis. The recipe for this is very simple: choose the coefficients a, b, c, d of the matrix to be real. If we also assume that $ad - bc = 1$, we get a set of matrices called $SL(2, \mathbb{R})$, where 'S' indicates that the determinant is 1.

The product of two such matrices still has real entries so $SL(2, \mathbb{R})$ forms a group, in fact a **subgroup** of $SL(2, \mathbb{C})$.[1] If you apply such a matrix to a real number x, the result $T(x) = (ax + b)/cx + d$ is another real number (or ∞ if $x = -d/c$). According to our rules for operating with ∞, $T(\infty)$ is also a real number a/c.

This can be expressed more neatly if we think in terms of the Riemann sphere. Just as we adjoined ∞ to $\mathbb{C}$ to make the Riemann sphere, frequently denoted $\widehat{\mathbb{C}}$, so we can adjoin ∞ to $\mathbb{R}$ to obtain what we call the **extended real line** $\widehat{\mathbb{R}}$. This makes sense if we think of $\widehat{\mathbb{R}}$ as a circle through ∞; stereographically projected on to the sphere, $\widehat{\mathbb{R}}$ is just a great circle through the North Pole. The calculation we have just done shows that *the transformations in $SL(2, \mathbb{R})$ map the extended real line $\widehat{\mathbb{R}}$ to itself.*

The extended real line divides the complex plane into two halves: the upper half plane $\mathbb{H}$ in which $\text{Im}\, z > 0$, and the lower half plane for which $\text{Im}\, z < 0$. As you would expect, every T in $SL(2, \mathbb{R})$ maps each of these half planes to themselves. But here's a twist: the map $z \mapsto 1/z$, which is given by the real matrix $\begin{pmatrix} 0 & 1 \\ 1 & 0 \end{pmatrix}$, carries i to $1/i = -i$ so it swaps the two half planes around! We are saved by the fact that the determinant of this matrix is -1 so it is not in $SL(2, \mathbb{R})$: see Note 3.6. For this reason, we sometimes refer to $SL(2, \mathbb{R})$ as the **upper half plane group**: it is the group of linear fractional transformations which preserve the upper half plane.

Matrices in $SL(2, \mathbb{R})$ have a real trace, so by the classification on p. 82, they may be either elliptic, parabolic or hyperbolic, but never strictly loxodromic. In other words, for matrices in $SL(2, \mathbb{R})$, you can never have a spiral twist. Another useful fact about these maps is that they carry circles C with centres on the real axis to other circles $T(C)$ also with centres on the real axis. We mustn't think this is because T maps the centre of C to the centre of $T(C)$ – that's not how this non-linear world works.

[1]The symbols $\mathbb{R}$ and $\mathbb{C}$ are standard mathematical notation for the real and complex numbers respectively. We shan't have much cause to use them but they are convenient here.

Note 3.6: Why $SL(2, \mathbb{R})$ **maps the upper half plane** $\mathbb{H}$ **to itself**

Let $T = \begin{pmatrix} a & b \\ c & d \end{pmatrix}$ be a real matrix with determinant 1. We want to show that $T(\mathbb{H}) = \mathbb{H}$. Let's calculate the imaginary part of $(az + b)/(cz + d)$. Multiplying top and bottom by $c\bar{z} + d$, the denominator becomes the real number $|cz + d|^2$. Then we can compare imaginary parts to get the elegant identity

$$\text{Im}\left(\frac{az + b}{cz + d}\right) = \frac{(ad - bc)}{|cz + d|^2}\,\text{Im}\, z = \frac{\text{Im}\, z}{|cz + d|^2}.$$

Thus $\text{Im}\, z > 0$ implies $\text{Im}\,(az + b)/(cz + d) > 0$.

The reason is that circles whose centres are on $\mathbb{R}$ are the same as circles which cross $\mathbb{R}$ twice at right angles. Remember that T preserves angles, so if C crosses $\mathbb{R}$ at right angles, so does $T(C)$. We should also include vertical lines perpendicular to $\mathbb{R}$: these are really limiting cases of circles meeting $\mathbb{R}$ at right angles, going through one finite point and ∞. If we think of these as 'extended circles' orthogonal to the extended real line $\hat{\mathbb{R}}$, the correct statement is that T maps any extended circle C orthogonal to $\hat{\mathbb{R}}$ to another circle of the same kind.

We have already seen lots of examples of maps in $SL(2, \mathbb{R})$ – except that some of the pictures need to be rotated around! The lower right frame of Figure 3.9 shows a typical hyperbolic with real entries, while you get a typical elliptic by turning the lower left frame on its side. You can tell you have to do this because the elliptic map in the figure takes the imaginary axis to itself, so to get a map which preserves the real axis, you have to make a 90° turn. The simplest parabolics are just the horizontal translations $z \mapsto z + a$ where a is a real number. Another useful parabolic map was shown in Figure 3.6. That map has fixed point 0 and takes the imaginary axis to itself, so rotating through 90° gives another map in $SL(2, \mathbb{R})$.

If T is any hyperbolic transformation in $SL(2, \mathbb{R})$, then its sink and source are automatically real. Geometrically, this is because orbit points $T^n(z)$ approach the sink when n tends to ∞ and approach the source when n tends to $-\infty$. Since when z is real, $T^n(z)$ stays real, the sink and the source had better be real as well. For the same reason, if T is a parabolic in $SL(2, \mathbb{R})$, then its one fixed point, being both a sink and a source, has to be real. Elliptic transformations in $SL(2, \mathbb{R})$ must have one fixed point in the upper half plane and one in the lower. The second fixed point is always the complex conjugate of the first. In fact, if an elliptic transformation had a real fixed point a, then T could not possibly be a rotation near a, because then it would not map the real axis through a to itself.

Recipe II: A map which carries the real axis to the unit circle.

A nice easy recipe for when company drops in unexpectedly, the Cayley map K is both important and pretty. It is given by the formula

$$K(z) = (z - i)/(z + i).$$

It is easy to calculate a few values:

$$K(0) = -1, \; K(\infty) = 1, \; K(-1) = i, \; K(1) = -i, \; K(i) = 0, \; K(-i) = \infty.$$

Note that 4 points on the extended real line, namely $-1, 0, 1, \infty$, are mapped to four points on the unit circle, namely $i, -1, -i, 1$. This shows

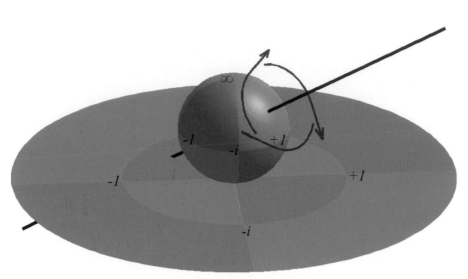

Figure 3.11. The Cayley map. The Cayley map on the Riemann sphere is a rotation through 120° around the centre of the octant with vertices $\infty, 1, -i$. On the sphere, the unit disk becomes the southern hemisphere, that is the bottom half, and the upper half plane is the back of the sphere – from the viewpoint of the figure. Can you visualize how this rotation takes the back half of the sphere to the bottom half?

that the image of the real line is a circle which meets the unit circle in more than two points: the only possibility is that these two circles are, in fact, the same. Thus K maps the extended real line to the unit circle. Since it maps the point i in the upper half plane $\mathbb{H}$ to the point 0 inside the unit circle, it follows that K maps the entire region $\mathbb{H}$ exactly to the region inside the unit circle, usually called the **unit disk** denoted by $\mathbb{D}$.

It is a curious fact that K^3 is the identity map: if you trace through what happens to each of the six points above under K^3, you will find they all come back to their starting points. The action of K can be seen three-dimensionally using the octahedral tiling of the sphere shown in Figure 3.11. It is a rotation by 120° about an axis through the centre of the sphere and the midpoints of the two octants with vertices $(\infty, 1, -i)$ $(-1, i, 0)$. This checks with our calculations that $K(\infty) = 1, K(1) = -i$ and $K(-i) = \infty$. As we have drawn it in Figure 3.11, the 'lift' of the unit disk to the sphere is the southern hemisphere, while the upper half plane lifts to the back.

Recipe III: Maps which carry the unit circle to itself.
The third recipe we shall need is for Möbius maps which carry some circle to itself. If we restrict ourselves to the unit circle, then using the Cayley map K and conjugation will do the trick. In fact, since K maps the upper half plane $\mathbb{H}$ to the unit disk $\mathbb{D}$, then if we conjugate a map which carries the upper half plane to itself by K, the result will be a map which carries the unit disk to

itself. This needs a bit of algebra to work out but the result is nice: after careful calculation you find that the conjugate $\hat{T} = KTK^{-1}$ of *any* transformation T in $SL(2, \mathbb{R})$ has the special form

$$\hat{T} = \begin{pmatrix} u & v \\ \bar{v} & \bar{u} \end{pmatrix}$$

where u and v are complex numbers such that $|u|^2 - |v|^2 = 1$. Everything works because, since $T(\mathbb{H}) = \mathbb{H}$, then

$$\hat{T}(\mathbb{D}) = KTK^{-1}(\mathbb{D}) = KT(\mathbb{H}) = K(\mathbb{H}) = \mathbb{D}.$$

In other words, since T is in the upper half plane group $SL(2, \mathbb{R})$, its conjugate $\hat{T} = KTK^{-1}$ maps $\mathbb{D}$ to $\mathbb{D}$. The set of all such matrices form a group, the **unit circle group**, denoted in mathematical books by $SU(1,1)$.[1] Note that the trace of any such T is twice the real part of u and has no imaginary part, so that once again T is either hyperbolic, elliptic or parabolic but not loxodromic.

We can get lots of examples by choosing suitable values for u and v. If $u = e^{i\theta}$ and $v = 0$, then $T = \begin{pmatrix} e^{i\theta} & 0 \\ 0 & e^{-i\theta} \end{pmatrix}$ is the map $z \mapsto e^{2i\theta}z$, which rotates by the angle 2θ about 0. These are the simplest elliptic maps which obviously map the unit disk to itself. If, on the other hand, u is real and greater than 1, then we get the hyperbolic map

$$T = \begin{pmatrix} u & \sqrt{u^2 - 1} \\ \sqrt{u^2 - 1} & u \end{pmatrix}$$

which is also in $SL(2, \mathbb{R})$. An example is the map in the lower right frame of Figure 3.9. As you can check, T stretches $\mathbb{D}$ symmetrically away from the source at -1 towards the sink at 1. Maps like this, which we sometimes call **special stretch maps**, will come in useful later.

Geometrically, another way to make examples is to take any elliptic, hyperbolic or parabolic Möbius map, find a circle which the map leaves fixed and conjugate this circle to the unit circle. Such a circle might be any one of the blue circles in Figure 3.9 or Figure 3.6.

Recipe IV: Maps which pair circles.

The top right frame of Figure 3.9 showed a loxodromic transformation T with a sequence of coloured circular disks nesting down to each of its fixed points. These disks were mapped one to the other by T: let D be the red disk on the left and let D' be the one on the right. Then T maps the region inside D to region outside D' and the region outside D to the region inside D'.

Sometimes we want to reverse this process: starting from two disjoint circles C and C', suppose we can construct a Möbius transformation which maps C to C', taking the inside (respectively outside) of C to the outside

[1] In this notation, S stands for 'special', meaning the determinant is 1 and U stands for 'unitary', meaning that when these matrices act on 2-vectors they preserve a strange kind of squared length, namely $|x|^2 - |y|^2$.

(respectively inside) of C'. Iterating T on C' will lead to a sequence of disks nesting down on the sink $\text{Fix}^+ T$, while iterating the inverse T^{-1} on C yields a sequence nesting down on the source $\text{Fix}^- T$. We say that a map T like this **pairs** the circles C and C'. Such maps will be our main building blocks in the next chapter.

Here is a recipe for finding a map which pairs two circles. Suppose that C has centre P and radius r, while C' has centre Q and radius s. The 'proof by pictures' in Figure 3.12 explains why the map $z \mapsto rs/(z - P) + Q$ will work. (Remember that P and Q are complex numbers as well as points, which explains why we can use them in an algebraic formula like this.)

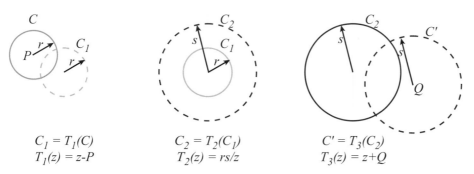

$$C_1 = T_1(C) \qquad\qquad C_2 = T_2(C_1) \qquad\qquad C' = T_3(C_2)$$
$$T_1(z) = z - P \qquad\qquad T_2(z) = rs/z \qquad\qquad T_3(z) = z + Q$$

Figure 3.12. The map $z \mapsto rs/(z - P) + Q$ carries the green circle C to the blue circle C'. Broken down into 3 steps, $z \mapsto z - P$ translates C to the yellow circle C_1 with centre 0; $z \mapsto rs/z$ inverts it taking the inside of C_1 to the outside of the red circle C_2 with radius s (note that $|rs/z| = rs/|z|$, hence $|z| \leq r$ implies $|rs/z| \geq s$); and finally $z \mapsto z + Q$ translates the outside of C_2 onto the outside of circle C'.

Sometimes we need more freedom in choosing our maps. Could there have been other transformations which pair the same circles? We can split the middle step $T_2(z) = rs/z$ into three parts: $z \mapsto z/r$, $z \mapsto 1/z$ and $z \mapsto sz$. The first scales C_1 to the unit circle, the second inverts in the unit circle and the third scales the unit circle up to C_2. Why not insert a transformation which takes the unit circle to itself in between? Our previous recipe told us exactly what these maps are: they all look like $z \mapsto \frac{uz+v}{\bar{v}z+\bar{u}}$ where $|u|^2 - |v|^2 = 1$. Inserting this into the diagram and calculating all the compositions carefully, we come up with the transformation

$$z \mapsto s\frac{\bar{v}(z - P) + r\bar{u}}{u(z - P) + rv} + Q.$$

By judiciously choosing the values of u and v, you can move 0 to any point in the unit disk: first stretch by choosing u and v real so that 0 gets moved to a chosen point v/u inside the unit disk and on the real axis, and then rotate about 0 until your point reaches the required position in $\mathbb{D}$. Combining this with the rest of the maps in the diagram you can, for example, find a transformation carrying the inside of any circle C to the outside of any other circle C', with the additional requirement that some given point inside C maps to the point ∞ outside C'.

Recipe V: The image of a circle by a Möbius map. To draw any of the figures in this chapter we need to implement one of the most frequent requirements of our graphics: to calculate the image of a specific circle C under a specific Möbius map T. The best thing is to make a computer subroutine which does this. We shall call it `mobius_on_circle`. The input data will be the centre P and radius r of the circle C, and the matrix $\begin{pmatrix} a & b \\ c & d \end{pmatrix}$ representing the map T. The output will be the centre Q and radius s of the image circle $T(C)$.

The reader who tries this will find it is not at all obvious how to do the calculation in a relatively painless way. Remember, even though Möbius maps carry circles to circles, they do not (usually) carry centres to centres so it will *not* usually be true that $T(P) = Q$. A subroutine for calculating the image of a circle under a Möbius transformation is presented in Box 10. The reason it works has to do with inversion, and you can find Grandma's explanation in Note 3.7.

Box 10: `mobius_on_circle`

This routine takes as input a circle with centre P and radius r, and a transformation $T = \begin{pmatrix} a & b \\ c & d \end{pmatrix}$. The program calculates and outputs the centre Q and radius s of the circle $T(C)$. If:

(1) $\quad z = P - \left(r^2 \big/ \overline{(d/c + P)} \right)$

(2) $\quad Q = \dfrac{az + b}{cz + d}$

(3) $\quad s = \left| Q - \dfrac{a(P + r) + b}{c(P + r) + d} \right|$

then the output circle has centre Q and radius s.

```
structure circle {complex cen, real rad}
```
A new datatype containing the centre `cen` = P and radius `rad` = r of a given circle C.

```
circle mobius_on_circle(matrix T, circle C){
   circle D;
   complex z;
   z = C.cen - (C.rad*C.rad)/cxconj(T.d/T.c + C.cen);
   D.cen = mobius_on_point(T,z);
   D.rad = cxabs(D.cen - mobius_on_point(T, C.cen+C.rad));
   return( D );
}
```

Projects

3.1: More about spirals

Here are a few questions about maps similar to the one in Figure 3.1.

(1) If $T(z) = kz$, then we can calculate its inverse by using the map $J(z) = 1/z$ which interchanges its two fixed points. Calculate JTJ^{-1} and show it has a source at $J^{-1}(0) = \infty$ and a sink at $J^{-1}(\infty) = 0$. Now verify that $T^{-1} = JTJ^{-1}$. This is a special case of a general identity: try proving that if T is any Möbius transformation with two fixed points and J is any Möbius transformation which interchanges them, then $T^{-1} = JTJ^{-1}$.

(2) What happens to $T(z) = 2z$ if we lift to the sphere and then rotate by our map R (see p. 65) before projecting back onto the plane? The latitude and longitude circles become the yellow and green families you can see in Figure 3.4. The yellow circles are flowlines. Where are the source and the sink? Work out the formula for the action of this map.

(3) In several pictures we drew continuous spirals linking up the orbit points $T^n(z)$. If $T(z) = kz$, then $T^n(z) = k^n z$, so the spiral consists of points $k^t z$ for all real numbers t. Since k is complex, we had better be careful about what this means. Using the polar form $k = |k|(\cos\theta + i\sin\theta)$ set $k^t = |k|^t(\cos(t\theta) + i\sin(t\theta))$. Now: what is the formula for the double spiral curves used in Dr. Stickler's walk? If his head was at the point 0.25 when it crossed the real axis in Figure 3.3, how far was it from the origin when it exploded on the left?

Note 3.7: Grandma's explanation of `mobius_on_circle`.

Remember the formula for inverting in a circle centre P and radius r: the point z is mapped to $P + r^2/(\overline{(z-P)})$. In particular, the centre P inverts into the point at infinity ∞.

We know that Möbius maps preserve all properties of points which can be described in terms of angles and circles. The magic of conjugation tells us that if z and w are inverse with respect to a circle C, then $T(z)$ and $T(w)$ are inverse with respect to the circle $T(C)$. In the present case, let us apply this fact to the circle C' and the inverse T^{-1} of the map T. Since Q and ∞ are inverse with respect to C', we see that $T^{-1}(\infty)$ and $T^{-1}(Q)$ are inverse with respect to $T^{-1}(C') = C$.

We can work this out more specifically: if $T = \begin{pmatrix} a & b \\ c & d \end{pmatrix}$ has determinant 1, then as we know,

$T^{-1} = \begin{pmatrix} d & -b \\ -c & a \end{pmatrix}$ so $T^{-1}(\infty) = -d/c$. This means that $w_0 = T^{-1}(Q)$ is the inverse of $-d/c$ in the circle C. Applying the appropriate formula for inverting gives $w_0 = T^{-1}(Q) = P + r^2/(\overline{-d/c - P})$. This explains the first output of Grandma's recipe:

$$Q = T(T^{-1}(Q)) = T(w_0) = T\left(P + r^2/(\overline{-d/c - P})\right).$$

You can compute the radius s by taking the distance from Q to any point on $C' = T(C)$. Now $P + r$ certainly lies on the first circle C, and so $w = \frac{a(P+r)+b}{c(P+r)+d}$ is on C'. This explains why we can find the new radius from the formula $s = |w - Q|$. There are a few annoying exceptions to all this, concerning the cases in which either C or C' (or both) are straight lines. We explore how to deal with these in Project 3.7.

3.2: Möbius maps and triples of points

The aim of this project is to show that there is exactly one Möbius map which carries a given triple of points P, Q, R to another triple P', Q', R'.

(1)   Show that the only Möbius map which fixes all three points $0, 1, \infty$ is the identity transformation $\begin{pmatrix} 1 & 0 \\ 0 & 1 \end{pmatrix}$.

(2)   Verify that the map

$$S(z) = \frac{(z - P)(Q - R)}{(z - R)(Q - P)}$$

carries the three given points P, Q, R to the three special points $0, 1, \infty$ in that order.

(3)   Put these results together to show that if T fixes each of P, Q and R, then T is the identity. Hint: conjugate T by S.

(4)   Put everything together to show that if S and T are both Möbius maps which carry P to P', Q to Q' and R to R', then $S = T$.

3.3: A special transformation for later use

Suppose that $0 < s < t < 1$ and let T be the map

$$T(z) = \frac{(s + t)z - 2st}{-2z + (s + t)}.$$

What is the normalized matrix form of T with determinant 1? What is its trace and, hence, what kind of Möbius map is T? Why would you expect that T has both its fixed points on the real line? Why does T map the extended real line to itself? Show by direct calculation that the fixed points of T are $\pm\sqrt{st}$. Which is the sink and which is the source?

Now let C be the circle with centre on the real axis which meets the real axis in the points s and t. Let C' be the reflection of C in the imaginary axis, meeting the real axis at $-t$ and $-s$. Check that $T(s) = -s$ and $T(t) = -t$. Using the fact that T maps the extended real axis to itself, show that in fact $T(C) = C'$. Then check that T maps the *outside* of C to the *inside* of C'. (This shows that T pairs C to C' as described in Recipe IV.) You can do this by checking that just one point outside C maps to a point inside C', so work out $T(\infty)$ and $T(0)$ and show they both are between $-s$ and $-t$.

3.4: Making two circles concentric

The problem here is to show that if C and C' are any two disjoint circles, then there is a Möbius transformation which maps C to the unit circle and C' to a concentric circle inside. Here are some hints. Using just affine transformations (that is, by translating and scaling), you can map C to the unit circle. If needed, use $T(z) = 1/z$ to place C' inside C, and then use a rotation to get its centre on the real axis. Now the tricky part: use one of the special stretch transformations $\begin{pmatrix} u & \sqrt{u^2 - 1} \\ \sqrt{u^2 - 1} & u \end{pmatrix}$ from Recipe III to shift C' so that its centre moves to 0 *without shifting C.*

3.5: Radii of circles

This is a harder project which contains the essence of what makes our fractal pictures in the next chapter work, see Project 4.5. Take any pair of concentric disks $D \subset E$.[1] Apply a Möbius transformation T which doesn't turn D and E inside out, so that $T(D) \subset T(E)$ and ∞ is outside $T(E)$. Prove that

[1]The notation $D \subset E$ means that D is inside E.

$$\frac{\text{radius of } T(E)}{\text{radius of } T(D)} \geq \frac{\text{radius of } E}{\text{radius of } D} > 1.$$

Method 1: First try this in the special case in which E is the unit circle and T is a special stretch map. If D has radius r, use the formula for distortion of distance on p. 77 applied to $T(r)$ and $T(-r)$ to show that $\text{rad}(T(D)) = r/((u^2 - 1)(1 - r^2) + 1)$. Now explain why you may as well replace T by any map ST for which S is either a scaling or translation. Pick such an S for which $ST(E) = E$. This means we just have to show that if $U(E) = E$ then $\text{rad}(D)/\text{rad}(U(D)) \geq 1$. Find a W such that W is a stretch followed by a rotation and such that $W(D) = U(D)$, then work with W instead of U.

Method 2: Use the recipe `mobius_on_circle` in Box 10. With the notation in the box, if a circle D has centre 0 and radius r, show that

$$\text{rad}(T(D)) = |r - z|/(|cz + d||cr + d|) = r/(|d|^2 - |c|^2 r^2).$$

(You will need to use that $|\bar{c}r + \bar{d}| = |cr + d|$.) Use the fact that $T(D)$ does not contain ∞ to deduce that $|d/c| > r$. Now use a similar formula for E to compare the radii of $T(D)$ and $T(E)$.

3.6: Pairing tangent circles

In Recipe IV we saw how to find a Möbius map T taking the inside of one circle C to the outside of another disjoint circle C' and vice versa. Is it possible to do the same thing if the circles are tangent? The simplest case can be seen in Figure 3.6 where T is parabolic and C and C' are the large red circles. But there are other possibilities.

First, assume T is parabolic and show that if such circles can be found, then their point of tangency must be the fixed point of T. (The fixed point has to be inside both C and C'; why?) You can find a suitable T and pair of circles by using the method of conjugation and studying the case where T is a translation.[2] However, as you can see on the left in Figure 3.13, it is not necessary for the orbits of T to move along circles perpendicular to C and C'. Can you find formulas which implement this?

[2]As discussed in Note 2.5, you have to be careful when you do this, because what was the *inside* of a disk or half plane before conjugation may well be the *outside* afterwards.

If T is hyperbolic, try finding a suitable pair of circles by conjugating T to a scaling map $T(z) = kz$. Choose the point of tangency to be one of the fixed points.[3] This is illustrated on the right in Figure 3.13. The trouble is that one disk contains both fixed points. If the tangency point is $\text{Fix}^+ T$, then as n tends to ∞, the backward images $T^{-n}(C')$ won't nest down to $\text{Fix}^- T$.

[3]You have to think about insides and outsides carefully to do this: in this configuration you will only see what you want by recognizing that the second of the paired disks D' is the region which looks as if it is *outside* your image circle C'.

3.7: Code for lines as well as circles

In Recipe V, we dealt at length with calculating the image of a circle under a Möbius transformation. However, as we have seen, we also need to include lines as a special type of circle. Several steps are involved.

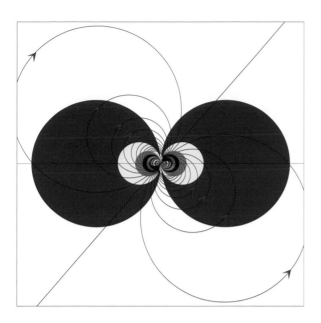 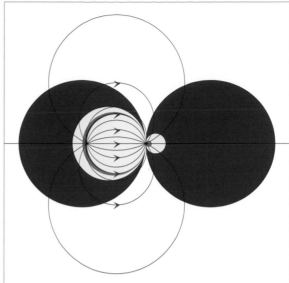

Figure 3.13. Two surprising examples of Möbius transformations pairing tangent disks by taking the outside of the first to the inside of the second.

The first is that we need a data structure which describes a line. We could use the coefficients a, b of a linear equation for a line $y = ax + b$, but this would leave out vertical lines. The simplest way to handle all lines at once is to write them in the form

$$\cos \theta \cdot x + \sin \theta \cdot y = d,$$

with a **direction** θ (with $0 \leq \theta < 2\pi$) and an **offset** $d \geq 0$.

Armed with a way of describing lines as well as circles, we can go back to Grandma's recipe in Box 10 and single out the case in which $T(C)$ is a line for separate treatment. Show that $T(C)$ is a line if and only if $|d/c + P| = r$. Then work out formulas for direction and offset in this special case. You will also need to work out the centre and radius of the circle $T(L)$ if L is a given line.

An alternative is to handle circles and lines simultaneously by using the family of equations:

$$a \cdot |z|^2 - 2 \operatorname{Re}(z \cdot (b - ic)) + d = 0.$$

Here a, b, c, d are all real numbers which are required to satisfy $b^2 + c^2 > ad$. Reminiscent of the way in which 2×2 matrices describe Möbius maps, if t is any non-zero number, then the equation with coefficients ta, tb, tc, td has the same solutions. So if $a \neq 0$, we can modify the equation by dividing by a. Show that in this case the equation describes a circle with centre $(b + ic)/a$ and radius $\sqrt{(b^2 + c^2)/a - d}$. On the other hand, show that if $a = 0$, then the equation describes a line with direction $\arg(b + ic)$ and offset $d/(2\sqrt{b^2 + c^2})$.

CHAPTER 4

The Schottky dance

So, Nat'ralists observe, a Flea
Hath smaller Fleas that on him prey;
And these hath smaller Fleas to bite 'em;
And so proceed ad infinitum.

Jonathan Swift, *On Poetry*

Having learnt about the dynamics of a single Möbius map in the last chapter, we now embark on the topic which will occupy almost all the rest of the book: what patterns are simultaneously symmetrical under *two* Möbius maps? This turns out to be a very fruitful question, because two transformations can interact in several very different ways. They will try their best to dance together – but they do not always succeed. Like two friends making music, they may perform in simple harmony, they may be elaborately contrapuntal, or the result may be total dissonance. This chapter studies one of the simplest possible arrangements. As we go further, the full range of complexity from relative order to total chaos will gradually unfold.

In Chapter 1, we studied the collection of all possible transformations obtained by composing (that is, iterating) two initial transformations and their inverses in any order whatsoever. At that point, we were talking about Euclidean symmetries which preserved a pattern of tiles on a bathroom floor, but there is no reason why what we said there should not equally be applied to Möbius transformations which, as we have seen, can be thought of as symmetries of patterns on the Riemann sphere.

Our basic building blocks will be transformations which pair circles as in the two righthand frames on p. 81. We described the set-up in detail in Recipe IV on p. 89. Start with a single loxodromic transformation T and a pair of non-intersecting disks D and D', chosen so that T maps the *outside* of D to the *inside* of D'. In this situation, T is said to **pair** the disks D and D'. The repelling fixed point or source is inside D and the attracting fixed

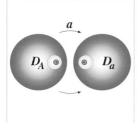

Box 11: **Notation for circle pairing**

A single loxodromic transformation a is said to **pair** the disks D_A and D_a if:

(1) the disks D_A and D_a do not overlap;

(2) the *outside* of the disk D_A is mapped by a to the *inside* of the disk D_a and the inside of the disk D_A is mapped to the outside of the disk D_a;

(3) the circle C_A which bounds the disk D_A is mapped by a to the circle C_a which bounds the disk D_a;

(4) the attracting fixed point $\mathrm{Fix}^+(a)$ of a is inside D_a and the repelling fixed point $\mathrm{Fix}^-(a)$ of a is inside D_A;

(5) successive powers of a shrink the disk D_a to smaller and smaller disks containing $\mathrm{Fix}^+(a)$, while successive powers of a^{-1} shrink D_A to smaller and smaller disks containing $\mathrm{Fix}^-(a)$.

The inverse of a transformation symbolised by a lower case letter like a or b, is denoted by the corresponding upper case letter like A or B. Thus $a^{-1} = A$ and $A^{-1} = a$. Using this convention, for any upper or lower case letter x, the disk D_x contains the attracting fixed point of x.

---

point or sink is inside D'. Successive images of D and D' nest down on the source and the sink.

For historical reasons, instead of denoting transformations with letters like S and T, we are going to change our notation and systematically begin calling them a and b. We shall want names for their inverses, and from now on a^{-1} will be denoted by A, and b^{-1} by B.[1] The disks which go with a will be called D_a and D_A, chosen so that a maps the *outside* of D_A to the *inside* of D_a and vice versa. The circles which bound these disks will be called C_a and C_A. To remember which way round it is, notice we have chosen the labels in such a way that D_a contains the attracting fixed point of a. Notice the distinction here between the *circle C_a* (that is, the circumference traced out by drawing with a pair of compasses) and the coloured area or *disk D_a* which it bounds.

Figure 4.1 shows a picture of two transformations a and b arranged so that the four disks D_a, D_A, D_b and D_B do not overlap. One pair of disks is pink and the other is blue. The fun begins when we start following one of the maps A or a by one of the maps B or b. Figure 4.2 shows the result. All we have done is repeatedly applied all the four transformations a, A, b and B to the initial configuration of 4 disjoint circles! The wonderful feature of this picture is that no matter how far you 'zoom in', the essentials

[1] An aside to our mathematical colleagues: we adopted this rather non-standard 'change-of-case convention' notation out of necessity. When writing programs, you can imagine that storing something like a^-1 in a string variable is a trifle more laborious than storing A. Once having made the change, we never wanted to go back.

of the pattern repeat, as you can see from Figure 4.3. Like big fleas bitten by lesser fleas bitten by yet lesser fleas, each disk contains three further smaller disks, which in turn each contain another three. In Figure 4.2, we have only shown the disks to five levels. In Figure 4.3, we zoomed in with high magnification[1] on the tiny black rectangle marked on the first picture. Allowing for colour-cycling, the appearance of the zoom is not very different from the whole picture. As we penetrate further and further, we find ourselves looking at minuscule disks which nest down inside other disks, themselves in turn inside other disks, each disk yet more tiny than the ones which came before.

The collection of all possible compositions of the Möbius transformations a and b and their inverses A and B which were used to make this picture is a good example of a group. A group, remember, is a collection of symmetries with the two properties:

(1)   if S and T are in the collection, then so is the composition ST;

(2)   if S is in the collection, then so is the inverse S^{-1}.

In this particular group, each possible composition can be thought of as a '**word**' written in an alphabet containing only the four letters a, A, b and B. A typical member would be the transformation $aBBaAab$. The collection of all these compositions (in any order whatsoever) is called the **Schottky group** on the **generators** a and b, in honour of Friedrich Schottky who first manufactured groups of Möbius maps in this way. Of course, you can also have Schottky groups made with more than two generators.

Although each word written in this alphabet represents a transformation, different words don't necessarily represent different transformations. For example, the sequence aA cancels to give the identity transformation 1, so $aBBaAab$ represents the same map as the shorter word $aBBab$. And be careful: in a Schottky group, ab is not the same as ba so we can't, for example, simplify Bab first to Bba hence to a. In fact, no simplifications other than cancelling out adjacent a's and A's and adjacent b's and B's are possible. A group in which these are the only cancellations which happen is called **free**.

We expect groups to be associated with symmetrical objects. We shall shortly find a tiling hidden in our picture, a tiling which is symmetrical or **invariant** under the transformations in the Schottky group. But there is another symmetrical object hidden here as well. Repetition of similar structures (like for example all these nested circles) at all scales means there is likely to be some sort of fractal object lurking in the depths. In

[1] Jonathan Swift (1667–1745) apparently wrote his famous doggerel in response to the invention of the microscope. Food for thought!

Figure 4.1. The four initial Schottky disks. The transformation a maps the exterior of D_A onto the interior of D_a, and b maps the exterior of D_B onto the interior of D_b.

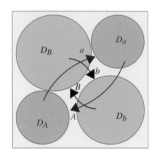

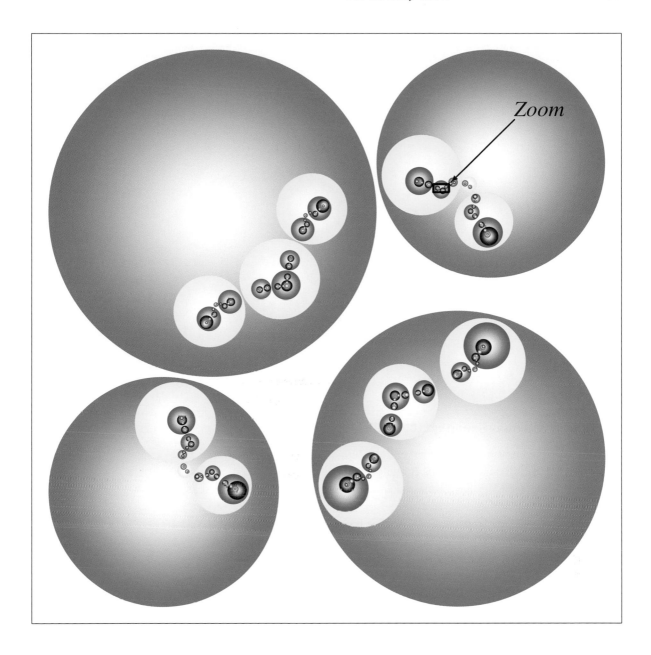

Figure 4.2. The Schottky array. Images of the initial four Schottky disks under iterations of the transformations a, A, b and B. You can see disks within disks nesting down on what look like specks of dust.

fact there appears to be something almost ethereal at the very limit of all these microscopic circles, a smattering of constantly fragmenting dust. As you can see from our quotation on p. xvi, Klein's imagination failed him when he tried to conceive the appearance of this dust. This dust is nothing but *the collection of those points which belong to disks at every single level*. There is exactly one particle of dust at the end point of each infinite chain of nested disks.

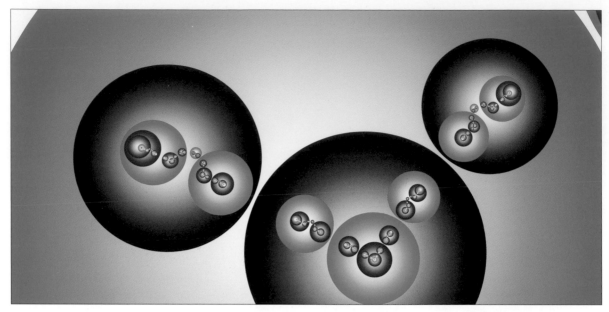

Figure 4.3. A zoom showing the small black rectangle in the picture on the previous page.

Friedrich Hermann Schottky, 1851–1935

Schottky was born in Breslau (then part of Germany but now in Poland) and received his doctorate in 1875 in Berlin working with both Weierstrass and Helmholtz. He seems to have been rather a dreamer, having been arrested as a student because he had forgotten to register in time for his military service. Fortunately he proved so useless as a soldier he was discharged after only six weeks. However he was a talented mathematician and the groups introduced here appeared in his thesis. These examples intrigued Klein, who pointed them out to Poincaré. Schottky also discovered the correct number of parameters needed to describe all possible Schottky groups; see p. 383. Later, he worked on the theory of functions of a complex variable, teaching at the University of Berlin until 1926.

The points of dust are at the limit of the nesting circles, so the dust is called the **limit set** of the group. Because of the way we created it, *each transformation in the Schottky group will leave the dust-like limit set unchanged.* We have created our first really intricate pattern – not a tiling but a chaotic 'fractal' – which is symmetrical or invariant under the transformations in a group. We shall have more to say about fractals and limit

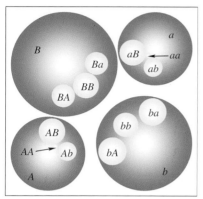
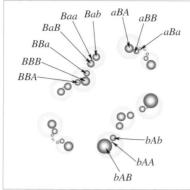
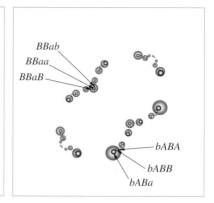

Figure 4.4. The first Schottky disks. The left frame shows images (in yellow) of the four initial red disks under combinations of two of a, A, b, B. The centre frame shows level 3 disks (in green) nested inside the level 2 disks. On the right are the level 4 disks.

points in the next chapter. For the present, though, let's look in a bit more detail at how the creation of new disks works.

To analyse how the Schottky group builds up its array of nesting circles, we need to look carefully at Figure 4.4, which shows the first steps in the Schottky dance. The initial four disks D_a, D_b, D_A and D_B, sometimes called the **initial Schottky disks**, are shown in red. Since Möbius transformations always map circles to circles, when any of the four transformations a, b, A, B is applied to any the four red disks, all four images are disks themselves. In the figure, these image disks are all coloured yellow.

Let's look more closely at the effect of the transformation a. Since a maps the outside of D_A onto D_a, this disk is in fact nothing new. What about the other three new disks $a(D_a), a(D_b)$ and $a(D_B)$? Well, since D_a, D_b and D_B are all *outside* D_A, and since a maps the *outside* of D_A to the

---

Note 4.1: **Alphabet soup**

The collection of all possible words which can be written using the four-letter alphabet a, A, b and B forms a group. For example, if $aBBaBab$ and $BBaBab$ are both members of the group, then so is their composition $aBBaBab\ BBaBab = aBBaBaBaBab$, where we have cancelled out the sequence bB at the join. The *inverse* of the transformation $aBBab$ is the sequence $BAbbA$, which we got by writing the word in reverse order and replacing each symbol (for example A) by its inverse (in this

case a). (This basic rule was discussed in Chapter 1.) That $BAbbA$ is indeed the inverse transformation can easily be checked by the calculation

$$aBBab\ BAbbA = aBBa\ AbbA = aBB\ bbA$$
$$= aB\ bA = a\ A = nothing.$$

In each case we cancelled a pair aA, bB, Bb or Aa until we destroyed it all. What is the identity in this group? It is *nothing*, or the *empty word*.

inside of D_a, the image disks must land inside D_a. They are the three yellow disks inside D_a in the figure. In just the same way, applying the transformation $A = a^{-1}$ creates three new yellow disks

$$A(D_A), \quad A(D_b), \quad A(D_B)$$

all of which are inside the disk D_A. Applying b creates new yellow disks

$$b(D_a), \quad b(D_A), \quad b(D_b)$$

inside D_b, and applying B creates further yellow disks

$$B(D_a), \quad B(D_A), \quad B(D_B)$$

inside D_B. These are the disks you can see with their labels on the left.

It is not hard to imagine that as we go on applying the transformations a, A, b, B, more levels of smaller disks will appear. For example, there are in total $3 \times 3 = 9$ yellow disks outside D_A. Applying a, we would by the same reasoning expect to find in total 9 disks inside D_A. Three of these should be inside the yellow disk $a(D_a)$, three inside $a(D_b)$ and three inside $a(D_B)$. You can see the disks at this level in green in the centre frame. Thus circles within circles continue *ad infinitum*. We call the entire collection of all the smaller and smaller image disks which are created by this process, the **Schottky array**.

The pattern in which the four Schottky circles map under the four generators will recur repeatedly in our constructions. To recapitulate: we have 4 disjoint disks D_a, D_A, D_b and D_B and the region outside them all, which we shall call F. The map a carries the outside of D_A to the inside of D_a, and the inside of D_A expands to swallow up everything except D_a. In symbols:

$$a(D_A) = D_A \cup D_b \cup D_B \cup F,$$

where $S \cup T$ means all the points which are either in the set S or in the set T or in both. The other generators have an analogous effect. We call this arrangement of disks and maps the **Schottky dynamics**. As we shall see, you can have Schottky dynamics even if some of the circles touch (but not cross). In the final chapters, we shall find that it is even possible to make a similar construction by pairing shapes which aren't circles. As long as the paired shapes are disjoint, any group like this is nowadays called a **(general) Schottky group**. Groups in which the shapes paired are circles are sometimes called **classical Schottky groups**, to distinguish them from their more exotic cousins.

From the computer pictures, it seems obvious the diameters of a sequence of nesting disks always shrink to 0. But we must tread with caution: what seems obvious at first sight may, in slightly differing circumstances, be false. We saw examples in Project 3.6 in which two of the original circles to be paired are tangent rather than disjoint, and in which the diameters of the nesting circles never get near to 0. Fortunately for our programs, the fact is that in the situation of pairing *disjoint* circles, the diameters *do* always shrink to 0. In fact, for any Schottky group, there are constants $k > 1$ and C such that the radii of all circles of level l are at most Ck^{-l}. One says that the circles shrink to zero **exponentially fast**. You can work through why this is so in Project 4.5.

Labels and trees

There is a nice way of recording the pattern of circles nested within circles, which mathematicians and computer programmers alike call a **tree**. By this, they mean a collection of vertices or **nodes**, connected by edges or arrows. Each node has a level, 0, 1, 2, and so on. There is exactly one level 0 node, out of which emanate a number of arrows connecting to the nodes at level 1. From the level 1 nodes, in turn, emanate arrows connecting to certain of the nodes at level 2. The pattern continues in the same way: each node at level 2 is connected by one backwards pointing arrow to a node at level 1 and by a number of forward pointing arrows to certain of the nodes at level 3, and so on.

The tree in Figure 4.5 is a schematic way of showing the arrangement of nested Schottky disks out to level 3. The tree has one initial point or node labelled 1, which we call the node at level 0. (This inconsistency is a bit annoying, but we ask the reader to bear with it; if we did not choose the labels this way the problem would pop out elsewhere and cause more trouble later on.) From the initial level 0 node emanate four arrows pointing to four nodes labelled a, b, A and B, representing the four initial disks D_a, D_b, D_A and D_B. These are the nodes at level 1. From each of these in turn emanate three arrows pointing to nodes at level 2. From the level 1 node a the three outward pointing arrows go to nodes aa, aB, ab, which are the three possible words with initial letter a and no cancellations like aA or Bb. Likewise, from the node A, three outward arrows point to the level 2 nodes AA, AB and Ab.

We can, of course, extend this tree to further levels, and the pattern repeats. Each level N node is reached by one arrow arriving from the level $N - 1$ node 'behind' it in the tree; while emanating from it are three further arrows pointing to three new nodes of level $N + 1$. You get the labels of the three new nodes by adding each of the 3 letters which don't cancel with the final (rightmost) letter of the word to the right end of the label. The fourth letter cancels, reversing the direction of the arrow and bringing us back to the word directly behind us at one level lower. For example,

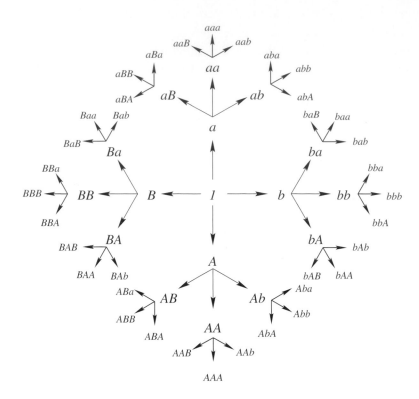

Figure 4.5. The tree of words to third level. The four arrows meeting each node represent *right* multiplication by each of the four generators *a*, *b*, *A* and *B*. At each of the nodes of level 1 or higher, there is one inward arrow and three outward arrows. If we multiply a word on the right by the *inverse* of its very last letter, we cancel that letter and go back down the tree along the inward arrow.

at the node *Ab*, the final letter is *b*, which cancels with *B*. Therefore, the three forward arrows go to *Aba*, *AbA* and *Abb*, while the previous node in the tree is *A* (which you might also think of as *AbB*). Notice that none of the labels on the nodes will ever contain an inverse pair such as *aA* or *Bb*. Words like this are called **reduced**. Each reduced word in our four-letter alphabet occurs at exactly one node on this **word tree**, in other words, the word tree is a perfect catalogue of the reduced words in the group.

The images of the Schottky disks are labelled in exactly the same way as the nodes on the word tree. To do this properly, when we write down a label, we have to be sure we know exactly which image disk we mean. Keep an eye on the labels in Figures 4.4. Each new disk was formed by applying a certain reduced word to one of the initial four disks D_a, D_A, D_b, D_B. The name of each new disk should record both the original disk and the transformation which was used. Our convention is that we write D_{aa} for the disk $a(D_a)$, D_{ab} for the disk $a(D_b)$ and D_{aB} for the disk $a(D_B)$. Doing this for the other three Schottky disks, we get disks $D_{Ab} = A(D_b)$, $D_{AA} = A(D_A)$, and so on. In fact, there is a distinct disk for

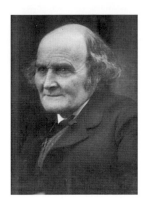

Arthur Cayley, 1821–1895

The work of the eminent English mathematician Arthur Cayley on algebraic geometry in the mid-nineteenth century, was, in Klein's opinion, of 'outstanding importance'. In the 1850's, Cayley wrote three important and influential papers formulating and developing the idea of a group as an abstract set of symbols which could be combined mechanically in accordance with definite rules. Later he originated the idea of a graph with elements of a group as nodes. The edges joining the nodes really represent multiplication by the generating elements of the group. The resulting graph is called the **Cayley Graph** and is now a fundamental construction in many parts of mathematics.

Cayley was born in Surrey, England but spent his childhood in St. Petersburg where his father was a merchant. After his family returned to England, he studied at Cambridge where he was Senior Wrangler in 1842. He started his career as a lawyer, spending 14 years practising law and devoting his spare time to his real love, mathematics. His energy may be guessed by the fact that he published 250 papers in this period! He was appointed to a professorship in Cambridge in 1863 at a considerable reduction in salary, where he spent the rest of his life. He continued to be very prolific and was characterized by one of his friends by 'the active glance of his grey eyes and his peculiar boyish smile'.

Cayley's work touches our story at another point through his 1859 paper *A sixth memoir on quantics* which showed how the measurements of distance and angles in both Euclidean and spherical geometry could be deduced as a special case of projective geometry. This was a beautiful example of the use of quantities invariant under a group of transformations and strongly influenced Klein's early work, inspiring his thinking about groups, symmetry and geometry, and leading to his interest in non-Euclidean geometry and hence to all the topics of our book.

each of the twelve possible two letter sequences of the symbols a, A, b, B which do not form the identity sequence $aA = Aa = bB = Bb = 1$.

The general rule is that for any word $xyz \ldots st$, the notation $D_{xyz\ldots st}$ means the disk $xyz \ldots s(D_t)$ obtained by transforming D_t by the map $xyz \ldots s$. Let's look at some examples. For instance, D_{abA} is $ab(D_A)$. So we also get $D_{abA} = ab(D_A) = a(b(D_A)) = a(D_{bA})$, which shows how the disks with level 3 labels are obtained by applying the generators to the disks of level 2. In the same way, the name $D_{aBaBBBab}$ indicates the disk which is the image of D_b under the word $aBaBBBa$. Alternatively, it is the image of D_{BBab} under the word $aBaB$.

Suppose we want to locate the disk D_{abA}. First of all, we know that $D_{bA} = b(D_A)$ is inside D_b. Therefore $D_{abA} = a(D_{bA})$ must be inside $a(D_b) = D_{ab}$. This gives us a chain of nested disks: D_{abA} is inside D_{ab} which in turn is inside D_a. In set theory notation,

$$D_a \supset D_{ab} \supset D_{abA}.$$

(The symbolism $S \supset T$ means that every point in the set T also is a point in the set S, but not necessarily the other way round: it says the set S contains the set T.)

In exactly the same way, for any reduced word $xyz\ldots st$ in our alphabet a, A, b, B, there is a chain of nested disks

$$D_x \supset D_{xy} \supset D_{xyz} \supset \ldots \supset D_{xyz\ldots s} \supset D_{xyz\ldots st}.$$

Note the order in which things are done. *You are warned* that this may not be quite what you think. Even though you will almost certainly want to say so, D_{ab} is the image of D_b under a and *not* an image of the disk D_a under any of the four transformations a, A, b or B. Virtually everyone first studying this subject has fallen into this trap. In fact, the four disks $D_a, D_{ab}, D_{abA}, D_{abAB}$ are images (under which transformations?) of the four *different* original Schottky disks D_a, D_b, D_A and D_B. In Figure 4.6 we have coloured all the images of the same disk in the same colour. See how this is quite different from our earlier colouring scheme.

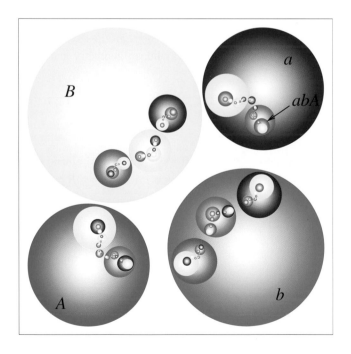

Figure 4.6. The Schottky code. The rule for colouring this picture is that all the images of the same disk are the same colour: all images of D_a are red, all images of D_b are blue, and so on. For example D_{abA} is the green disk inside a larger blue disk inside a still larger red disk.

You can also see from our pictures that the word tree has more significance than just being a convenient way of recording successive images of the Schottky disks D_a, D_A, D_b, D_B. The roughly circular array of the vertices at a given level in the tree is reflected in the roughly circular array of the image disks. When we come to write actual programs, this ordering will be a key to making good plots. We chose the labelling so that disks corresponding to two vertices which are close on the tree, are also reasonably close on the plane. There is one respect, however, in which the comparison between the tree and the disks is not at all exact: you can see clearly that there are marked differences in size among disks of the same level. Thus in Figure 4.2, on level 1, the diameters of the original Schottky disks are comparable. However, by level 3, the sizes of the image disks vary by almost two orders of magnitude. This is a major complication if you want to make good programs: you can see it already caused us trouble when we tried to write in the labels of all the disks in Figure 4.4.

Tiles and pretzels

Thus far, our symmetry groups have always gone hand in hand with a tiling: either the bathroom floor tiles in the picture on p. 17 in Chapter 1, or the ring-shaped tiles associated to the group generated by a single loxodromic transformation in Chapter 3. As we already hinted, there is still a set of symmetrical tiles hidden amid the circles in the Schottky array. It is a bit harder to spot them, because their shape is rather more exotic than those we have met thus far. Perhaps a volunteer could help us. Dr. Stickler, would you do us the favour of guiding us around?

Figure 4.7 shows a time elapsed photograph of Dr. Stickler's journey. Starting boldly at the centre of the picture, you see him shrinking and turning as he progresses ever more deeply into the Schottky array. A new Dr. Stickler appears for each transformation in the group. The most striking feature of the picture is that the arrows exactly reflect the pattern in the word tree. Each Dr. Stickler is at a node, and if you copied over the labelling from the word tree you would find each copy labelled by the exact transformation which gets him from his central position to the given location. Of course, we only show a finite number of levels, but it is not hard to imagine the whole infinite tree of words extending outwards, the edges getting ever shorter until we eventually reach a limit point at the infinite end of the branch. The picture confirms our intuition that the limit set is nothing other than the boundary of the word tree.

The largest Dr. Stickler at the root of the tree is outside all four

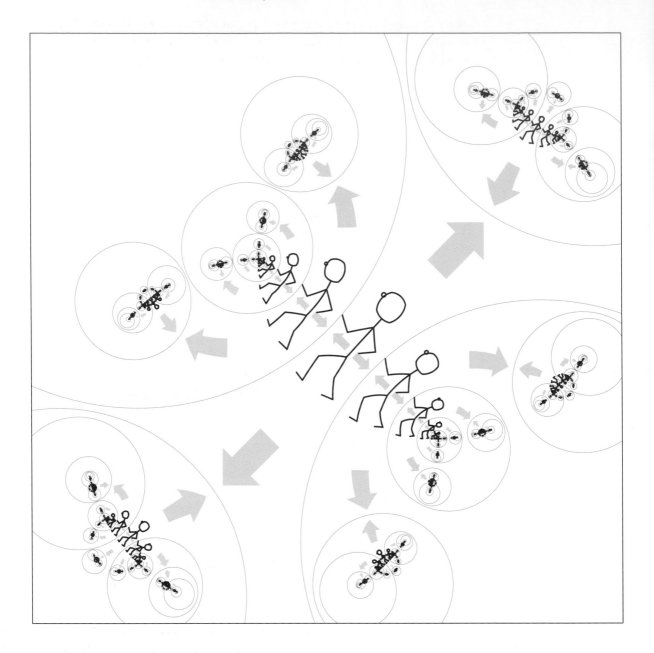

original Schottky disks, in the region which forms the white background in Figure 4.2. This 'holey' region looks rather like a big chunk of swiss cheese we name F (for '*fromage*'). The generating transformations a, b, A, and B all copy F into vaguely similar pieces of swiss cheese in each of their respective Schottky disks. It is a bit easier to see the four new regions in Figure 4.2. There, the new regions are red, one for each level 1 vertex of the word tree. They are still circular disks from which three other

Figure 4.7. Our Schottky group takes Dr. Stickler for a walk. The picture cements the connection between the abstract tree of words, the nested Schottky disks and the Swiss cheese tiles.

smaller circular disks have been removed, but the proportions and relative positions of the missing disks have changed. Notice how the four new Dr. Stickler's have dramatically different sizes and shapes. This is because, as we have seen, Möbius transformations distort not only size but also shape.

The next level of swiss cheese regions in Figure 4.2 are yellow; you can count 12 of these corresponding to the 12 reduced words of length 2. In a similar way, Figure 4.7 is made up by applying longer and longer words to the original Dr. Stickler and the original cheese F. In Figure 4.2 the 108 fourth-level images of F are blue. (The new copies do not look exactly the same because of the usual non-linear distortions.) The image cheeses fit exactly along their edges, and it appears that they gradually fill up the whole Riemann sphere. The only points we shall never reach are the limit points at the ends of infinite chains of nesting disks in the Schottky array. (In just the same way, the concentric ring-shaped tiles for a single loxodromic transformation filled up the whole sphere except for the source and the sink.) In other words, the copies of F are moved around by our Schottky 'symmetry group' in such a way that they tile or tessellate everything except the limit set. This is the reason why it makes sense to call each copy of F a 'tile'. It was Fricke who first gave a name to the region filled up by the tiles: he called it the **set of discontinuity**, because it is the region of the plane where the images of Dr. Stickler under the group remain separated from each other and don't pile up. Nowadays, this set is more often called the **ordinary set** or **regular set** of the group. In the ordinary set, Dr Stickler's walk is relatively easy to understand. The limit set is what is left after you have punched out all the tiles.

In Chapter 1, Dr. Stickler cut out one of the squares from the symmetrical world of identical fields and farms in Figure 1.1, and glued it up matching two points whenever there was a transformation in the symmetry group which carried one into the other. The result was the doughnut illustrated in Figure 1.14. On p. 85 in Chapter 3, we saw him making a similar construction, gluing together points on the two boundaries of the now ring-shaped tile.

Can we make a similar surface by gluing the edges of the Schottky swiss cheese tile F? We start by making a rubber version F, as depicted in the first frame of Figure 4.8. It looks like the plane with four circular holes punched out. Strictly speaking, of course, we should be imagining the tile on the Riemann sphere, explaining why Dr. Stickler is pursing together the part of the plane stretching out to infinity and adding one extra point '∞'.

Following the recipe used before, he now has to glue together the edges

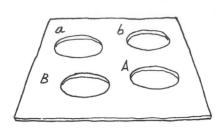

Figure 4.8. Dr. Stickler creates a pretzel by gluing up a slice of swiss cheese!

BEGIN WITH A PLANE MINUS FOUR DISKS, WITH CIRCLE a TO BE GLUED TO A AND b TO B.

ADD THE POINT AT INFINITY TO THE PLANE TO FORM A CLOSED SURFACE.

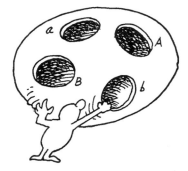

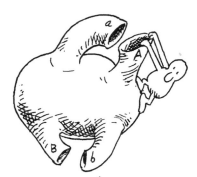

STRETCH SURFACE TO MAKE CORRESPONDING CIRCLES SIDE-BY-SIDE.

PULL OUT REGIONS SURROUNDING THE CIRCLES TO BRING THEM TOGETHER.

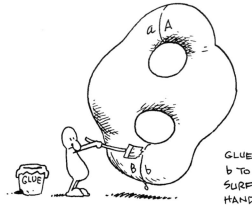

GLUE a TO A AND b TO B, FORMING A SURFACE WITH TWO HANDLES.

of this tile as dictated by the symmetries of the group. The generating transformations a and b match the boundaries of the missing disks in pairs, so he must match up each point P on the boundary circle C_a with its symmetrical point $A(P)$ on the boundary circle C_A, and similarly with points on circles C_b and C_B. As we see, he first stretches the surface so as to arrange the circles which are to be glued side by side. Then, pulling the disks outwards in pairs and bringing them together, he carefully flattens C_a against C_A and C_b against C_B. Gluing the circles together, he finally gets the double doughnut or 'pretzel' shown in the last frame. The surface has two holes or handles, so in mathematical language, he has created a **surface of genus two**.

Plotting the Schottky array

The pictures we have just been admiring are our first real glimpse of the type of images we want to create. How are we going to write programs to make pictures like these? Armed with the word tree, it is not as hard as it might at first appear. Assuming that by now the reader has a reasonable set-up for programming including matrices, complex arithmetic and making basic graphic plots, let's start with some basic features common to all the graphics algorithms we are going to write.

Set-up of generators. At the outset, we have to specify the group of transformations we want to use. Usually this just means defining the generators a, b, and so on, which will all be 2×2 complex matrices. The entries in the matrices will be specified as algebraic expressions or formulas which depend on certain **parameters**. We can choose the parameters when running the program but the formulas will be fixed. At this point, we won't go into any discussion of what the parameters and formulas are. You can find some simple examples to get started in Projects 4.1 and 4.2, and we shall be meeting many more recipes as we go along.

Let's assume we have just two generators a and b, together with their inverses A and B. These four transformations should be stored as a list of 2×2 matrices with complex entries

$$\texttt{gens[1]} = a, \quad \texttt{gens[2]} = b, \quad \texttt{gens[3]} = A, \quad \texttt{gens[4]} = B.$$

The exact ordering of this list is not very important but one thing we will always need to know is which generator is the inverse of which, in order to make sure we generate only reduced words and don't needlessly multiply a matrix by its inverse. We will do this using an array of 4 integers `inv`, where `inv[i]=j` means that `gens[j]` is the inverse of `gens[i]`. With the generators stored as above, we would get `inv[1]=3`, `inv[2]=4`, `inv[3]=1`, `inv[4]=2`.

Plotting objects. During the course of the program, we shall issue instructions to plot various things. For example, we may want to plot images of Dr. Stickler, or images of the initial four Schottky circles C_a, C_A, C_b, C_B. We must set up these **seed figures** at the beginning of the program. Our basic graphic operation will be to apply a word W made up of letters in the alphabet a, A, b, B to one of these seed figures and plot the result with the command `plot W applied to seed[i]`. The plots are of two basic types:

Tiling plots This plots something every time we come to a new node in the group. We are laying down something, for example Dr. Stickler, for each element of the group. Drawing the circles for our Schottky groups is like this, because each circle such as C_a or C_{Abba} is attached to a word a or $Abba$ and so on, with the exception that there is no circle attached to the identity map.

Limit set plots In these plots we do not actually plot objects for all the words which we calculate, but only those associated to the longest words which we get to in our search algorithm (sometimes called the 'leaves' of the search tree). Ideally, to make pictures of limit sets we want to plot objects for which the associated word is 'infinite'. Of course this is impossible, so we plot the objects which correspond to the longest words in our search.

Enumeration of words. The heart of the program is to catalogue the words in the group. We have already learnt how to do this with the aid of the word tree. Each word (including the **trivial** or **empty** word 1) is a node. Each reduced word W is connected by forward arrows to three new words Wx, where x can be any of the generators a, A, b, B *except* the inverse of the last letter in W. (There is an exception for the empty or trivial word which is connected by four arrows to the four generating words at level one.)

Search algorithms. We have not yet specified any particular order for running through the words. It is not even clear whether considering the words in order of length (that is, level) is the best approach. There are several competing methods of enumeration, which in the language of computer science are known as **search algorithms**, as if we were exploring the tree in search of something. (Which, in fact, we are. Our quest is for *visual accuracy*.) Different algorithms give different ways of organizing the search. The three algorithms we shall describe are:

Breadth-First Search or BFS This searches the tree level-by-level, not leaving a level until we have explored every word at that level.

Depth-First Search or DFS This follows a branch of the tree all the way to a node that activates a termination procedure, at which point we back up until we find an new unexplored branch.

Random Search or IFS This uses some method of 'randomly' selecting words in the group. Generally, the randomness is fake, simulated by some software that produces what appears to be a sequence of random numbers. IFS stands for Iterated Function System, a term coined by Michael Barnsley for a large class of algorithms of this type.

Termination procedures. Ideally, we would like to consider every single word in the group. In reality of course we can't do this, because there are infinitely many of them. Therefore, we must adopt a **termination procedure** to decide when we have examined enough. The definitive procedure would ensure that we consider exactly as many words as are necessary to produce a visually correct picture, according to the resolution of our printing or previewing devices (one of which is of course our own eyes). For pictures like those in this chapter, we might decide to plot only those words for which the Schottky disk had radius less than 10^{-3} cm. (or 10 microns, roughly the size of a cell in your body). It will be harder to choose a termination procedure in later chapters with more complicated plots.

Breadth-first search

The simplest programs use breadth-first search. Although not necessarily the most efficient or best method, this way will get us started quickly. Breadth-first search works by recording all the words up to a given level, working level by level. Here are the data structures we need:

`lev` The current level we are at.

`levmax` The maximum level we wish to search. Therefore, `lev` is not greater than `levmax`.

`group` A growing list of words in the group that we have generated. It starts out as a huge empty part of memory, space enough for (in our code box) one million 2×2 complex matrices (32 megs. in single precision). Having room for an array of such gargantuan dimensions is a defect of the breadth-first search. The array is gradually filled up as the program churns along computing more and more words in the group.

tag This is a list of integers of exactly the same length as group. As we generate words in the group, we record here the index (an integer between 1 and 4) of the last (rightmost) generator used in the word. Thus if we multiply on the right by gens[3] and put the result in group[56732], we also set tag[56732]=3.

num We shall need to keep track of the number of words enumerated at each level. In the present case, this is actually $4 \cdot 3^{lev-1}$. But in more complicated cases it might not be so easy to determine in advance and it is natural to keep track of this by setting num[lev] equal to the place in the array group where the first word of level lev is recorded.

Box 12 contains pseudo-code of a program for doing all this. Referring to the box to follow each step, the code starts with data type declarations which may or may not be necessary in the chosen programming system. Many languages require maximum array dimensions to be explicitly given at the start. In our example, we stipulate a maximum of one million words.[1] We have assumed there is some way of telling the compiler that each entry in group is a 2×2 complex matrix: we indicate this in the pseudo-code by giving the data type first, then the name of the array of this data type. The same holds for the simpler arrays whose entries are circles or integers.

[1] Setting limits in advance is reassuring to the computer, although it can be a bit nerve-wracking for the humans.

Next, we just indicate with INITIALIZE whatever is needed to choose the actual generators of the group and the four Schottky circles, and to specify which generators are the inverses of which. Note that gens[4] in this part of the code means we are declaring that gens will contain four matrices, not that we are accessing its fourth entry.

The procedural part of the code is divided into three sections, the first dealing with the words of length 1, the second with words of all lengths 2 through lev-1 and the last part with the plotting. The middle section is set up as a for-loop over all intermediate levels.

The first step just sets group[i] = gens[i] for each $i = 1, 2, 3, 4$ and plots the four initial circles. Remember that our numbering convention is that gens[1] = a, gens[2] = b, gens[3] = A, and gens[4] = B. We set num[1]=1, num[2]=5 to indicate that level 1 begins at slot 1, level 2 at slot 5.

At all intermediate levels, we need to add to our list of words. This is done with a for...do control loop from lev=2 to levmax-1.

What is done at each level is the stuff enclosed in the braces. These commands assume that the list group contains all words at level lev-1 in the slots beginning at num[lev-1] and ending at num[lev]-1. Thus the first empty place in the array is num[lev]. The inner core of the algorithm is labelled the *Main Enumeration Loop*. This loop calculates all the elements of the group out to a large level. It builds this list level by level, creating the next level by applying all the generators to the list of group elements in the current level.

Box 12: Breadth-first search

```
matrix gens[4], group[1000000];                    Declare the arrays needed.
circle circ[4], newcirc;
integer lev, levmax, tag[1000000], inv[4], num[levmax];

INITIALIZE gens[4], circ[4], inv[4];               Performed elsewhere.

for i from 1 to 4 do {                              Put the 4 generators in group
   group[i] = gens[i];                              and give them tag i.
   tag[i]=i;
   plot circ[i]; }                                  Plot the 4 initial circles.

num[1] = 1;                                         The index where level 1 begins.
num[2] = 5;                                         The index where level 2 begins.

for lev from 2 to levmax-1 do {                     MAIN ENUMERATION LOOP
   inew = num[lev];                                 Initialize the counter.
   for iold from num[lev-1] to num[lev]-1 do {      Run over words in the last level.
   for j from 1 to 4 do {                           Run over generators
      if j = inv[tag[iold]] then next;              but avoid cancellations.
      group[inew] = group[iold] * gens[j];          Add a new word to the list
      tag[inew] = j;                                and tag with last generator used.
      inew = inew+1;                                Advance the counter.
   } }
   num[lev+1] = inew;                               Record the end of this level.
}                                                   END ENUMERATION LOOP

for i from 1 to num[levmax]-1 do {                  OUTPUT LOOP
for j from 1 to 4 do {                              Run over circles
   if j = inv[tag[i]] then next                     but avoid cancellations.
   newcirc =                                        Compute the new circle
      mobius_on_circ(group[i], circ[j]);
   plot newcirc;                                    and plot it.
} }                                                 END OUTPUT LOOP
```

At each point, `iold` points to a spot in `group` occupied by a word of level `lev-1` and `inew` points to the first empty spot in the array `group`. So we take the word `group[iold]` and try to multiply it on the right by each of the four generators `gens[j]`. We have to avoid multiplying by the one generator which cancels the rightmost generator in this word. The generator to avoid is thus `inv[tag[iold]]`. In our innermost loop, we check whether we are about to apply the forbidden generator, and if so we use the command `next` to break off that pass of the loop and skip to the next one. In this way, we move slowly, left to right, through all the words of level `lev-1` and multiply each one by the three generators which produce a reduced word of size `lev`. When we're done, we record in `num[lev+1]` the next empty slot where words of the next level will begin. Here is a 'trace' of the words in `group` as the level ranges from 1 to 3:

$$a \longrightarrow b \longrightarrow A \longrightarrow B \longrightarrow$$
$$aa \longrightarrow ab \longrightarrow aB \longrightarrow ba \longrightarrow bb \longrightarrow bA \longrightarrow$$
$$Ab \longrightarrow AA \longrightarrow AB \longrightarrow Ba \longrightarrow BA \longrightarrow BB \longrightarrow$$
$$aaa \longrightarrow aab \longrightarrow aaB \longrightarrow aba \longrightarrow abb \longrightarrow abA \longrightarrow$$
$$aBa \longrightarrow aBA \longrightarrow aBB \longrightarrow baa \longrightarrow bab \longrightarrow baB \longrightarrow$$
$$bba \longrightarrow bbb \longrightarrow bbA \longrightarrow \ldots \longrightarrow BBa \longrightarrow BBA \longrightarrow BBB.$$

Except that we got tired of writing and skipped the 20 words between *bbA* and *BBa*, we have taken care to list the words above exactly in the order that would be computed by our BFS algorithm if we set `gens[1]=a`, `gens[2]=b`, `gens[3]=A`, `gens[4]=B`. The order does not match the order witnessed going around in the word tree in Figure 4.5. In fact, the algorithm we have presented lists the words in **dictionary ordering**, with '*a*' considered the first letter, '*b*' the second, '*A*' the third and '*B*' the fourth.[1] A big advantage of the DFS algorithm is that, as explained in the next chapter, it will relate our programs to the ordering in the word tree, calculating nearby points at roughly the same time.

This particular code is for a tiling type plot, so in the last part, for each word *W* which we have calculated, we plot those three of the four circles WC_a, WC_b, WC_A, WC_B for which the last generator in *W* doesn't cancel the label of the circle. (The circle which gets omitted is of a lower level and has already been plotted.) We use a list of circles

$$\text{circ}[1] = C_a, \quad \text{circ}[2] = C_b, \quad \text{circ}[3] = C_A, \quad \text{circ}[4] = C_B$$

following the same order as in `gens`. The formula needed for plotting the image of a circle by a Möbius transformation is on p. 91.

[1] In the dictionary ordering, the word *bAba* comes before *bAbB* and after *baBB*, and so on.

Notice that the word lists grow extremely rapidly, multiplying by 3 with each augmentation. Already at level 10, the length is more than a million, which would exhaust the defined limit of our arrays above. Many of our illustrations require consideration of words up to level 1000 or more in order to obtain visual accuracy. As it turns out, most of the words of high level correspond to visually undetectable features, so that we should be discarding rather than preserving them from step to step. Exploring such questions is a very interesting problem, but we shall shortly leave breadth-first methods and move on to the more efficient DFS algorithm we prefer.

If you want to see this code in action, you will probably want some formulas for actual matrices. As mentioned above, you can find some recipes in Projects 4.1 and 4.2.

Projects

4.1: Schottky groups with limit set on the real line

In Project 3.3 in the last chapter, we met the special Möbius transformation

$$z \mapsto \frac{(s+t)z - 2st}{-2z + (s+t)}.$$

Assuming $0 < s < t < 1$, it paired the circle with diameter on the segment between s and t with its reflection in the imaginary axis. If this transformation is called a, which of these circles should be called C_a and which C_A?

Now add another similar transformation

$$b(z) = \frac{(s+t)z + 2}{2stz + (s+t)}$$

to make a Schottky group. Find the normalized form of the matrix b, and check that b maps the outside of the circle C_B in Figure 4.9 onto the inside of circle C_b. What is the correct order of image Schottky circles at level 2 and level 3? (Your answer should *not* be the same as the order round the edge of Figure 4.5.) Why is the entire limit set of the group generated by a and b contained in the real line?

Figure 4.9. The arrangement of Schottky circles for the group described in Project 4.1.

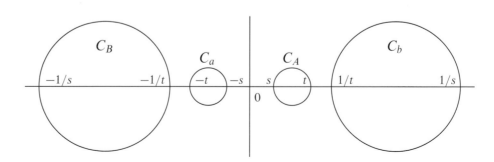

4.2: The θ-Schottky groups

Here are more concrete formulas for another special family of Schottky groups based on the symmetrical arrangement of circles in Figure 4.10. The formulas depend on an angle θ which measures the position of the circles, so we call them the θ-**Schottky groups**. The initial Schottky circles have been arranged orthogonally to the unit circle with centres at $r, ir, -r, -ir$ where $r = \sec\theta$.

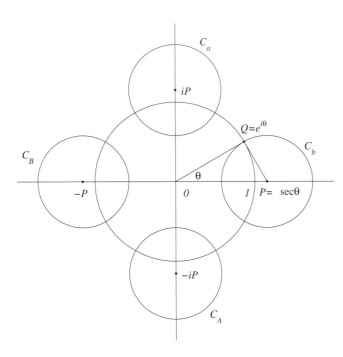

Figure 4.10. The Schottky circles for a θ-Schottky group. All four circles have the same radius, $\tan\theta$. The circle C_b has centre at $P = 1/\cos\theta$ on the horizontal real axis, meeting the unit circle at the point Q for which $\angle POQ = \theta$. In complex number notation, $Q = \cos\theta + i\sin\theta = e^{i\theta}$. Symmetrically, it also meets the unit circle at $Q' = e^{-i\theta}$. The centre of C_B is at $-1/\cos\theta$, while C_a and C_A have centres at $\pm i/\cos\theta$ on the vertical imaginary axis.

To get a group, we have to choose a, pairing C_a and C_A, and b, pairing C_b and C_B. It turns out this can be done with the transformations

$$b = \frac{1}{\sin\theta}\begin{pmatrix} 1 & \cos\theta \\ \cos\theta & 1 \end{pmatrix} \quad \text{and} \quad a = \frac{1}{\sin\theta}\begin{pmatrix} 1 & i\cos\theta \\ -i\cos\theta & 1 \end{pmatrix}.$$

Check that b is one of the stretch transformations we met on p. 89. It pushes circles orthogonal to the unit circle rightwards along the x-axis into other circles of the same kind. Explain why this means that to check that b pairs C_B to C_b, we only have to check that $b(-\sec\theta + \tan\theta) = \sec\theta + \tan\theta$.

The easiest way to find a is to observe that if we rotated everything through $90°$, then a would look exactly like b. Indeed, anticlockwise rotation R through $90°$ maps C_B to C_A and C_b to C_a. This means that $a = RbR^{-1}$, which you can check by calculating

$$RbR^{-1}(C_A) = Rb(C_B) = R(C_b) = C_a.$$

Find R in normalized matrix form and use it to verify our formula for a.

Now conjugate this example to make a group whose limit set is contained in the real line and with Schottky circles arranged in the same order as in Figure 4.5.

4.3: Fuchsian Schottky groups

The groups in the last two projects are called **Fuchsian Schottky groups** because they have the special property that all their limit points are on either a line or a circle. Explain why if the initial four Schottky circles are all orthogonal to a given circle, then so are all their images. Why does this mean the limit set must be contained in the circle?

Can you manufacture a Fuchsian Schottky group whose 4 initial circles are all orthogonal to the unit circle but not all the same size?

4.4: Schottky groups with more than two generators

There is no reason why the Schottky construction has to start from only two Möbius maps. For instance, we could take six disjoint circles and construct three maps a, b, c which match them in pairs. The whole formalism of words extends with an alphabet of six letters a, b, c, A, B, C. A similar construction of nested circles extends too, giving us a fractal picture of nested circles corresponding to the new word tree. A good program which handles two-generator groups requires only the change of the dimension of an array to implement this. Figure 4.11 shows an example of what you may find.

4.5: Shrinking circles?

We want to show why for any Schottky group, there are constants $k > 1$ and C such that the radii of all circles of level l are at most Ck^{-l}. Remember that a circle at level l is at the end of chain of length l, as in the example:

$$D_{baBaB} \subset D_{baBa} \subset \cdots \subset D_b.$$

Let's look for a constant k such that at each step along the way, the radius shrinks at least by k, that is $\text{rad}(D_{baBaB}) < \text{rad}(D_{baBa})/k$, $\text{rad}(D_{baBa}) < \text{rad}(D_{baB})/k$ and so on. Now the pair $D_{baBaB} \subset D_{baBa}$ can be obtained by applying the map baB to the level 2 pair $D_{aB} \subset D_a$ and the same holds for any nested pair in our construction. There are twelve such pairs $D_{xy} \subset D_x$ to which we can apply maps in our group to get all nested pairs in our chains. Now we use the results of Projects 3.4 and 3.5. Take any one of these pairs $D_{xy} \subset D_x$ and call them D and E. Find a Möbius map S which carries them to a concentric pair $D_0 \subset E_0$. Let $k = \text{rad}(E_0)/\text{rad}(D_0)$. Obviously $k > 1$. Project 3.5 tells us that

$$\frac{\text{radius of } T(E_0)}{\text{radius of } T(D_0)} \geq k$$

for any Möbius map T which doesn't turn the circles inside out. So it must also be true that

$$\frac{\text{radius of } US^{-1}(E_0)}{\text{radius of } US^{-1}(D_0)} \geq k$$

CHAPTER 5

Fractal dust and infinite words

Even in a single pore are inconceivably many lands, countless as particles of dust... In every particle of dust in these lands, one also differentiates countless lands, some small, others large...

Avatamsaka Sutra[1]

Allowing our circle program to run for a long time, we find ourselves drawing great numbers of circles, circles within circles within circles. The disks of higher and higher levels seem to recede 'into the distance' as their diameters rapidly shrink to zero. As we learnt in the last chapter, there is a pot of gold at the end of each infinite sequence of nested circles, a very special point called a **limit point**. You can see these limit points glowing yellow in Figure 5.1. They are a smattering of points in the plane, what Mandelbrot calls 'fractal dust', known to the experts rather less colourfully as the **limit set** of the Schottky group generated by the transformations a and b. We shall be explaining more about what is meant by fractal dust as we go along.

In the last chapter we described a way of drawing all the Schottky circles. We can easily change it to draw only the limit set: run the BFS circle drawing program without plotting any circles until the words we are examining are extremely long. Eventually, all the circles will become extremely small, and the plot we get will be a very good approximation to the fractal dust.

In order to understand the limit set better, we are going to label each and every one of its infinitely many points in a distinctive way by thinking of them as the tips on an *infinite* tree of words. The procedure is not as mysterious as you might think: it is analogous to using infinite decimal expansions to locate points along a line. This analogy is both close and extremely useful, and will help us feel at home with different ways of describing limit points. Once set up, the labelling will enable us to dispense with the nesting circles and work directly with limit points

---

[1] From the *The Weaving of Mantra*, by Ryuîchi Abé, © Columbia University Press. Reproduced by permission of the publisher.

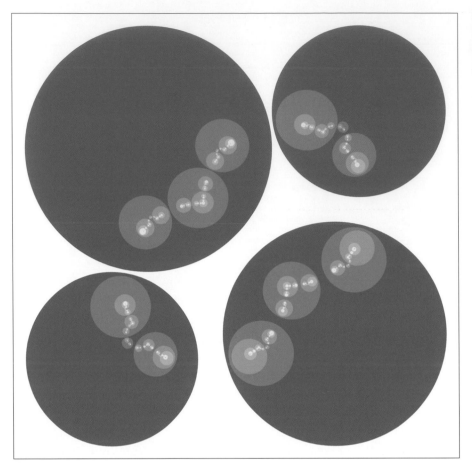

Figure 5.1. The Schottky dust. Mandelbrot coined an evocative name for sets like this which break into further pieces every time you zoom in: he called them *fractal dust*. The picture is the same as Figure 4.2, but we have chosen the colouring to highlight the limit points in glowing yellow.

themselves. If we are interested in seeing limit set pictures without all the Schottky circles obscuring the view, this will greatly speed things up. It will become indispensible when in later chapters we abandon the Schottky circles altogether and start drawing limit sets formed from groups of transformations created in quite different ways.

In the second part of the chapter, we shall be introducing a new limit set program based on the principle of **depth-first**, rather than breadth-first, search. DFS will be the algorithm which accompanies us from now on, and virtually all our limit set pictures were plotted by this method. (We should note however that some distinguished researchers have made very good plots by elaborations of breadth-first techniques.)

Limit points and infinite words

As we have seen, each limit point is at the end of an infinite sequence of nesting disks. Thus a natural way to 'name' it is by the infinite sequence of disks which nest down onto it. For example, the limit point which is at the end of the infinite chain of nesting disks

$$D_a \supset D_{aB} \supset D_{aBB} \supset \cdots \supset D_{aBBab} \supset \cdots$$

might just as well be represented by the infinite sequence of letters *aBBab* $\cdots$.

Since a finite string of letters makes a (finite) word, we call an infinite string of letters like this, an **infinite word**. Notice how we indicate that the word *aBBab* $\cdots$ is infinite. There is no concluding symbol, just a string of dots at the end. The dots indicate that the word goes on forever ... , as opposed to a finite word which we always indicate by inserting a final letter, such as *aBBabba* $\cdots$ *A*. This is familiar in writing the value of 'pi': one typically writes $\pi = 3.1415 \cdots$ meaning that its decimal expansion begins as written but continues forever. This seemingly pedantic distinction may appear to be a minor point, but it is easy to be tripped up.

It still makes perfectly good sense to talk about infinite words being reduced. A reduced infinite word is one which doesn't contain cancelling sequences like *aA* or *Bb*, so as to avoid needless doubling back. For example, *aBBaBBaBB* $\cdots$ is reduced while *aBbaBbaBb* $\cdots$ is not.

An infinite reduced word corresponds to an infinite path down the word tree. For example, the word *aBBaBB* $\cdots$ corresponds to the red path in Figure 5.2. Starting from the central node 1 (corresponding to the empty word), we follow the arrows to the nodes whose labels are *a*, *aB*, *aBB*, *aBBa*, *aBBaB*, . . . and so on. If the word wasn't reduced, we could still follow the arrows but we would find ourselves doubling back, a wasteful process we would do better to avoid.

Any infinite reduced word also specifies an infinite sequence of nested disks. Just as the finite reduced word *aBBaB* gives a nesting

$$D_a \supset D_{aB} \supset D_{aBB} \supset D_{aBBa} \supset D_{aBBaB},$$

so the *infinite* reduced word *aBBaBBaB* $\cdots$ gives an infinite nesting

$$D_a \supset D_{aB} \supset D_{aBB} \supset D_{aBBa} \supset D_{aBBaB} \supset \cdots \supset D_{aBBaBBaB} \supset \cdots.$$

Once again we use dots at the end of our formula, this time to indicate that the string of inclusions is to be continued without end. As we saw in Project 4.5, the diameters of such sequences of disks always shrink to zero, so that the disks nest down on a definite point which is the limit point represented by the infinite word. All words which begin with a certain string, for example *aaBABABBBA*, are located inside the same disk $D_{aaBABABBBA}$. If the string is long, then the corresponding disk will be small, so that all the words beginning with this particular string will be quite close.

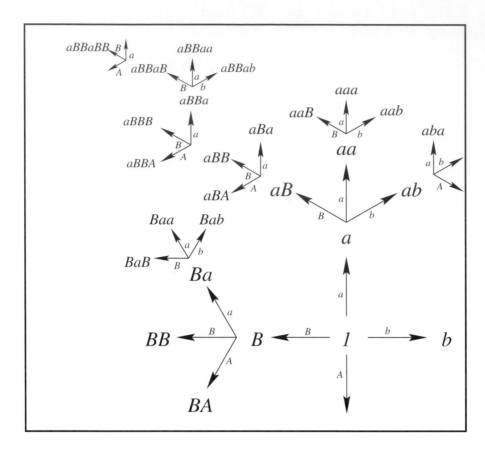

Figure 5.2. You can't go home again. This is part the same word tree as that in Figure 4.5, extended in the upper left corner. The red arrows and labels highlight an infinite path corresponding to the infinite word *aBBaBB*⋯. Following the arrows, the path keeps going forever, but, of course, we only had a finite amount of time to draw this picture.

There is one crucial difference between finite and infinite words, which will cause confusion if we don't get it straight. A finite word in the generators corresponds to a matrix product which can be computed explicitly given the generating transformations a and b. For example, putting in $a = \begin{pmatrix} 1 & 2 \\ 0 & 1 \end{pmatrix}$ and $b = \begin{pmatrix} 1 & 0 \\ 2 & 1 \end{pmatrix}$, we get

$$aBBaBA = \begin{pmatrix} 1 & 2 \\ 0 & 1 \end{pmatrix} \begin{pmatrix} 1 & 0 \\ -2 & 1 \end{pmatrix} \begin{pmatrix} 1 & 0 \\ -2 & 1 \end{pmatrix} \begin{pmatrix} 1 & 2 \\ 0 & 1 \end{pmatrix} \begin{pmatrix} 1 & 0 \\ -2 & 1 \end{pmatrix} \begin{pmatrix} 1 & -2 \\ 0 & 1 \end{pmatrix}$$

$$= \ldots \text{(after deleted expletives)} \ldots = \begin{pmatrix} -11 & 7 \\ -8 & 5 \end{pmatrix}.$$

If the word were infinite, however, what could this mean? It is not hard to see that, as we multiply more and more matrices together, that the entries tend, on the whole, to increase without bound. So multiplying together the infinite sequence of matrices corresponding to an infinite word doesn't seem as if it can make any sense. Fortunately, we never actually have to do this: remember all we are trying to do is to find the limit point represented by the infinite word. We shall never have to make sense of an infinite product of matrices. A finite word represents a composition of Möbius maps, but an infinite word represents a limit point of a group. To understand how these infinite words work, let's spend a bit of time exploring the analogy with decimals alluded to above.

Decimals

It may seem rather foreign to encode limit points by infinite reduced words, but it is not actually such a startlingly new concept as you might think. Intuitively familiar since schooldays, infinite decimals do exactly the same thing.

A terminating decimal is really the same as an infinite decimal whose expansion ends in an infinite string of 0's; you can think of 8.08 as short for 8.08000 . . . . The next easiest infinite expansions to handle are those in which a certain finite pattern repeats, such as

$$3.142857142857142857142857142857142857142857142857\cdots,$$

[1] At the time of writing, the digits of π have been calculated to one gazillion decimal places. Calculations are still going on. See *Talking About Pi* at www.cecm.sfu.ca/ ~jborwein/pi_cover. html.

which you may recognise as the decimal expansion of 22/7. On the other hand, the numbers

$$3.14159265358979323846264338327950288419716939937510\cdots$$

in the expansion for π go on forever without any discernable pattern at all.[1]

Note 5.1: **Infinite intersections**

We shall often have cause to talk about 'infinite nestings' or 'infinite intersections' of disks or other sets. There is no mysticism here: to say a point x is in an infinite intersection of sets $X_1, X_2, X_3, \ldots$ means simply that x is in the set X_n for each value $n = 1, 2, 3, \ldots$. Notice that as usual we use $\ldots$ at the end of a list to indicate that it goes on forever without end. If we meant the list to be finite, then we would have written $X_1, X_2, X_3, \ldots, X_n$. It is one of the main postulates or unproven assumptions in the mathematical theory of the real numbers that such an infinite nesting of non-empty bounded 'closed' sets always has at least one point x in its infinite intersection. The one twist is that one has to make sure that each X_n is 'closed'. This is a technical word which means that the limit of any sequence of points in X_n is itself in X_n, more intuitively, that X_n includes all its boundary. (Thus if X_n is a disk, it is closed only if we include its circumference.) This postulate is a theorem in some ways of developing the theory of real numbers: it depends on where you want to start. Good discussions of this circle of ideas can be found in many textbooks such as *Calculus* by Michael Spivak, Publish or Perish Press, 1994.

Repeating strings are especially useful, both for decimals and for our Schottky words. We shall call the repeating string the **repetend**.[1] In the United States, it is usual to indicate this by putting a bar over the repetend. On the other side of the Atlantic, it is more usual to use dots. Since lines are easier to read, we shall opt for the American convention and write, for example,

$$22/7 = 3.142857142857142857142857\cdots = 3.\overline{142857}$$

(3.1̇42857̇ on the old-fashioned side of the Atlantic.) The length of the repetend is called the **period**; in our example, the expansion of 22/7 has period 6. All fractions have repeating decimal expansions and repeating decimal expansions are always fractions: see Note 5.2. There may of course be some digits before the repetend, for example $3.14\overline{3} = 43/300$ and $0.08\overline{0} = 0.08 = 2/25$.

Even though two expansions may be infinite, there is an easy way to tell when the numbers they represent are close. We naturally group numbers whose decimal expansions begin in the same way. A programmer would say they have the same **prefix**, that is they both begin with the same finite string of digits. For example, 21.378892 and 21.37345 both have the same prefix 21.37. They agree up to the second decimal place, that is, they are within 10^{-2} of each other.

We make the analogy with Schottky groups by drawing a circle around the set of all numbers with a given prefix. This is shown in Figure 5.3, in which we have drawn some of the circles out to level 4. The large green circle contains exactly those numbers whose prefix before the decimal point is the digit 0, that is, all numbers between 0 and 1. The leftmost of the small labelled orange circles contains all numbers with the prefix 0.00, that is, numbers between 0.00 and 0.01. Each circle contains ten circles each of one-tenth the size, corresponding to the selection of the next decimal digit in the expansion. Due to the very rapid decrease in radius,

[1] The authors disclaim responsibility for this inelegant but useful word. One of us learned it during an otherwise somewhat sleepy seminar, and has thought that it deserved better publicity ever since.

Note 5.2: **Handling repeating decimals**

In some mythical golden age every schoolboy used to learn that repeating decimals correspond exactly to fractions, more formally rational numbers which can be written in the form p/q for integers p and q. For example, most people know that $0.3333\cdots$ is exactly equal to the fraction $1/3$. One way to work this out is the following trick.

Suppose $x = 0.\overline{abc}$. Multiplying both sides by 1000 gives $1000x = abc.\overline{abc}$. Subtracting, $999x = abc$ and so $x = abc/999$. You may like to apply this method to find the rational number whose expansion is $0.\overline{123}$.

The decimal expansion of an irrational number never settles down to an infinitely repeating string.

it is only possible to see a few generations of circles. In the picture we haven't been able to go beyond the fourth level, all numbers in one of the tiny blue circles having the same prefix of length 4.

Figure 5.3. Circles and decimal expansions. Each circle contains all real numbers with a prescribed prefix of digits. The tiny blue ones along the central line are too small to have their prefix printed.

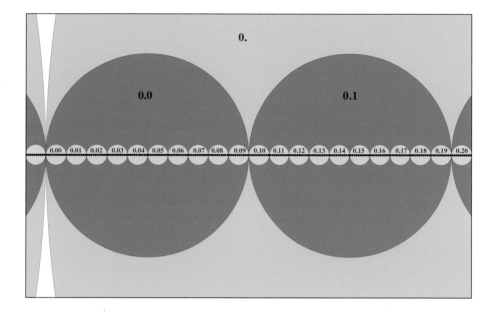

The nested circles should immediately remind you of the circles in the Schottky array. In fact, Figure 5.3 makes clear that each point on the real number line lies at the intersection of a rapidly shrinking infinite sequence of nested disks. The main difference is that many circles in the decimal picture are tangent, while all the circles in the Schottky array are disjoint.

The tangencies in Figure 5.3 are connected to the problem of infinitely repeating 9's.[1] They explain visually why for example

$$0.9999\overline{9} = 1.000\overline{0}.$$

[1] This is closely related to one of Zeno's paradoxes: how can the hare catch up to the tortoise when he needs to take an infinite number of steps, first to get to where the tortoise started, second to get where the tortoise was by the time he got there, and so on?

Whenever two circles are tangent, you can see two infinite sequences of nested disks, one shrinking down to the point of tangency from the left and one from the right. These points are exactly those which have two decimal expansions, one ending in all 0's and one ending in all 9's. For example, $0.07\overline{9}$ is the tangency point of the two orange circles whose three letter labels represent the prefixes 0.07 and 0.08. The tangency indicates that $0.07\overline{9}$ is equal to 0.08. Even for decimals such points are the exceptions: 'most' points lie inside only one infinite nest.

In the Schottky case, the initial circles are disjoint so tangent circles which create ambiguities do not occur. When in the next chapter we start to explore what happens when some of the Schottky circles become tangent, we shall have to deal with the phenomenon of the same limit point being represented by two different infinite words.

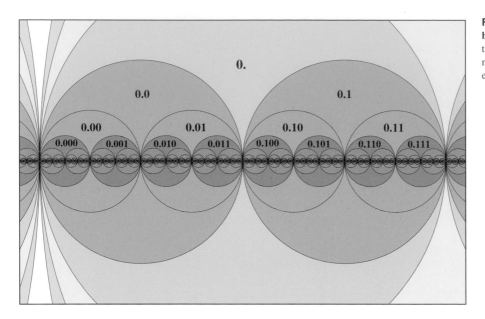

Figure 5.4. Circles and binary expansions. Because there are fewer of them, the nesting circles are much easier to see.

Infinite words do differ from decimals in one important respect. A decimal by its very definition always represents a definite point on the number line, whereas the actual point in the plane represented by an infinite word *depends on the specific choice of the matrices a and b.* In other words, for a fixed infinite word *aBBAabbA* · · · , as we change the coefficients in *a* and *b the actual points represented by aBBAabbA* · · · *will also change.* You can see this must be so by looking at the point whose infinite word

---

Note 5.3: **Two-fingered persons**

The exact number of symbols used to create a number expansion does not need to be 10. If we had evolved as two-fingered people, with one finger on each hand, then number expansions would probably be written with only two possible digits 0 and 1. This is **binary notation**, in which numbers are represented by infinite strings of 0's and 1's, their position in the string signifying increasing and decreasing powers of 2. Thus, 5 is written in binary as 101, which is short for $1 \cdot 2^2 + 0 \cdot 2^1 + 1 \cdot 2^0$, while 1.01 means $1 \cdot 2^0 + 0 \cdot 2^{-1} + 1 \cdot 2^{-2} = 5/4$. In **ternary notation**, we would use the three symbols $0, 1, 2$ and their position in the string would signify increasing and decreasing powers of 3. In the ternary system, 1.02 represents the number $1 \cdot 3^0 + 0 \cdot 3^{-1} + 2 \cdot 3^{-2} = 11/9$. Binary expansions are illustrated in Figure 5.4.

is just a string of repeating a's. As we are about to see in the next section, this particular string represents the attracting fixed point of a, otherwise written $\text{Fix}^+ a$. As we know from the fixed point formula, this will change depending on the entries in the matrix a.

Principles for working with infinite words

Box 13 displays three basic principles which are indispensible for working with limit points as infinite words. In this section, we shall be going into some detail about how and why these principles work. Later on, we shall see how they can be used to find several different ways of drawing as many of the limit points as you want. To save writing, we are going to copy the decimal notation for repeating ends, so that for example $aBBa\overline{B\vphantom{b}Abb}$ will be shorthand for $aBBaBAbbAbbAbb\cdots$.

The first principle is perhaps the most important of all: limit points are close if they begin with a long common string. Points are compared by scanning their infinite words letter by letter, starting from the left, until we arrive at the first pair of letters that differ, when we stop. This gives us their largest common prefix W, a finite reduced word in the four symbols a, b, A, B. Both points will be in the disk D_W, which, if W is long, is likely

---

Note 5.4: Counting

Just how many limit points are there? This is not such a silly question as it seems: of course, there are infinitely many, but perhaps if we arranged them cleverly they could all be written down in a list. The German mathematician Georg Cantor (1845–1918) gave a famous argument to show that while you can indeed make such a list of all the rational numbers, infinite decimals are quite a different kettle of fish.

Suppose you could write down all possible infinite decimals between 0 and 1:

$$
\begin{array}{llllll}
0. & \boxed{a_1} & a_2 & a_3 & a_4 & \cdots \\
0. & b_1 & \boxed{b_2} & b_3 & b_4 & \cdots \\
0. & c_1 & c_2 & \boxed{c_3} & c_4 & \cdots \\
0. & d_1 & d_2 & d_3 & \boxed{d_4} & \cdots
\end{array}
$$

and so on. This is impossible, Cantor reasoned, because if we choose digits A_1 different from a_1, B_2 different from b_2, and so on, then even if we go writing for infinitely long time, we shall never have listed the number $0.A_1B_2C_3D_4\cdots$. The square boxes suggest why this is called *Cantor's diagonal argument*.

If you can list a collection of things then you can count them, so a set of things which can be listed is called **countable**. Sets which can't be listed, on the other hand, are called **uncountable**. Cantor's argument shows that the set of infinite decimals is uncountable, and you can prove that the limit set is uncountable in exactly the same way.

We shall be meeting another of Cantor's ideas on p. 136.

Box 13: Principles for working with infinite words

Rule 1 Infinite words with a long common prefix define limit points which are close.

Rule 2 Finite words act on limit points in exactly the way you would expect. If P is a limit point, then $W(P)$ has the infinite word found by sticking W in front of the infinite word for P and making all necessary cancellations at the join.

Rule 3 Infinite words with repetends correspond to fixed points of transformations in the group. Suppose W and V are finite reduced words. Then:

- If the initial and final letters of W are not inverses of each other, then $\overline{W} = \text{Fix}^+(W)$.
- If the initial and final letters of W are inverses of each other, then $\text{Fix}^+(W)$ is represented by the infinite reduced word formed from $\overline{W}$ by making all possible cancellations at the joins between successive W's.
- Having made cancellations if necessary, then $V\overline{W} = \text{Fix}^+(VWV^{-1}) = V(\text{Fix}^+(W))$.

---

to be small. For example, suppose we want to compare

$$bbaaBBAAbaba\cdots \quad \text{and} \quad bbaaBBAABBaB\cdots.$$

After scanning to the ninth place we detect a difference b versus B. The two words are in the same eighth level disk $D_{bbaaBBAA}$, which means that they are almost certainly extremely close.

The second of our three principles comes in when we want to know the effect of a word W (thought of as a Möbius map) on a limit point P. There is an obvious guess for the answer: write W in front of the infinite word for P and make all the cancellations you need at the join. In other words, $W(P)$ should be found by the simplest possible recipe: take the finite word W and 'multiply' it into the lefthand end of the infinite word for P. For example, if $W = aBBAbA$ and $P = aBA\overline{AbA}$, then you would guess that $W(P) = aBBAA\overline{AbA}$. We just took W and multiplied it into the lefthand end of the infinite word for P, cancelling the string $bAaB$ at the join. The result is obviously another infinite reduced word, so certainly represents another limit point of the group.

To see why this works, let's consider a point P whose infinite expansion is $aabaB\cdots$. To make life simple we'll take W to be the sequence

BaB: that way, since *BaB* doesn't end in *A*, there is no cancellation at the join between *W* and *P*.

We know that *P* is at the end of the infinite shrinking nest

$$D_a \supset D_{aa} \supset D_{aab} \supset D_{aaba} \supset D_{aabaB} \supset \cdots,$$

and therefore $W(P)$ must be at the end of the nesting sequence

$$W(D_a) \supset W(D_{aa}) \supset W(D_{aab}) \supset W(D_{aaba}) \supset W(D_{aabaB}) \supset \cdots.$$

Putting in $W = BaB$ we work out

$$BaB(D_a) = D_{BaBa}, \ BaB(D_{aa}) = D_{BaBaa},$$
$$BaB(D_{aab}) = D_{BaBaab}, \ BaB(D_{aaba}) = D_{BaBaaba}$$

and so on, which means that $W(P)$ is at the end of the nesting sequence

$$D_{BaBa} \supset D_{BaBaa} \supset D_{BaBaab} \supset D_{BaBaaba} \supset \cdots.$$

This proves what we suspected, that $W(P)$ must have the expansion *BaBaabaB* $\cdots$. The case in which *W* ends with *a* so you get cancellations is discussed in Note 5.5.

The third principle has to do with the meaning of periodic or repeating words. The simplest case is $\bar{a}$, the sequence of infinitely repeating *a*'s. Because the sequence of nested disks

$$D_a \supset D_{aa} \supset D_{aaa} \supset \cdots$$

shrink down to the attractive fixed point $\text{Fix}^+(a)$, we find $\bar{a} = \text{Fix}^+(a)$.

Note 5.5: **Cancellations aren't a big deal**

What would happen if *W* ended with *A*, so that it cancelled into the expansion *aabaB* $\cdots$? Let's try: take $W = bAA$, so that $W(P)$ is at the end of the nesting sequence

$$bAA(D_a) \supset bAA(D_{aa}) \supset$$
$$bAA(D_{aab}) \supset bAA(D_{aaba}) \supset \cdots.$$

The first two terms are a bit tricky, so let's leave them on one side and go to the third term $bAA(D_{aab})$. The rules for working with nesting disks from the last chapter tell us that $bAA(D_{aab}) = D_{bAAaab}$. The *a*'s and *A*'s cancel out and we are left with D_{bb}. Further down the nest it gets easier:

$$bAA(D_{aaba}) = D_{bba}, \ bAA(D_{aabaB}) = D_{bbaB}$$

and so on. We conclude that $bAA(P)$ is at the end of the nesting sequence

$$D_{bb} \supset D_{bba} \supset D_{bbaB} \cdots.$$

In other words, $bAA(P)$ lies in the limit set and is represented by the infinite word *bbaB* $\cdots$. The fact that we ignored the first terms doesn't matter, since we are only interested what happens infinitely far inside the nest: funny things going on at the beginning make no difference at all.

By similar reasoning you might like to try to convince yourself that if $W = aaBBBAAA$, then $W(\bar{a})$ has the expansion $aaBBB\bar{a}$. In other words, there may be a lot of cancellation but we always get a sensible answer: since *W* is finite the cancellation will always be done in a finite amount of time.

As another illustration of Rule 3, suppose for example we want the fixed point of the transformation $W = aBB$. Think about the infinite word $\overline{W}$ which is $WWW \cdots$ (continuing with W forever), that is, $aBBaBBa\overline{BB}$. This is reduced, so it represents a point P in the limit set. By Rule 2, $W(P)$ is the infinite word formed by prepending[1] W to the word for P and making all possible cancellations at the join. But there is no cancellation, so $W(P) = W\overline{W} = \overline{W} = P$! We conclude that P is fixed by W, exactly as predicted by Rule 3.

[1] i.e. sticking in front of

There is one more point to check: why is P the *attracting* fixed point of aBB? Well, P is the intersection of the chain of circles

$$D_W \supset D_{WW} \supset D_{WWW} \supset \cdots .$$

The effect of W is to push each of the nesting disks inside the next: $W(D_W) = D_{WW}, W(D_{WW}) = D_{WWW}$ and so on. Thus W maps each disk to the next smaller disk, shrinking as it goes, so that P must be the attracting point of $W = aBB$. In a similar way, the repelling point of W must be the infinite word $\overline{W^{-1}}$. As above, $\overline{W^{-1}} = W^{-1}\overline{W^{-1}}$, so $\overline{W^{-1}}$ is the *attractive* fixed point of W^{-1}.

Notice incidentally that this shows that W has *two* fixed points. This proves that none of the transformations in a Schottky group can be parabolic. Perhaps even more importantly, none of the reduced words W can be the identity, because as we have seen, all of the disks D_W are different. As mentioned earlier, see p. 98, a group with this property is called free. We shall be meeting a few non-free groups of Möbius maps in Chapter 11, when we explore what happens when we allow the Schottky disks to cross.

Drawing the limit set: what should we actually plot?

Although in an ideal world we might want to draw every single one of uncountably many points in the limit set, this is obviously a task which even the most powerful computer will struggle to perform. It is also unnecessary, because both our printer and our eye have a finite visibility threshold beyond which finer detail simply will not be seen.[2]

[2] Herein lies one of the advantages of making your own plots: you will be able to zoom in and replot the picture to whatever level of resolution you please.

There is actually not one, but a whole multitude of ways of plotting a finite approximation to the limit set of our group. Going back to our original program for plotting Schottky circles, the most obvious thing to do would be to calculate all the image Schottky circles $W(C_a), W(C_b), W(C_A), W(C_B)$ where W runs over all reduced words of level `levmax`, taking care that the last letter of W is not the inverse of the label of the circle to which it is applied. Another possibility, avoiding the need for calculating lots of image circles, is to pick any seed point P

in the initial Schottky Swiss cheese tile F (see p. 108). Any word W maps the tile F and with it the seed point P, so that $W(P)$ is inside the image tile $W(F)$. As we apply longer and longer words W, the disks D_W get ever smaller, so if P and Q were both inside F then their images $W(P)$ and $W(Q)$ must be very close. The question of which seed point to choose gets less and less relevant and if we only plot $W(P)$ for very long words W, we shall again get a plausible rendering of the limit set. This phenomenon, called **independence of starting point**, lends a wonderful stability to limit points. It is closely connected to **chaotic motion**, as explained in Note 5.6.

One disadvantage of this method is that if the seed point P is not in the limit set, then neither are the image points $W(P)$. As W gets longer and longer they get nearer and nearer to limit points but nevertheless they are not quite there. We are faced with the awkward problem of how long to go on calculating before actually plotting the point $W(P)$.

Fortunately, Rule 3 in the previous section gives us an effective method of computing points in the limit set itself. We can use fixed points of transformations W in our group because Rule 3 shows that these are nothing but the limit points given by the infinite word $\overline{W}$ (making cancellations if necessary) and that, conversely, any limit point given by an infinite word which is eventually periodic is such a fixed point. We can get as close as we want to *all* limit points in this way. Suppose for example we want to plot the limit point P associated to an infinite word $aBaBaaaBBBa\cdots$. Using Rule 1, we know that P will be pretty close to the limit point

---

Note 5.6: **The butterfly effect**

Independence of starting point is actually the flip side of the famous **butterfly effect** in chaos theory: the idea that a butterfly flapping its wings in Coventry can cause tornadoes in Oklahoma (possibly setting off blizzards in Providence along the way). The point of this analogy is to explain the feature of chaos called **sensitive dependence on initial conditions** (alias starting points). In our context, we can explain the butterfly effect like this. If P and Q are nearby limit points, then their infinite words begin with a long common string $W = aBBaBABABaaa$ say. Now apply in order the maps A, b, b, steadily unravelling W. The pairs of points $A(P), A(Q)$; $bA(P), bA(Q)$; and so on will get steadily further apart (because their common strings are shorter and shorter), until eventually we reach the pair $W^{-1}(P)$ and $W^{-1}(Q)$ which must be in different initial Schottky disks (for example D_A and D_B), because they have no common string at all. In other words, no matter how close P is to Q, if we run the above process for long enough it will be completely unpredictable where we end up. Limit sets, it seems, are closely connected to chaos.

represented by any other point which begins with the same string, for example $aBaBaaaBBB\bar{a}$. And by Rule 3, this equals $aBaBaaaBBB(\text{Fix}^+ a)$.

We can convert this idea into an algorithm: fix a level `levmax` and plot all limit points corresponding to all infinite reduced words which, after the initial prefix of length `levmax`, consist of infinite repetition of just one letter. If you think for a moment, you will see this is the same as plotting all possible points $W(\text{Fix}^+ a), W(\text{Fix}^+ A), W(\text{Fix}^+ b)$ and $W(\text{Fix}^+ B)$ where W runs over all choices of reduced words of length at most `levmax` *whose final letter does not cancel with the label of the fixed point to which it is applied.* The choice of words W we must use will therefore depend on which of the four fixed points we are handling at the time. To take a simple example, if `levmax` = 2 then we choose W out of the collection

$$aa, \ ab, \ aB, \ ba, \ bb, \ bA, \ AA, \ Ab, \ AB, \ Ba, \ BA, \ BB$$

and we plot the 36 points

$$aa(\bar{a}), \ ab(\bar{a}), \ aB(\bar{a}), \ ba(\bar{a}), \ \ldots, \ AB(\bar{B}), \ Ba(\bar{B}), \ BA(\bar{B}), \ BB(\bar{B}).$$

There is nothing very special about running through the images of the four fixed points $\bar{a}, \bar{b}, \bar{A}$ and $\bar{B}$. We could just as well run through all words ending with the repetends $\bar{a}, \bar{b}, \overline{ab}$. This would mean we were plotting all points $W(\text{Fix}^+ a), W(\text{Fix}^+ b)$ and $W(\text{Fix}^+ ab)$, once again taking care never to allow cancellations between the beginning of the repetend in question and the end of W. Later on, we shall find it useful to use other repetends as well. For example, in Chapter 6, a lot of attention will be focussed on the repetend $abAB$. To get the very convoluted plots towards the end of the book, we had to make some extremely intricate choices of the repetends indeed.

In Box 14 we have listed five different ways of making a limit set plot. Most of them should be self-explanatory: which one we choose is a matter of the level of accuracy we want, convenience, and taste.

Methods (1) and (2) plot points not actually in the limit set, but they work fine if the level N is chosen large enough. If the seed point P is a limit point, then all the points plotted by Method (2) are also limit points. Methods (3)–(5) also plot only limit points, giving better accuracy. Which limit points to plot is a question of some importance for obtaining the most accurate pictures. For now, we shall adopt Method (3) for simplicity; in subsequent chapters, we will examine the need for extending it to Method (4). Method (5) is worth experimenting with, but it is not selective enough in the points it plots to be very efficient. Before we go ahead

---

Box 14: **All roads lead to Rome**

The limit set of a Schottky group can be plotted in many different ways. Here is a list of possible things we might do:

(1)   Calculate all the image Schottky circles $W(C_a)$, $W(C_b)$, $W(C_A)$, $W(C_B)$ for all words W of level N for some very large number N such that the last generator in W isn't the inverse of the generator associated to the circle. Plot all of these circles.

(2)   Pick any seed point P in the central tile F and plot $W(P)$ for all words W of some very large length N.

(3)   Plot all limit points which begin with any finite string of length at most N and continue with an infinite string of a's, A's, b's, or B's. That is, plot all points $W(\mathrm{Fix}^+(x))$, where x is any one of the four generators and W is any reduced word.

(4)   Do the same as in (3) but replacing the repetends $\overline{a}, \overline{b}, \overline{A}, \overline{B}$ by any other finite set of repetends. Equivalently, plot all points $W(\mathrm{Fix}^+(V))$, where V is any one of a fixed finite choice of reduced words.

(5)   Plot all fixed points of words of length at most N.

---

with the details of our program, though, let us pause to reflect a bit more on the nature of fractal dust.

Weighing the dust

Dust is really rather dry and prone to being swept under the rug. Later our dust will congeal and be harder to ignore, but even now it turns out to have some weight – provided you use the right set of scales! The thing about weighing dust is that you need to allow for the fact that it is more substantial than a set of isolated points but less substantial than a thin length of hair. Hair in its turn has less substance than that useful commodity variously called cling film or plastic wrap, which in its turn has much less substance than a bottle of milk. What we have in mind is that a few isolated points are zero-dimensional, hair is one-dimensional and plastic wrap is two-dimensional. Milk is the only truly three-dimensional volume-occupying thing in our list.[1] Maybe our dust ought to have a 'fractional' dimension a bit bigger than 0!

What does dimension have to do with weights and measures? Well, one-dimensional objects have length, two-dimensional objects have area

[1] In the real world of course, hair has non-zero diameter and plastic wrap has some thickness. Following Euclid, we are imposing a more perfect geometric vision on the world.

and three-dimensional objects have volume. What these quantities have in common is that if you cut the object into pieces, then the measure of the whole is the sum of the measures of the parts. Where they part company is in the way in which they behave when you expand or shrink the object. When one hair is twice as big as another, its length is exactly twice as great. When one piece of wrap is twice as big as another, (in all directions of course), it has four times the area. Similarly, a bottle which is double the size of another holds eight times as much milk. As we learn in high school, the rule is that if the dimension of the object is d and if you expand (or contract) by a factor k, then its measure (that is its length, area or volume as the case may be) changes by the factor k^d.

Everything thus far goes back way before Euclid, but it was not until 1919 that Felix Hausdorff pointed out that there could be other types of measurements for which the same rules held but for which d was *not* an integer! He invented a way of making a 'd-dimensional measurement' of things for every positive real number d. This d-dimensional measure has two basic properties. Firstly, when you break an object apart into pieces, the measure of the whole is the sum of the measures of the parts. Secondly, if you expand or contract it by a factor k, its 'd-dimensional measure' changes by a factor k^d. If $d = 1$, Hausdorff's measure is of course just length; if $d = 2$, it is area; and if $d = 3$, it is volume.

Now Hausdorff went further, and noticed that to measure objects you don't have to know their dimension in advance. For example, we say a hair is 1-dimensional because it has zero 2-dimensional measure or area. A piece of cling film is 2-dimensional because it has zero 3-dimensional measure (that is, volume) but also infinite 1-dimensional measure (that is, length). Both the 1-dimensional measure or 'length' and the 2-dimensional measure or 'area' of the contents of a milk bottle on the other hand (should you be so foolish as to try to measure them) should surely be considered infinite, because they must be greater than that of any of the infinitely vast amounts of idealised hair or cling film you could stuff into the bottle. Hausdorff said that a thing's dimension should be the special number D for which its 'D-dimensional measure' is neither 0 nor ∞.[1] The number D is called its **Hausdorff dimension**. A fuller explanation of Hausdorff's measures and dimensions is given in Note 5.7.

Before we go on to see how this applies to our Schottky dust, let's look at a simplified example – what the physicists would call a 'toy model'. Examples of this type were designed, once again, by Cantor. Nowadays, any fractal dust formed by roughly self-similar repetition is called a **Cantor set**. Instead of the plane, start with the unit interval, that is, all the real

[1] This is not strictly accurate: for example the complex plane is a two-dimensional thing with infinite area. Strictly speaking, you can tell that the dimension is D by checking that the d-dimensional measure is ∞ if $d < D$ and 0 if $d > D$.

numbers between 0 and 1 laid out on a line. Let's imitate the game we have been playing with Schottky circles but make its pieces as simple and regular as possible. Instead of four circles, let's take the two line segments, one from 0 to 1/3 and the other from 2/3 to 1. The next step in building the Schottky array was to construct three smaller circles inside each of the initial ones: let's do something simpler and construct two line segments inside each of the initial line segments. So we construct the two smaller segments from 0 to 1/9 and from 2/9 to 1/3 inside the segment from 0 to 1/3: see Figure 5.5. Likewise, throw out the middle of the segment from 2/3 to 1 and get two smaller ones from 2/3 to 7/9 and from 8/9 to 1. The pattern should be clear: at each stage, we will have lots of ever tinier line segments and we throw out their middles, getting twice as many segments, each with one third the length. The fractal dust is the set of points which never get thrown out, but are in one of the segments at every stage.

Figure 5.5. Cantor's dust. The first seven stages in the construction of Cantor's dust. At each stage, the middle third is thrown out of every segment from the previous stage. This one-dimensional configuration is stretched for visibility.

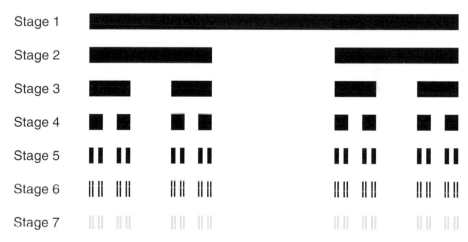

Let's try applying Hausdorff's definition. It is pretty clear that there must be equal amounts of dust in each level N segment, so let's call this amount x_N.[1] The dust in each segment fragments into the dust in the 2 subsegments at the next level, so $x_N = 2x_{N+1}$. On the other hand, if we expand a level $N + 1$ subsegment by a factor 3, then its dust exactly fits over the dust in an level N segment. According to Hausdorff's definitions, when we do this, the d-dimensional measure of this part of the dust expands by the factor 3^d. So we get another equation: $x_N = 3^d x_{N+1}$. Comparing the two, we see that $3^d = 2$. Thus $d = \log(2)/\log(3) \approx .63$ is the true dimension of the dust this construction creates!

A very similar method works for measuring the Schottky dust. Instead

[1] It's important that x_N is the amount of *dust* in the segment, not the size of the segment itself.

of segments fragmenting, we have a collection of disks: four at the first stage, dividing into three forever thereafter. The role of the expansion factor 3 which got us from segments of level $N + 1$ to level N is played by our four Möbius maps a, A, b and B. The difference is only in the detail, because the fragmentation at each level isn't quite as nice and regular. Still, it's not too bad either: let W be a reduced word which we suppose for example doesn't end with B. If we let L_W be that part of the limit set (dust) contained in the disk D_W, then we can express the splitting into finer levels by saying that

$$L_W = L_{Wa} \cup L_{WA} \cup L_{WB}.$$

Let's suppose also that W doesn't begin with A. Then we can expand

---

Note 5.7: **Hausdorff's measures and dimensions**

Suppose you want to measure the area of a complicated set X in the plane. One way to do it is to 'cover' X by a sequence of small disks $D_1, D_2, D_3, \cdots$ in such a way that every point of X is in at least one of the sets D_i. The area will be at most

$$\pi/4 \left(\text{diam}(D_1)^2 + \text{diam}(D_2)^2 + \cdots \right).$$

Of course this will probably give too large an answer, not only because the disks overlap, but also because they will probably cover some parts of the plane which are not actually in X. So we look at all possible ways to cover, insisting that all of the disks must be allowed to get as small as we please. The two-dimensional area of X is the lowest limiting value of sums of disk areas you can get.

What happens in this process if X is just a line segment, say of length 1? In this case, you can cover X exactly by N disks each of diameter $1/N$. So according to our definition, the area of X is at most $\pi/4 \times N \times (1/N)^2 = \text{constant}/N$. By taking N extremely large, we have to conclude that the line X has two-dimensional area 0. To find the one-dimensional length of X on the other hand, we have to do just the same as before, covering X by disks

but evaluating

$$\text{diam}(D_1) + \text{diam}(D_2) + \cdots .$$

To get the Hausdorff d-dimensional measure is just the same: up to a constant factor it is the least value you can get for

$$\text{diam}(D_1)^d + \text{diam}(D_2)^d + \cdots .$$

In practice (though not in theory) we usually imagine all these disks are roughly the same size. So if each disk has diameter some small number ϵ, and if $N(\epsilon)$ of them are needed to cover X, then roughly speaking, the d-dimensional measure is proportional to the limiting value of $N(\epsilon)\epsilon^d$.

Usually there will be exactly one value for d for which this limit is neither 0 nor ∞. That will be the number D for which $N(\epsilon) \propto 1/\epsilon^D$. This special value D is the Hausdorff dimension of X.

For example, if X is a relatively smooth curve of total length L, then we need roughly L/ϵ disks of diameter ϵ to cover it, in other words, $N(\epsilon) \propto 1/\epsilon$ and so $D = 1$. If, on the other hand, X more or less solidly fills in part of the plane with area A, then the number $N(\epsilon)$ of disks of diameter ϵ and hence area ϵ^2 needed to cover it is roughly A/ϵ^2, so $N(\epsilon) \propto 1/\epsilon^2$ and $D = 2$.

everything using a to get

$$a(L_W) = L_{aW} = a(L_{Wa}) \cup a(L_{WA}) \cup a(L_{WB}).$$

So if we can work out the expansion factors, we get an equation which relates the measure of the whole piece to the sum of the measures of its parts.

Because the dust isn't so regular, instead of one number x_N, we now have to deal with the measures x_W of each piece L_W, where W is a reduced word of length N. And because the Möbius maps a, b, A and B expand and shrink by varying amounts, we don't get exact equations between the x_W but only approximations which get more and more accurate as we use higher levels. It turns out that these equations only have a solution (other than $x_W = 0$ or $x_W = \infty$) for one special value of d. Curt McMullen showed how this idea can be applied to work out Hausdorff dimensions of our Schottky limit sets. For example, the limit set in Figure 5.6, for which we used this algorithm, has dimension $0.817265\cdots$. Since you might like to actually try using it, we have explained the gory details in Project 5.5.

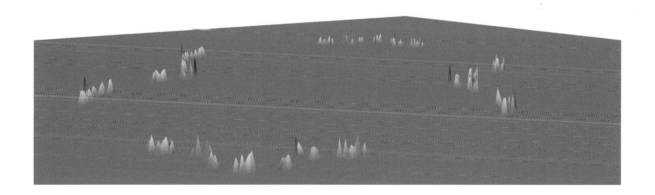

So what is a fractal?

We have used the word 'fractal' quite a few times by now, but thus far have avoided an explanation of exactly what it means. The word was coined by Benoit Mandelbrot, who pointed out that objects with non-integer Hausdorff dimension, far from being the mathematical curiosities they had been taken for until the mid-1970's[1], actually abound in nature – objects like the outlines of trees or craggy moonscapes or meandering water channels or the crests of Hokusai's *Great Wave* illustrated on p. 371. What all these have in common is that just about however much you 'zoom in' to magnify them, they still look more or less the same. If you see a picture of the moon's surface, unless there is an astronaut standing by to give you a sense of scale or you are a lunar geologist, it is really pretty hard to tell if you are looking at a mountain range, a few smallish mansized rocks, or a magnified image of some grains of sand.

This is a rather vague definition, but there are lots of objects like this in mathematics too. Our Schottky dust is a case in point. When you zoom in on a small part of the picture, you see yet more fragments inside the whole. Moreover, the pattern in which the pieces fragment is not arbitrary, but remain at least roughly regular at all levels. The lines from the Avatamsaka Sutra at the head of this chapter are an almost perfect description!

Mandelbrot renamed Hausdorff dimension as **fractal dimension** and popularised his ideas in his now famous book *The Fractal Geometry of Nature*.[1] Fractals are crinkled, fragmented or convoluted shapes which, no matter how you 'zoom in' to magnify them, still look (roughly) just as much crinkled, fragmented or convoluted as before. When you magnify a conventional curve or piece of surface, on the other hand, it looks roughly straight or flat[2]. A fractal can be viewed on many different scales, from macroscopic to microscopic, and on each different scale the same intricate structure persists. There is no universally agreed definition of exactly what we should mean by a fractal but two points are central. First, it should be an object with some type of non-integer dimension, such as Hausdorff dimension.[3] Secondly, it should be approximately (or 'statistically') 'self-similar', meaning that even minutely small pieces should blow up into large pieces which are approximately or statistically similar to the whole shape.

Nowadays, fractals are being used more and more by mathematicians and others seeking to model real life phenomena like turbulence or the flow of traffic on the Internet. They are closely connected with 'chaos', another widely used term which seems to defy a precise mathematical definition. In fact the limit set is what is sometimes known as a 'strange attractor' for the group's dynamics, because as we continue to iterate our maps, then, like poor Dr. Stickler in the picture on p. 108, the orbit of any point will get sucked in ever closer to the limit set. A point in the limit set, on the other hand, hops around chaotically under the dynamics of the group, see Note 5.6. This is closely connected to the random or IFS algorithm we shall be describing on p. 152. Curt McMullen likes to call the limit set the **chaotic set**, and truly this would be a better name.

Even though fractals are a 'new' invention, it has long been understood

[1] W. H. Freeman, 1982. First published as *Fractals: Form, Chance and Dimension*, 1977.

[2] This is of course the main principle of differential calculus!

[3] There are actually quite a few slightly varying different types of fractal dimension. At least for the reasonably regular objects we shall be creating in this book, most of these differing definitions turn out to be the same.

that Cantor's sets are closely connected to models of complicated dynamics. In fact our exact coding of limit points for a Fuchsian Schottky group was used by Hedlund in the 1930's to create a model for chaotic motion in which the trajectories of a particle come back infinitely many times closer and closer to their starting point without ever exactly repeating their motion.

Depth-first search: how to extricate oneself from a maze

Now that we have a good picture of the limit set, we turn to the intricacies of depth-first search. For the reader who prefers not to get involved in the programming details, we begin by explaining the difference between the breadth-first and depth-first approaches with a brief vignette on the pros and cons of solving mazes by the two methods.

The breadth-first algorithm in the last chapter worked by enumerating all the words of the group out to a given preset level lev, then calculating an object (usually an image circle) corresponding to each word. We found all the circles at one level before going on to the next. We chose a definite termination point by specifying in advance the maximum level levmax to which the program would run. The problem with this is that you need to store all the $4 \cdot 3^{\text{lev}-1}$ matrices at each level at once, because they will all be needed to go to the next level. Each matrix is specified by four complex or eight real numbers and typically a single real number is stored in 8 bytes. This means that if the level is 15, the storage requirement will be about $8 \times 8 \times 4 \times 3^{14} \doteq 1.2$ gigabytes.[1] Of course, these days, that is small change, but why store all these matrices anyway? For each word, all we need is to multiply its constituents. This suggests it would be an improvement to follow one branch of the tree for very many levels before changing the initial letters in the word string. This is the approach of depth-first search. The principle is encapsulated in the words of the hero in Borges' story *The garden of forking paths*[2]: "The instructions to turn always to the left reminded me that such was the common procedure for discovering the central point of certain labyrinths ... ". The DFS algorithm is a bit trickier to program than the BFS one, but for our particular goal of accurate rendition of limit sets it has some distinct advantages, especially when, from the next chapter, the limit sets become curves.

Figure 5.7 highlights the difference between breadth-first and depth-first searches by suggesting two different methods of extricating oneself from a maze. On the left, we have the lone hero Theseus, aided only by Ariadne's golden thread, which he systematically follows in his search for the exit to the Labyrinth. The thread is an exact record of the path he has

[1] The symbol $\doteq$ means 'is a close numerical approximation to'.

[2] Extract from LABYRINTHS by Jorge Luis Borges. Copyright © New Directions Publishing Corporations, 1962, 1964, used by permission of The Wylie Agency (UK) Limited.

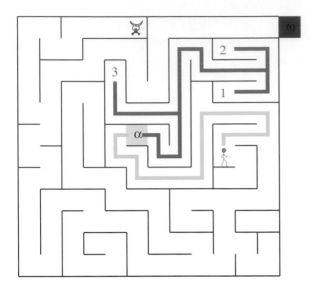
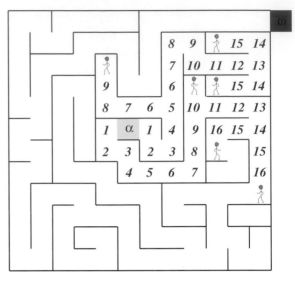

currently followed from his starting point. Theseus has to be meticulous about always turning right when he reaches a junction he has not previously encountered. (Borges' hero does the same thing, but turns always to the left.) When he returns from a dead end, he goes back to the first junction at which there were alternative turnings and takes the next unexplored route immediately to the left of his path thus far. Some of his failed paths are shown on the maze, numbered according to this right-hand rule. Let's hope he gets past the Minotaur! You may wish to investigate whether, blindly following the rules of this algorithm, he does indeed escape.

On the right, a vast army with unlimited resources is crashing through the maze. At every junction, plenty of soldiers are available to send down every possible turn. Victory is inevitable! The steps taken by the soldiers are numbered. From steps 1 to 5 only two soldiers are needed. At step 6, a third soldier must be sent out to follow a new route. By the 17th step, six soldiers were needed to follow all routes, although only two remain active, since the other four have all run into dead ends. Perhaps they should go back and make themselves useful!

You can see how the two algorithms may appeal to different computational resources: the lone processor versus the massively parallel machine with huge amounts of memory. Of course, to truly take advantage of the breadth-first algorithm we need one of those new state-of-the-art machines that are capable of creating new processors on demand. (Calm down, you technophiles. There really isn't such a machine; we were just kidding. Or were we?)

Figure 5.7. Solving mazes systematically. On the left you can see Theseus (looking suspiciously like Dr. Stickler) attached to his golden thread as he searches the depths of the maze. He started at the point labelled α. His first efforts, turning right whenever possible, are shown by the green paths ending at 1, 2 and 3. He is currently on his fourth attempt. On the right, a veritable *Hydra* of soldiers (also curiously resembling Dr. Stickler) have marched 17 steps through the maze. At each junction, new soldiers are created to follow each possible path. Some of the soldiers have become stuck in dead ends; they, however, are expendable.

Box 15: **Cyclic permutation**

Cyclic permutation is a simple operation we shall often want to perform on a string of symbols. Suppose the string is $uvw\cdots z$, and imagine these symbols written around a circle so the string has no particular beginning or end. A **cyclic permutation** is any other string you can get by reading the same sequence round the circle in the same direction, with any starting point you choose. All these different strings are said to have the same **cyclic order**.

The simplest cyclic permutation is made by removing the first symbol u in the string and adding it on at the end. Thus the string $uvwxyz$ cycles to $vwxyzu$, then to $wxyzuv$. We can list all the cyclic permutations: they are $uvwxyz$, $vwxyzu$, $wxyzuv$, $xyzuvw$, $yzuvwx$, $zuvwxy$. This is shown in the picture below with a diagram of the letters around a wheel. As the wheel rotates, the position of the letters changes cyclically.

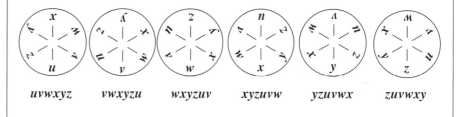

$uvwxyz$ $vwxyzu$ $wxyzuv$ $xyzuvw$ $yzuvwx$ $zuvwxy$

When we implement depth-first search in a Schottky group, the maze is nothing other than our word tree. What does 'turning right' mean in this context? To sort it out properly we need to use the idea of **cyclic permutation**, explained in Box 15. The four generators should be treated in the **cyclic order** a,B,A,b, in other words we are interested in all four cyclic permutations a,B,A,b and B,A,b,a and A,b,a,B and b,a,B,A at the same time. Think of them arranged anticlockwise around a circle: a at 12 o'clock, B at 9 o'clock, A at 6 o'clock and b at 3 o'clock. If we are only interested in cyclic order, then we are allowed to rotate the clock. Thus a could be anywhere but the sequence a,B,A,b must follow on anticlockwise round the clock face.

You can see this cyclic order if you examine Figure 5.2. Look at the incoming and outgoing arrows at each vertex. The labels on the outgoing arrows are always in the same cyclic order; three of the four symbols from the sequence a,B,A,b. The fourth incoming arrow is labelled by the inverse of the fourth missing symbol, so if we replace this by an outgoing arrow carrying the inverse symbol, we see that the *the four outgoing arrows are labelled by the four generators in the same cyclic order at each vertex of the word tree.*

Now here is how to systematically find left and right branches in depth-first search. Suppose you follow a path down the word tree, for example the red path in Figure 5.2. You arrive at a vertex with the three outgoing branches arranged in their fixed cyclic order taken from the sequence a, B, A, b. Suppose the vertex is labelled by a word W ending in B, as is the situation in Figure 5.8. Then the last incoming arrow on your path must have been labelled B, so that the reversed backwards arrow must be labelled b. Now b, a, B, A is a cyclic permutation of a, B, A, b, so the three forward directions emanating from W are, in anticlockwise cyclic order, a, B, A. This establishes your sense of direction: from your viewpoint walking forwards along the path, A is the left-most turn, B is straight ahead and a is the right-most turn. In other words, left and right on the word tree will always mean left and right as on Figure 5.2, relative to the path from which you arrived starting at 1. Try finding incoming arrows labelled B in the figure and checking this out.

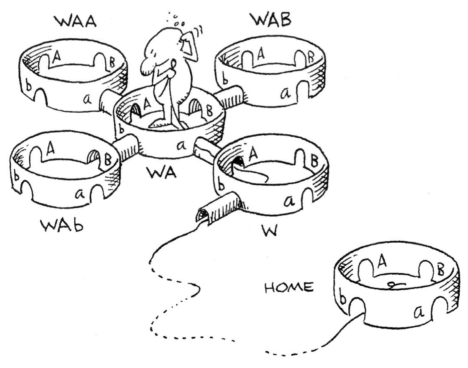

Figure 5.8. Keeping track of left and right. Here Theseus (alias Dr. Stickler) is having a hard time finding his way through this latter day maze. To find the minotaur, he had better keep a tight hold on his thread and make sure he follows the instructions to turn always to the right.

The left-right part of the DFS algorithm can be cleverly implemented using a variant of addition called **modulo 4 arithmetic**. Start by assigning a number to each generator by describing a as 1, b as 2, A as 3 and

B as 4. (Note this is not the same as the cyclic order we were using up above; the order here is a convention and we shall stick with this numbering in our program in the next section). Now let these numbers cycle $\cdots, 4, 1, 2, 3, 4, 1, 2, \cdots$. In modulo 4 arithmetic, if j is a number in this list, then $j + 1$ always refers to the next number and $j - 1$ refers to the number before. So $1 + 1 = 2$, $2 + 1 = 3$, $3 + 1 = 4$ *and* $4 + 1 = 1$; moreover $4 - 1 = 3$, $3 - 1 = 2$, $2 - 1 = 1$ *and* $1 - 1 = 4$. Now here's the trick: with this convention, the generator attached to $j + 2$ (which is always equal to $j - 2$) is the inverse of generator j. And the generators which can follow generator j without cancellation are numbered $j - 1, j, j + 1$ interpreted modulo 4! Therefore in our description of paths down the word tree, if the incoming path is labelled j, then the rightmost outgoing branch is always numbered $j + 1$, while the leftmost is $j - 1$. The trick works because we arranged that generator $j \pm 2$ always corresponds to the inverse of the incoming generator j, and hence is the one direction we have to avoid.

Depth-first search in code

Assuming the technophobes have left the room, we now settle down to some serious business with depth-first search algorithm. If you aren't interested in the miry details, skip ahead to p. 152. (You may find yourself returning to this section later though; the way the DFS algorithm works will impinge on our theory later in the book.)

At any given time, we shall be working with only one word, but it will be necessary to keep track of how it was formed by successive multiplications by either a, b, A or B. Therefore, the word will be stored successively as the entries word[i] of an array word where i ranges from 1 to the length or level lev of the word. Each entry word[i] is the 2×2 matrix corresponding to the product of the i left-most generators in the word. For example, suppose our current word is *aabAB*, which has length 5. Then our array word would contain word[1] = a, word[2] = aa, word[3] = aab, word[4] = $aabA$, and word[5] = $aabAB$.

Remember the entries in word are the actual matrices that would be computed (so the array word truthfully has dimensions levmax $\times$ 4, as explained before). We also need to know exactly how these matrices were computed, and to that end we introduce an array tags of integers. We assign the standard numbering $a \leftrightarrow 1$, $b \leftrightarrow 2$, $A \leftrightarrow 3$, $B \leftrightarrow 4$ to the generators and put the generators in an array gens of four complex 2×2 matrices and their attractive fixed points in an array fix of 4 complex numbers. Then each entry tags[i] is the number of the *i*th generator in the word, so that the following equation holds:

$$\texttt{word[i] = word[i-1] * gens[tags[i]]}.$$

That is, the partial word `word[i]` is obtained by multiplying `word[i-1]` on the right by the generator with number `tags[i]`. In case $i = 1$, we just set `words[1]` equal to `gens[tags[1]]`. For our sample word *aabAB*, the array `tags` has entries 1, 1, 2, 3, 4, in order.

The maximum level `levmax` still represents the maximum possible depth to which we may search; however, we allow ourselves the freedom *not* to go all the way out to that level if we so choose. The set-up of the generators remains the same as in the breadth-first search, but we will now plot only the limit points using Method (3) in Box 14 on p. 135.

Before introducing pseudo-code for the heart of the algorithm, let us describe the major phases in general terms. Remember that we are about to venture into a maze of tunnels.

Go Forward Go forward one step in the tree of words. If there is a choice of branches to follow, always choose the 'right-most'. This choice corresponds to always cycling round the anticlockwise cyclic order a, B, A, b as illustrated in Figures 5.2 (on p. 124) and 5.8. As we move one step deeper, we increment `lev` by 1.

Termination of Branch Check Check to see if we have come to the end of a tunnel. The exact criterion for termination is one of the subtlest parts of the algorithm. The simplest requirement is that `lev` should be equal to `levmax`, but more elaborate criteria are better. For example, we might require the Schottky circle we are looking at to have radius less than a preset value, say 10^{-3}. If the criterion for termination is fulfilled, we first plot a limit point and then turn to the 'go backward' phase. If the criterion is unfulfilled, we return to the 'go forward' phase.

Go Backward Go backward along the route we have currently traced out in the tree until we come to the first node where there is a tunnel which we have not yet explored. In the backwards phase, we decrease `lev` by 1 for each node backwards that we travel. If we go all the way back to `lev = 0` and there are no more available turns, then our work is done and the tree search is finished. If we come to a node where there is an available turn, we point ourselves down the right-most available tunnel and go to the 'go forward' phase.

There are several branching points in the above algorithm, so in Figure 5.9 we have dusted off the venerable and still valuable programming tool of a **flowchart** which describes how it runs.

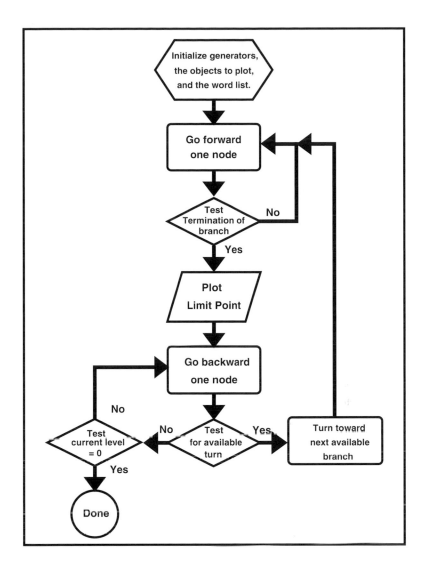

Figure 5.9. A flowchart for the depth-first search algorithm.

With this immaculate organization in hand, we now reveal the program in glorious pseudo-code in Box 16. This time, we have written the code as a 'main loop' calling five subroutines, followed by the details of each subroutine: modular programming style.

The `do` *procedure* `until` *test* construction means repeatedly execute the named procedure and after each execution check to see whether the test is true or false. If it comes out true, then stop; otherwise continue to execute. There are different variations of this construction in different languages, and even within one language we could vary the construction to test for termination before or after one execution of the procedure. The `until` *test* `do` *procedure* is used to indicate that the test should

Box 16: **Depth-first search**

```
matrix gens[4], word[10000]
complex fix[4], newpoint
circle circ[4], newcirc
integer lev, levmax, tags[10000]
real epsilon
Initialize Generators, Objects to Plot, Maximum level
lev = 1;                                              Initialization of word tree.
tags[1] = 1;
word[1] = gens[1];

until lev=0 and tags[1] = 2 do {                      BEGIN MAIN LOOP
                                                     Go down the tree until you are ready
    until branch_termination do go_forward;          to plot.
    do go_backward until lev = 0 or available_turn;  After plotting, climb back.
    turn_and_go_forward;                             When possible, take another branch.
}                                                    END MAIN LOOP

go_forward{
    lev = lev + 1                                    Advance level by one.
    tags[lev] = tags[lev-1] + 1 mod 4                Adopt right-most branch.
    word[lev] = word[lev-1] * gens[ tags[lev] ] }    Compute new word.

go_backward{
    lev = lev - 1 }                                  This is easy!

available_turn{
                                                     Have we taken left-most turn
    if tags[lev+1] - 1 = tags[lev] + 2 mod 4 then    already?
        RETURN(FALSE)                                If so, no more turns are available;
    else RETURN(TRUE) }                              if not, there is one.

turn_and_go_forward{
    tags[lev+1] = tags[lev+1] - 1 mod 4              Take the next left branch.
    if lev = 0 then word[1] = gens[tags[1]]
    else word[lev+1] = word[lev] * gens[tags[lev+1]] Compute the new word.
    lev = lev +1 }                                   Go forward one level.

branch_termination{
    newcirc=mob_on_circ(word[lev-1],circ[tags[lev]]) The new Schottky circle:
    if lev = levmax or radius(newcirc) < epsilon then  if it is small enough,
        newpoint=mob_on_point(word[lev],fix[tags[lev]]) compute a fixed point,
        plot newpoint and RETURN(TRUE)               and plot it.
    else RETURN FALSE. }
```

be performed before each execution. We carry out the first step forward before the main loop. Note that the program ends when we reach level 0 and there are no more possible routes to follow, which happens if the last route travelled began with the last of the four generators. With our conventions, the four branches of the tree of words will be explored in this order: first the *a*-branch, `tags[1]=1`, second the *B*-branch, `tags[1]=4`, third the *A*-branch, `tags[1]=3` and lastly the *b*-branch, `tags[1]=2`. (To check this is the right order, look at Figure 5.2.) So when you come down to `lev=0` and find `tags[1]=2`, you are finished.

This sort of programming is *so easy!* Unfortunately, the little bits `go_forward`, `branch_termination`, and so on represent subroutines we still have to write out. The work is always in the detail! Here are the tasks assigned to these subroutines:

go_forward Go one level deeper in the tree on the right-most branch available. (A branch is available if it hasn't been travelled before.)

go_backward Go one level up (or backward) in the tree along the route we have already travelled.

branch_termination Check to see whether we should stop going forward along this branch. This will happen either if we have reached the maximum level `levmax`, or if the circle in question is smaller than some small positive number `epsilon`, for example 10^{-3}, specified at the outset. Plot a limit point and return *true* if we should stop.

available_turn Check to see if there is another forward route leading from the current node that we have not previously travelled. If so, return *true*.

turn_and_go_forward Knowing there is a possible turn, make the turn and go forward one node.

The small number `epsilon` represents our visual threshold. We decide in advance not to plot detail below this level. The Greek letter ϵ, pronounced 'epsilon', is always used in mathematics to denote a very small, but positive, quantity.

When programming with subroutines, it is necessary to be very clear about the **scope** of the variables, meaning the parts of the program from which the variables in question are accessible. It is generally safer to use variables whose scope is limited to the procedure in which they appear; that way no other procedure can affect their values. In the program we are writing now, though, the variables `lev`, `word` and `gens` are all **global** variables accessible from any procedure. These variables represent our record of where we have got to in the word tree; like Ariadne's golden thread they ensure we shall never get lost.

When writing this code, it is very useful to be able to make the program output the words as they are enumerated. Here they are up to the maximum level 3:

$$a \longrightarrow ab \longrightarrow abA \longrightarrow abb \longrightarrow aba \longrightarrow aa \longrightarrow aab \longrightarrow aaa \longrightarrow$$
$$aaB \longrightarrow aB \longrightarrow aBa \longrightarrow aBB \longrightarrow aBA \longrightarrow$$
$$B \longrightarrow Ba \longrightarrow Bab \longrightarrow Baa \longrightarrow BaB \longrightarrow BB \longrightarrow BBa \longrightarrow$$
$$BBB \longrightarrow BBA \longrightarrow BA \longrightarrow BAB \longrightarrow BAA \longrightarrow BAb \longrightarrow$$
$$A \longrightarrow AB \longrightarrow ABa \longrightarrow ABB \longrightarrow ABA \longrightarrow AA \longrightarrow AAB \longrightarrow$$
$$AAA \longrightarrow AAb \longrightarrow Ab \longrightarrow AbA \longrightarrow Abb \longrightarrow Aba \longrightarrow$$
$$b \longrightarrow bA \longrightarrow bAB \longrightarrow bAA \longrightarrow bAb \longrightarrow bb \longrightarrow bbA \longrightarrow bbb \longrightarrow$$
$$bba \longrightarrow ba \longrightarrow bab \longrightarrow baa \longrightarrow baB.$$

We made this list by running through our DFS program exactly as recorded above and listing the words in the order that the program reaches them. Check it against Figure 5.2! The first word is the first generator a; going forward leads to the level 2 word ab. Note that b is generator 2, and since in modulo 4 arithmetic, 2 is one more than 1, the letter b is the next turn to the right. Since the maximum level hasn't been reached, we go on to the third word abA, appending generator number 3, again just one turn to the right. At that point we have to go backwards, and we find the first available turn is at level 2, giving the next word abb. Another turn at level 2 leads to the fifth word aba. At that point, we have to go back to level 1 to find an available turn. The process continues through 52 words in all. There are 36 words of level 3, and they are indeed exactly the same as the words listed in the level 3 breadth-first search algorithm. However, the ordering in which these words are encountered is substantially different.

A recursive implementation

Following Ariadne's thread seems a relatively straightforward idea and it is somewhat disconcerting that the flow chart in the last section should be so complicated with its three intertwined loops. Computer scientists, like mathematicians and physicists, believe that elegance is an important clue that you are doing things right and there is, indeed, a much more succinct way of writing the depth-first search program. This uses a basic programming language construction called **recursion**. Recursion was not allowed in FORTRAN until recently because it can lead to explosive calls on memory and might have considerable overhead. The issue led to an

[1] See the manifesto *Structure and Interpretation of Computer Programs*, by Abelson, Sussman and Sussman, MIT Press, 1996.

almost religious war between conservative numerical analysts using FOR-TRAN on one side and the artificial intelligence foreign legion speaking the many dialects of LISP on the other.[1] Recursion is, however, considered good programming style by most computer scientists.

A simple example is the recursive code for computing the **factorial function**. The symbol $n!$ (read 'n factorial') stands for the product of all positive integers from 1 to n: $n! = 1 \cdot 2 \cdot 3 \cdots n$. In traditional mathematics, it can be defined 'inductively' by the rules:

(1) $1! = 1$,

(2) if $n > 1$, then $n! = n \cdot (n - 1)!$

This can be translated into pseudo-code as:

```
fac(integer n){
   if n = 1 RETURN(1)
   else RETURN(n*fac(n-1))
}
```

Note that this code *calls itself*. Unless $n = 1$, when you run `fac`, to execute it you will have to run `fac` on smaller n's in order to evaluate $n!$.

Here is how this can be applied to depth-first searches of trees. Every tree starts at some 'root', attached to which are several subtrees, one for each edge emanating from the root. The idea is that a depth-first search of the whole tree is made up of depth-first searches of each of these subtrees, one at a time. This is accomplished by a piece of code which does the book-keeping needed to order the subtrees (in our case four, one for each generator) and 'calls' itself whenever it explores the subtree. The code is given in Box 17.

What happens is *exactly* the same as with the previous code. Before, as you ventured like Theseus further into the cave, you kept track of your trail with the two big arrays `words` and `tags`. This time, you keep calling the subroutine `Explore_tree`. The calling mechanism of your operating system places the data of the previous call, that is the word X and tag 1, on its 'stack', to be 'popped' when the present subroutine returns. Pretty much the same thing happens in the computer, but the programmer will have written a much shorter and hopefully more transparent piece of code.

The random search algorithm

As we explained in the last chapter, that there are three main methods of drawing the Schottky array. The first two, breadth-first and depth-first, we have now explored at some length. Just to complete our tour, we want to briefly explain the third method, the technique of random search. Mandelbrot was using a random search algorithm for his first explorations when he arrived in Harvard back in 1979. For the readers who have previous experience of iteration, this may well seem the most obvious method to choose.

When making either breadth- or depth-first searches, we always extended words on the right. This was because for a reasonably long word W and seed point z_0, the points $Wx(z_0)$ and $W(z_0)$ are close, for any choice of letter x in the alphabet a, A, b, B. The longer the word W, the

---

Box 17: Recursive depth-first search

```
global matrix gens[4];

global complex fix[4];
global real epsilon;


MAIN{
   integer k;
   INITIALIZE gens[4], fix[4];
   for k from 1 to 4 do {

      Explore_tree(gens[k],k);

   }
}



Explore_tree(matrix X, integer l) {
   matrix Y;
   integer k;

   for k from l-1 to l+1 do {

      Y = X * gens[k mod 4];

      if radius(circle(Y)) < epsilon then

         plot mobius_on_point(Y,fix[k mod 4]);
      else
         Explore_tree(Y,k);
   }
}
```

Make `gens`, `fixpt` and `epsilon` available to *all* procedures.

Performed elsewhere.

Start word with any of the 4 generators and explore this part of the tree.

This routine explores all words beginning with X. The integer l is the last generator used.

Loop over all generators except the inverse of `gen[l]`.

Make a bigger word with this generator.

If the circle is very small, terminate search and plot.

Otherwise, go forwards.

Go backwards automatically upon return.

better it works. Quite the opposite is true if we extend words on the left. The points $W(z_0)$ and $xW(z_0)$ can be very far apart. Left multiplication tends to scramble everything up. Following a path traced by repeated left multiplication takes us on a pretty chaotic tour.

Chaotic tours do have the advantage that in the course of time, you expect to get pretty near to every point.[1] This idea forms the basis of a completely different but in many cases very rapid method of generating and compactly storing all sorts of fractal images looking like mountains, trees, ferns and clouds. In the hands of Michael Barnsley and his co-workers, it has been elevated to a highly sophisticated technique. Commercially, it is used for making fantastic landscapes of the kind you see in Star Wars.

In Barnsley's Iterated Function System algorithm[2], or IFS for short, we start with a collection of transformations, in our case just the generators a, b, A and B. We also choose a seed point z_0. The algorithm plots a sequence of iterates $z_0, z_1 = x(z_0), z_2 = y(z_1), z_3 = z(z_2)$ and so on. The transformations $x, y, z, \ldots$ are picked 'randomly' from the initial collection a, b, A, and B.

As long as the choice is 'random', the points $z_0, z_1, z_2 \ldots$ joyfully spring all over the screen. If the initial data has been chosen cleverly enough (you can learn how in Barnsley's book), then out of this bewildering cloud of points emerges, miraculously, some wonderful shape like a fractal fern. In our case, the best choice is to set the program running with z_0 a limit point of the group. The jumping point never escapes the limit set and you can see it gradually tracing out the whole dust on the page. Unlike the depth-first algorithm, there is no very obvious stopping point: program termination may happen when the user tires and hits *Control-C*, or when some preset number of points has been plotted.

The IFS algorithm has a strong advantage in that it has virtually no overhead. There is no need to remember all the points that have been visited, nor the sequence of letters used to travel to along the path. On the other hand, in our situation, it does not produce terribly good results because, like breadth-first search, it doesn't know which long words are essential to get various parts of the limit set. Moreover with purely random choices we are likely to spend about a quarter of our time needlessly doubling back. For example, if we randomly choose generator a, the algorithm jumps us from the current point z to the new point $a(z)$. On the next pass, if we choose the inverse transformation A, we retrace our step to $A(a(z)) = z$, which is a waste of time. It would be better if, after having moved $z \rightarrow a(z)$, we had zero probability of choosing the

[1] This principle is known as the *Poincaré recurrence theorem*. A more precise version forms the basis of Boltzmann's famous *ergodic hypothesis*: for a sufficiently random path, the time spent in each region is proportional to its area. Such behaviour is always exhibited by paths which are chaotic in the strict mathematical sense.

[2] Beautifully described in Barnsley's book *Fractals Everywhere*, Academic Press, 1988, 1993. For further developments see *Fractals in Multimedia*, IMA vol. 132, Springer, 2002.

transformation A. Such a program, where the probabilities depend on what went immediately before, is called in the jargon a **Markov chain**[1]. It is plausible that one would do better by adjusting the probabilities of choosing the other three generators depending on the recent past travels of z. This would be an *adaptive* IFS algorithm, where on every pass of the main loop, the probabilities assigned to the different choices of transformation change. Such algorithms have not yet been much studied, and may well be a fruitful area for a creative programmer or mathematician to explore.

[1] After the same Andrei Andreevich Markov (1856–1922) who gave his name to the trace equation we shall be meeting in Chapter 6.

Projects

5.1: Yet more ways to plot the limit set

Why do each of the following methods of plotting the limit set work?

(a)   Method (5) in Box 14.

(b)   Start with a point P in *any* of the Schottky tiles and plot $W(P)$ for all words W of some very long length N.

(c)   Start with any point P in the limit set and plot $W(P)$ for all words W up to some long length N.

How large do you have to make N for these ideas to work?

5.2: Möbius maps and decimals

It turns out there are some Möbius maps relating to decimals which behave very like a Schottky group lurking behind the scenes. The maps in question are the ten rather simple transformations

$$T_j(z) = \frac{z+j}{10}, \quad \text{for } j = 0, 1, 2, \dots, 9.$$

Note 5.8: Choosing random generators

To run the IFS algorithm, one has to address the crucial question of how to pick a 'random' generating transformation x. With four transformations, the easiest method might be to toss a coin twice. If it comes up heads–heads we choose a, heads–tails we choose b, tails–heads we choose A and tails–tails we choose B. Of course we could vary the probabilities: we could toss a dice and choose a if it came up 1 or 2, b if it came up 3 or 4, A if it came up 5 and B if it came up 6. In this case the probabilities would be $1/3, 1/3, 1/6, 1/6$. In reality of course one would use a so-called 'random number generator' which gives seemingly random integers in some huge range $0, 1, \dots, N$ for some extremely large number N. Then if we are given four desired frequencies p_a, p_b, p_A and p_B with $p_a + p_b + p_A + p_B = 1$, you choose a 'random' integer n in this range and select a if $0 \le n < p_a N$, select b if $p_a N \le n < (p_a + p_b)N$, select A if $(p_a + p_b)N \le n < (p_a + p_b + p_A)N$ and finally select B if $(p_a + p_b + p_A)N \le n < N$.

[1] This is the reciprocal of the **golden ratio** $(1 + \sqrt{5})/2$ which we shall meet again in Chapter 10.

Let's look at a specific example: we'll pick the magic number $(\sqrt{5} - 1)/2$ whose expansion begins $0.618\ldots$.[1] Referring back to Figure 5.3, the starting circle C_0 in the nest is the circle of radius $1/2$ and centre $1/2$, containing all real numbers from 0 to 1. You can easily check the next few levels of nested circles are $C_1 = T_6(C_0)$, $C_2 = T_6(T_1(C_0))$, $C_3 = T_6(T_1(T_8(C_0)))$, and so on. Observe how the new transformations are appended at the *righthand end*.

The transformations T_j show us why repeating decimals are fractions. Show that if $x = 0.\overline{a_1 a_2 a_3}$, then $T_{a_3}(x) = 0.a_3\overline{a_1 a_2 a_3}$, $T_{a_2}(T_{a_3}(x)) = 0.a_2 a_3\overline{a_1 a_2 a_3}$ and finally

$$T_{a_1}(T_{a_2}(T_{a_3}(x))) = 0.a_1 a_2 a_3\overline{a_1 a_2 a_3} = x.$$

Thus x is a *fixed point* of the product transformation $T_{a_1}T_{a_2}T_{a_3}$. Write down a fixed point equation for $T_{a_1}T_{a_2}T_{a_3}$ and solve directly to find x.

Finally, the transformations T_j illustrate the phenomenon of 'independence of starting point' we discussed in Note 5.6. Show that with any starting point z at all, the successive images

$$z, T_6(z), T_6(T_1(z)), T_6(T_1((T_8(z)))) \cdots$$

come ever closer to $(\sqrt{5} - 1)/2$. However good an approximation we had to $(\sqrt{5} - 1)/2$, if we ran the above process backwards for long enough it would be completely unpredictable where we would end up.

5.3: Counting the rational numbers

In Note 5.4 we explained Cantor's argument for showing that the set of real numbers between 0 and 1 is uncountable. What about the *rational* numbers p/q. Are they countable or not? To count them, you have to write them in a list. But what is the next rational number after 0? If you tell me it is p/q, then I say $p/2q$ is smaller. Cantor found an ingenious way to solve this problem too: arrange the rationals as part of a square array and then begin counting diagonally from the top left corner. Try it! Another way of counting is based on so called **Farey sequences** which we shall meet in Chapter 9. A Farey sequence lists all rational numbers (between 0 and 1) for which the denominator q is at most some given integer N.

5.4: How many visible circles are there?

After implementing breadth-first or depth-first search, you can count how many circles you found of different sizes. One question to ask is how many circles there are of radius at least c for various values of c. The circles get smaller and smaller as the level increases, though not at all in a regular way. So for any specific c, if your search is deep enough, you will find all circles of radius at least c. Let $N(c)$ be the number of these circles. Why might you expect that $N(c)$ should behave like $(1/c)^d$ where d is the Hausdorff dimension, somewhere between 0 and 2? You can try to check this by making a 'log-log' plot and drawing the graph of $\log N(c)$ against $\log(1/c)$ (here c is going to zero, so $\log(1/c)$ is getting large). If what we say is true, the graph should be roughly a straight line of slope d. The larger d, the more complicated the dust. Try playing with this, as we did years ago when we first wrote our programs.

5.5: Calculating Hausdorff dimension à la Curt McMullen

Here are the details of Curt's numerical algorithm for finding the Hausdorff dimension of a Schottky dust. It is rather harder mathematically than most of the things we have been explaining – but then dust is quite complicated.

As explained on p. 138, if W is a reduced word of length N which neither begins with A nor ends with B, then we get the identity:

$$L_{aW} = a(L_W) = a(L_{Wa}) \cup a(L_{WA}) \cup a(L_{WB}).$$

How can we find the factor by which a expands L_{Wa}? On p. 77 we wrote down a formula which essentially does this. If the entries in the bottom row of a are c and d, then a very small circle around a point z is mapped to a very small circle around $a(z)$ with radius multiplied by approximately the factor $1/|cz + d|^2$. The smaller the circle, the more exact the factor.

If W is long (so N is large) then D_{Wa} is small, so we can estimate this expansion factor by taking z to be any point in D_{Wa}, for example $W(\bar{a})$. Let's write this factor as $k_{W\bar{a}}$. (Of course $k_{W\bar{a}}$ depends on a too, but for the sake of readability we leave this out. If we know a, then $k_{W\bar{a}}$ is a number we can compute.) This gives

$$x_{aW} \cdot (k_{W\bar{a}})^d \approx x_{Wa} + (k_{W\bar{A}})^d \cdot x_{WA} + (k_{W\bar{B}})^d \cdot x_{WB}.$$

Now make $\{x_W\}$ into a big column vector $\vec{x}$ with $4 \cdot 3^{N-1}$ entries. Then these equations look like $\vec{x} = M_d \vec{x}$ where M_d is a *huge* $4 \cdot 3^{N-1} \times 4 \cdot 3^{N-1}$ matrix. Luckily, most of the entries are zero; matrices like this are often called 'sparse'. We kept the subscript d because the number d is encoded in M_d. Obviously, we can solve these equations by either setting all the x_W's equal to 0 or ∞. We want to see if there is any other solution, which we expect will happen for exactly one value of d. We might as well solve for d and all the x_W's at the same time. What a nice idea!

We shall solve the equations by successive iteration. At any stage, we shall have a guess for d, a range from $d_{\min}$ to $d_{\max}$ which we are sure contains the true d, and a guess for $\vec{x}$. We also arrange that the sum of the components of $\vec{x}$ is 1, ensuring that the measure of the whole limit set is 1.

You can start with something dumb, for example $d_{\min} = 0, d_{\max} = 1, d = 1/2$ and all components of $\vec{x}$ equal to $1/(4 \cdot 3^{N-1})$. At each step, update $\vec{x}$ by (i) setting it equal to the matrix product $M_d \vec{x}$ and then (ii) dividing the new $\vec{x}$ by the sum of its components to make the updated sum still 1. If you repeat this 20 or 30 times, $\vec{x}$ will stop changing. Now notice what factor C you had to divide by in step (ii): if $C > 1$ then d is too large, so replace d by $(d + d_{\min})/2$ and $d_{\max}$ by the old value of d; if $C < 1$, then d is too small, so replace d by $(d + d_{\max})/2$ and $d_{\min}$ by the old value of d. Then go back to steps (i) and (ii). After a while, you'll converge on both the true value of d and the true answer for $\vec{x}$.

CHAPTER 6

Indra's necklace

Beautiful is thy wristlet, decked with stars and cunningly wrought with myriad-coloured jewels.

Gitanjali, Rabindranath Tagore[1]

The picture in Figure 6.1 shows our first truly impressive fractal construction created by the simultaneous symmetry of two Möbius maps. We made it by introducing what seems like only a small change in the set-up of the previous two chapters: we just let the four circles come together until they were tangent, forming, as we shall say, a kissing chain. In a beautiful way, this causes the limit set to coalesce from dust into a 'necklace'. If you look carefully, you will see how the nested Schottky circles form smaller and smaller chains of tangent circles. Instead of nesting down on fractal dust, these chains shrink down onto a glowing and rather curiously crinkled loop. The nesting circles seem to condense near the tangency points of the circles, highlighted by our colour coding like brilliant blue jewels. Except for the fact that the circles are tangent, this picture is entirely similar to the one in Figure 5.1.

To see better what is going on, in Figure 6.2 we zoomed in to the small black rectangle marked in the large scale picture. Now you can see more clearly the intricate complexity with which this marvellous necklace has been fashioned. Near the bottom you get a fairly clear view of some of the inner levels of chains of kissing circles, while near the main tangency point, nested tangent circles pile up creating a wonderful jewelled effect only hinted at in Figure 6.1. You can see more clearly here what it means to be a fractal – the same complicated patterns are repeating at ever decreasing scales.

This array of nesting tangent circles was created by exactly the same method as in the last chapter. Just as before, we paired opposite circles to each other, in the pattern shown in Figure 6.3. To distinguish our new groups from Schottky groups made by pairing disjoint circles, we shall call

---

[1] From *Collected Poems and Plays of Rabindranath Tagore*, Macmillan, London.

157

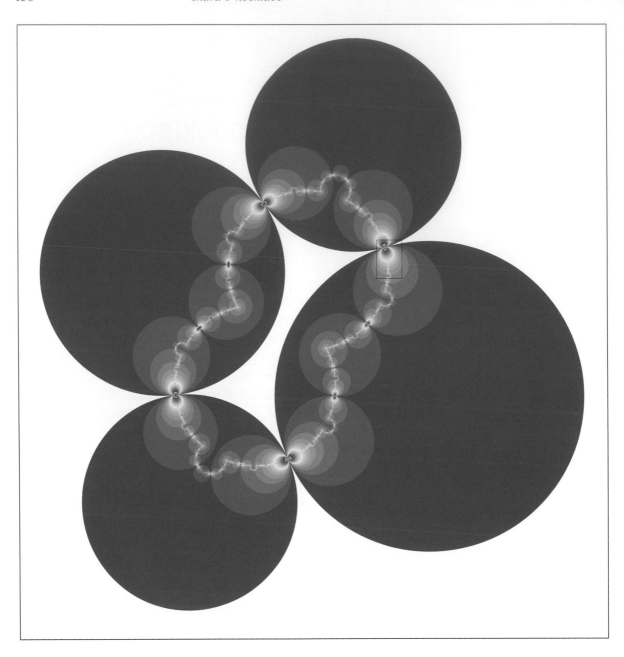

groups made by pairing tangent circles **kissing Schottky groups**, since the initial circles touch or 'kiss'. As you can see, inside each circle there are three further circles, tangent to each other and touching the outer circle at the point which it meets its neighbours in the initial chain of four. This pattern repeats from level to level, so that we get a succession of chains each contained within each other as the circles shrink. The chain at a given level is like a bead necklace. To create the next level, we insert

Figure 6.1. Indra's necklace. A limit set formed by a tangent circle chain. As usual, we have coloured the circles according to the 'level of nesting'. The limit set snakes its way through all the tangent circles, highlighted with a yellow glow.

Figure 6.2. A detail of Indra's necklace. This is a *zoom* into the small black rectangle marked in Figure 6.1. We are now seeing much deeper into the nesting and some of the colours used for very high levels become prominent, explaining why parts of the limit curve look bluish green.

into each bead three further beads, touching each other and the outer bead because they are all strung on the same string. The circles shrink from level to level, at each stage forming a more intricately convoluted chain made with three times as many smaller and more delicate beads. In the

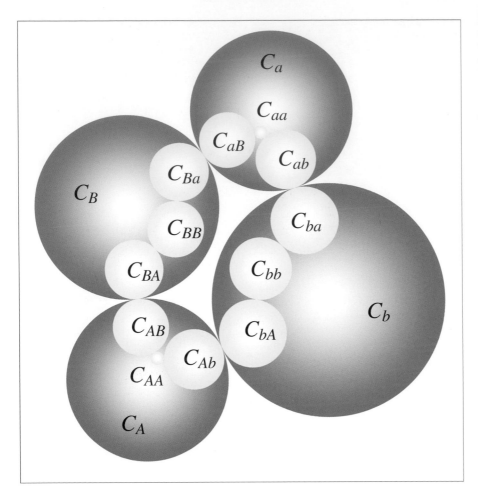

Figure 6.3. The tangent circle chain out to level 2 of the nesting. The four original circles are red, while the level two circles are yellow. The circles are labelled by their words in just the same way as in the last chapter, so that $C_{aa} = a(C_a)$, $C_{aB} = a(C_B)$, and so on. Notice the 'runts of the litter', the tiny circles C_{aa} and C_{AA}. They were so small that we had to put their labels off to their sides.

previous chapter we saw that individual nests of circles nested down to limit points. This time, the whole chain shrinks to a limiting curve, which surely deserves to be named for our title *Indra's pearls*.

As we have seen, starting from disjoint circles, the limit set is just a spray of fractal dust. You may wonder how it is that by just making the circles tangent, we get a limit set which is a **connected** loop, a loop, that is, which can be drawn continuously without lifting your pen from the page. Actually we have seen an example rather like this before, when we made the comparison between limit points and decimals. Limit points for a Schottky group form a fractal dust, whereas decimals represent all the points on that most basic of connected curves, the real line.

Our method of creating limit points is highly repetitive, so the mechanisms which determine the final outcome are set in motion at the initial stage. The limit set of a Schottky group, the Schottky dust, is contained inside the initial four Schottky circles, which split the limit set into four separate chunks. This initial separation propagates, so that the level-one Schottky disks subdivide each chunk of limit set into three further parts. Subdivision repeats at every level, so that, no matter how high the magnification, each visible speck of limit set fragments into further microscopic bits. For decimal circles on the other hand, the initial tangencies propagate to the smaller image disks, resulting in a tangent circle chain at each level which still contains the line. As a result, there is a nest of circles shrinking down to every single point, which explains why the 'limit set' for the decimal circles is the whole real line.

The pearls in Indra's necklace are strung in exactly the same way as the decimal circles, the only difference being that the tangent chains are now somewhat kinked. This effect is already quite visible even in the first few levels in Figure 6.3. Regularly proliferating kinks at microscopically small scales are the hallmark of a fractal curve. Anywhere you zoom in, the pattern of initial bends proliferates and repeats. No matter how much you magnify, the kinks will never straighten out. We have created our first limit set which is a fractal loop!

Limit sets which are loops like this are called **quasicircles**. They look vaguely circular, but all crinkled up. A group of Möbius transformations whose limit set is a quasicircle is called a **quasifuchsian group**. This is because a group whose limit set is actually a circle, as in Figure 6.5, is called **Fuchsian** after Lazarus Fuchs, about whom more on p. 179. The limit set of a quasifuchsian group is a connected curve which can be drawn in one continuous piece. In addition, the path never meets or crosses over itself until it finally gets back to the point at which it began. Whenever you have a group of Möbius transformations, that part of the plane which is not the limit set is called the **ordinary set** (sometimes the **regular set**) of the group. As you can see, the limit set of a quasifuchsian group always separates the ordinary set into two parts, the 'inside' and the 'outside'.

Quasifuchsian groups made by pairing Schottky circles, which we call kissing Schottky groups, feature in the third volume by Fricke and Klein. Figure 6.4 was drawn by one of Klein's students, by a happy chance a gifted draftsman, whose beautiful pictures were not to be improved until the advent of modern computer graphics a century later.

Tiling the inside and outside of the necklace

To get from Figure 5.1 to Figure 6.1, all we had to do was to bring the four Schottky circles together until they touched. As you can see

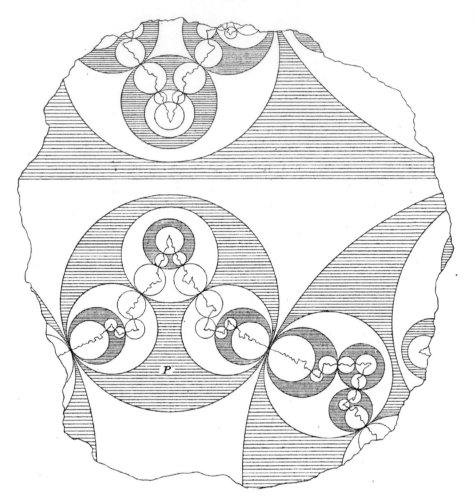

Figure 6.4. A reproduction of Figure 145 in Fricke and Klein, *Vorlesungen über die Theorie der Automorphen Functionen*, Leipzig, 1897, Vol. 1, p. 418, showing the limit set of a Schottky group made by pairing a chain of tangent circles. **Figure**

comparing the pictures, the white walls between the holes in the large white swiss cheese tile which surrounds the four red Schottky disks in Figure 5.1 have been pinched together at the tangency points of the red disks in Figure 6.1, breaking the white tile into two halves.

The two white half-tiles can be seen in Figure 6.1. The inner one is easy: it has four circular arc sides bounded by the parts of the four red Schottky circles inside the limit set. The outside white background is the other tile. It takes a little imagination to see that this is actually one piece with four circular arc sides. As usual, the trick is to imagine it on the Riemann sphere (we have shown this, with different colours, in the top left panel of Figure 6.10). Cut out the four red tangent circles, and the remainder of the sphere falls apart into two white parts, each bounded by four circular arcs. The part containing the North Pole projects to the white outer region of the plane.

Figure 6.5. (*Opposite*) **Dr. Stickler swept around by a Fuchsian group.** This group was made by pairing four tangent circles all orthogonal to another circle which passes through all four tangency points. This fifth circle is the limit set of the group. The red Dr. Sticklers are exploring the inner part of the ordinary set while the blue ones are swept forever around the outside. This group belongs to the family on p. 170 with parameters $u = \operatorname{Tr} a/2 = \sqrt{2}$, $y = \operatorname{Tr} b/2 = \sqrt{2}$, and $k = 1$.

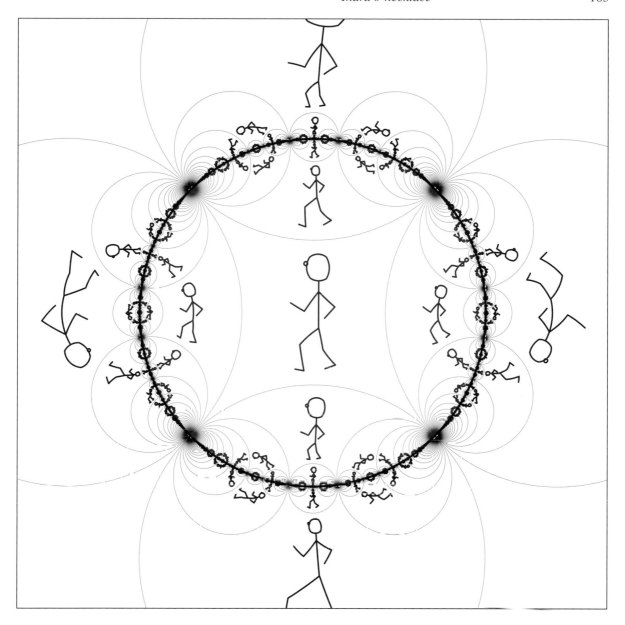

At the next level there are eight half-tiles: the four red regions inside the limit set, each bounded by four circular arcs, and another four outside, each of which (if we think of them on the Riemann sphere) has exactly the same shape. The pattern repeats at each level so that for example there are twelve half-tiles inside the limit set and twelve outside.

To get a picture of the whole tiling, let's take a very simple example starting with four symmetrically arranged circles of the same radius. The result, shown in Figure 6.5, is a Fuchsian group, recognisable because the limit set is just a rather boring circle. Actually the pinched off half-tiles look a bit more like the kind of tiles you might choose for your standard bathroom floor. Straight edged and equal sized they may not be, but at least they have four sides and no holes.

The limit set may be a circle, but the effect of all this on Dr. Stickler is not boring at all! We have drawn one Stickler in each half-tile. The one inside the limit circle is red and the one outside (perhaps because in places he has been flipped upside down) is blue. There is one full tile, and hence two half-tiles, for each element of the group. The inner one of this pair of half-tiles contains a red Stickler and the outer one contains a blue Stickler. You can find a symmetry in the group getting you from any red Stickler to any other one, but there is no symmetry which will ever get you from a red one to a blue one. This explains why the red Stickler are stuck inside the limit set, never escaping across it to the 'outside'. Smaller and smaller Sticklers pile up on the limit circle from both inside and outside. Their shape and size varies quite dramatically, even at the same level. Notice how rapidly they get smaller as you move towards the limit circle, and how many nooks and crannies they have to explore.[1]

One obvious feature of the picture is the vertical line of red Sticklers being pushed out from the source of a at the bottom centre and getting swallowed up as they head towards its sink at the top. You may be able to see a horizontal line of red Sticklers striding boldly along from the source of B on the extreme right to its sink on the extreme left, rapidly shrinking in size as they move along.[2]

In Figure 6.6 we sent Dr. Stickler to explore a rather more crinkled group. For a change, we started him on the boundary of the central tile, the righthand inner arc of the circle C_b. This means that each and every Dr. Stickler in the picture is also on the boundary of one of the tiles. Once again, smaller and smaller Dr. Sticklers pile up on the boundary of the inside half of the ordinary set. However far he gets whisked about, he is trapped forever in the part of the ordinary set enclosed inside the black limit curve, because each of the four generating transformations a, b, A and B transport this inner region to itself. We say the inner region is symmetrical or invariant under the group generated by a, b, A and B.

Note how every feature of Dr. Stickler's walk is repeated in miniature everywhere you look. At the fixed points of a and A, there are large kinks in the limit set, and on the boundary of every image of this circular path, you see another kink. At the fixed points of b and B, on the other hand, the limit set bulges out. Going further and further out you can see smaller and smaller bulges appearing in their turn. In each and every kink and

[1] This is because Dr. Stickler is really exploring **non-Euclidean** or **hyperbolic** space which is much 'bigger' than Euclidean space. In fact the inside of the circle is really a picture of the universe of two dimensional non-Euclidean geometry, in which all the copies of Dr. Stickler have the same size. You can learn more about this in the final chapter.

[2] The matrices we used to generate Figure 6.5 can be found at the bottom of p. 171.

Figure 6.6. Dr. Stickler taking a quasifuchsian stroll. Here Dr. Stickler is exploring the inside part of the ordinary set. This group belongs to the family on p. 170 with parameters $u = \text{Tr}\, a/2 \doteq 1.143$, $y = \text{Tr}\, b/2 \doteq 1.684$. In the picture we have thrown away everything outside the limit set, so you can see clearly all the half-tiles which remain. You should compare this picture with Figure 4.7.

each and every bulge you can see the same pattern of Dr. Sticklers striding along, the angles of the kinks and bulges always the same. This repetition of structure at finer and finer scales is a sure sign that the limit set is truly a fractal curve.

Matching the disks

How can we write down formulas for pairs of matrices a and b which will generate a group like the one in Figure 6.1? When we made Schottky groups, the Schottky disks were disjoint, and you could take any maps you choose which matched them in pairs. Now the disks are tangent, we have to be more careful what happens to the tangency points in order to get the beads linking properly in the second level chain.

First requirement. The problem can be seen in Figure 6.7, in which we have labelled the four tangency points, (in anticlockwise order starting from the tangency point of C_a and C_b), by P, Q, R and S. The map a

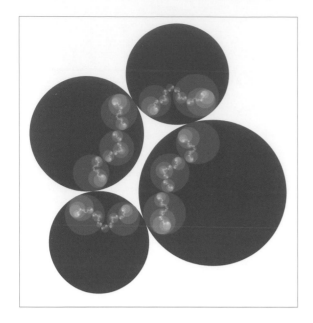

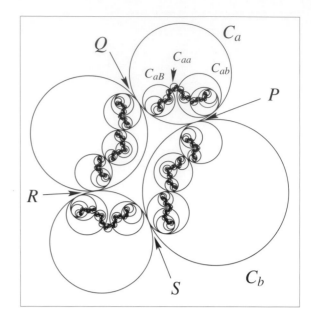

sends the outside of C_A to the inside of C_a, carrying the chain of three tangent circles C_b, C_B and C_a onto a chain of three tangent circles inside C_a. Since C_b and C_B are both tangent to C_A, their images $a(C_b) = C_{ab}$ and $a(C_B) = C_{aB}$ are tangent to $C_a = a(C_A)$. Unfortunately, with no further stipulations, there is no reason why the nesting circles should nest down on a curve. The figure shows why. The image chain is indeed tangent to C_a, but since we paid no attention to what happened to the points R and S, the image points $a(R)$ and $a(S)$ are in the wrong place.

To get the inner chain to match up properly, the right place for $a(R)$ would have been Q and for $a(S)$ would have been P. If we don't organise this, beads at level two will not be properly strung, because the string they are supposed to hang on must pass through the tangency points at stage one. Thus a necessary requirement for the limit set to be a curve is that the tangent points be mapped according to the rules:

$$a(R) = Q \qquad\qquad b(R) = S$$
$$a(S) = P \qquad\qquad b(Q) = P.$$

These mapping relations do not say anything about what a does to P or Q, or what b does to P or S. However, the inverse transformations A and B just follow the reverse of the above rules:

$$A(Q) = R \qquad\qquad B(S) = R$$
$$A(P) = S \qquad\qquad B(P) = Q.$$

Figure 6.7. What went wrong? To make these figures, we used the same four Schottky circles as in Figure 6.1. However, someone has changed the transformations a and b so that, while they still map the circles according to the Schottky rules, the tangency points do not map to tangency points. The delicate limit curve of Figure 6.1 has been shattered into shards of fractal dust. Actually, you may be able to tell that only the transformation a has been tampered with; b is still well-behaved. In the right frame we show a small green circle which is the Schottky circle C_{abAB}. It should be tangent to C_b, but is, alas, a small distance off.

The rules nicely link all four points together:

$$A(P) = S \quad \text{and} \quad B(S) = R \quad \text{and} \quad a(R) = Q \quad \text{and} \quad b(Q) = P.$$

After applying four transformations, we end up back at our starting point P, in summary, $a(b(A(B(P)))) = P$. In other words, these rules for matching the tangent points imply that P is a *fixed point* for the word *abAB*. Figure 6.7 shows what happens if you try to make a group without matching tangency points in the correct way.

The word *abAB* is very special. If we had $ab = ba$, then substituting and cancelling, we would get $abAB = baAB = bB = I$, so in some sense it measures how far a and b are from commuting. For this reason it is called the **commutator** of a and b. We shall come back to commutators in the next section. Figure 6.7 also illustrates another aspect of the word *abAB*. The blue circles are the paired Schottky circles C_B and C_b. We have seen that *abAB* fixes the point P which lies on C_b. Using the rules for applying words to Schottky circles, we have

$$abAB(\text{outside of } C_b) = abA(\text{inside of } C_B)$$
$$= \text{inside of } C_{abAB}.$$

In the righthand frame of Figure 6.7, the resulting circle is marked in green. It is the image of the blue circle C_b under the special word *abAB*. Since *abAB* fixes P and P belongs to C_b, then P should also belong to the green circle $abAB(C_b)$. That would mean C_b and the green circle should share the point P. They don't, so something has gone wrong with our system of labelling, which is why the limit set shatters.

Second requirement. Unfortunately, matching the tangent points is not the only thing you have to do to create a necklace. Suppose we do keep the matching rules so that *abAB* fixes P and also maps C_b onto the circle C_{abAB}, which is now tangent to C_b at P. Looking at Figure 6.2, you can see how all the image Schottky circles nest down from both sides of the tangency points of the red disks. *The critical thing is that the nested circles shrink down to points.* This tells us that the tangency point P between the circles C_a and C_b must be a *parabolic* fixed point of *abAB*: if *abAB* were loxodromic, then it would have to have another fixed point P'. This would prevent the image Schottky circles from shrinking to P.

An example where the circles don't shrink was shown on p. 95. In Figure 3.13 we saw two red disks paired to each other by a single hyperbolic transformation T. As usual, T maps the outside of the disk on the left to the inside of the disk on the right. One of the fixed points of T is the tangency point of the two red disks. As you can see, the image disks

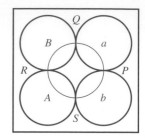

> ## Box 18: **The necklace conditions**
> To make a group whose limit set is a curve by pairing opposite disks in a tangent chain using two transformations a and b:
> - the tangency points must be correctly matched by the generators;
> - the transformation $abAB$ must be parabolic;
> - the trace of $abAB$ must be -2.

under T^{-1} fail to shrink down. Instead, they pile up on a limiting circle which goes through the two fixed points of T. The only way we can avoid something like this happening is to insist that T be parabolic, ensuring that the two fixed points of T coincide.

Third requirement. There is one final requirement which in the current examples will work out automatically, but which will be very important when we come to manufacture groups from a purely algebraic recipe in Chapter 8. Since $abAB$ is parabolic, its trace is either $+2$ or -2 (assuming, as always, that it has determinant 1). Changing the sign of all the entries in a matrix gives us the same Möbius map, so usually the sign of its trace is not important. However, it turns out that to get a quasifuchsian group with parabolic commutator, $\mathrm{Tr}\,abAB$ must *always* equal -2. In fact, changing the signs in a forces you to change the signs in its inverse A too, so the signs in $abAB$ don't change! The reasons behind this are explored in Project 6.1.

The conditions we have worked out in this section are very important facts for this and subsequent chapters, well worth recording in Box 18.

Commutators

We have just seen that to ensure our circle chain shrinks, we have to make sure that $abAB$ is parabolic. This so-called commutator plays an important role in any group. It measures the essential difference between the two simplest words of more than one letter: ab and ba. As we saw way back in Chapter 1, the order of multiplication (that is, composition) of transformations cannot be ignored. If ab and ba represent the same transformation in the group, they are said to commute. If they do not, it is the commutator $abAB$ which gets in the way, as you can see from the identity $(abAB)ba = ab$. In other words, $abAB$ is a measure of the difference between ab and ba.

In the last section, we arbitrarily made the choice of following the matching rules starting from the tangency point *P*. In consequence, we found that *P* was fixed by *abAB*, which had to be parabolic with trace −2. But wait a minute: if we had started from a different vertex, we would have found a different word, for example starting from *R* you find *ABab(R)* = *R* so that *ABab* should also be parabolic. It seems as if we have four different traces to contend with: the traces of *abAB*, *bABa*, *ABab* and *BabA*. Referring back to p. 143, we recognize that these are all cyclic permutations of each other. In addition, there were two choices of how to follow the matching rules starting at each of the four points: we could have gone in reverse as well and found *baBA(P)* = *P*, *BAba(R)* = *R*, and so on.

Fortunately the situation is not so complicated as it looks: the traces of all these transformations are in fact the same! Part of this we already knew, because the word *baBA* is the inverse of *abAB*, and certainly a matrix and its inverse have the same trace. In fact, cyclic permutation doesn't change traces either, because cyclic permutations are really conjugations in disguise. As we saw on p. 81, if *M* and *N* are matrices, then $\operatorname{Tr} NMN^{-1} = \operatorname{Tr} M$. Suppose, for example, starting from the string *abAB*, we wanted to move the symbol *a* to the back. The trick is to notice that $bABa = a^{-1}(abAB)a$. The rule about traces and conjugation shows that $\operatorname{Tr} bABa = \operatorname{Tr} a^{-1}(abAB)a = \operatorname{Tr} abAB$.

You can see in Figure 6.1 that the places where the pinch points of many tiles fit together, the kinks get extremely small and the limit set curve looks as if it is almost straight. The most prominent of such points are the meeting points of the four original Schottky circles. These are the fixed points of the parabolic commutator *abAB* and its cyclically permuted friends. Every other flat-looking point is the fixed point of a

---

Note 6.1: **The parabolic points on the limit set**

The limit set looks almost straight near the fixed points of parabolic transformations for two reasons: (a) the orbit of a point under the powers of a parabolic transformation is a subset of a circle through its fixed point; (b) the points of this orbit move towards the fixed point much more slowly than if it were loxodromic. We can see this if we conjugate and look at the simplest parabolic map $z \mapsto z + a$. In this case, (a) comes from the fact that the orbit of a point z is the set of all points $z + na$, which all lie on a straight line. As for (b), note that for the loxodromic map $z \mapsto kz$ with $|k| > 1$, the size of the orbit points $k^n z$ grows exponentially fast, whereas for the parabolic, the size (distance from 0) grows only linearly. In our pictures, this is reflected in the relative sizes of small nesting disks. Near a parabolic fixed point, diameters decrease much more slowly than they do elsewhere.

conjugate of the parabolic transformation *abAB*. The reason for this flatness is explained in Note 6.1.

Formulas for kissing Schottky groups

To draw pictures and examine Indra's necklace more closely, we need to write down formulas for both the initial circles and the Möbius maps *a* and *b* which pair them. Unfortunately, this is rather messy. Later, in Chapter 8, we shall introduce Grandma's ultimate recipe for generating loop shaped fractal limit sets which will allow us to drop the Schottky circles altogether, requiring only the input data of the traces of the matrices *a*, *b* and *ab*.[1] At this point, though, we want to explain in some detail some special examples of kissing Schottky groups which can be written down without too much pain. If you want to draw more varied chains of four tangent circles, look at Projects 6.3 and 6.4.

These special examples correspond to four circles arranged very symmetrically as in Figure 6.8. The formulas for the matrices which implement the pairing in this case are:

$$b = \begin{pmatrix} x & y \\ y & x \end{pmatrix} \quad \text{and} \quad a = \begin{pmatrix} u & ikv \\ -iv/k & u \end{pmatrix}.$$

You may recognize *b* as a stretch which has fixed points ± 1 and pushes points on the real axis from -1 to 1. It pairs the outside of circle C_B with centre $-x/y$ and radius $1/y$ to the inside of the circle C_b with centre x/y and the same radius. Likewise *a* pairs the outside of the circle C_A with centre $-iku/v$ and with radius k/v to the inside of the circle C_a centre iku/v and the same radius. The matrix entries can be any positive real numbers, provided they are related by the equations

$$x = \sqrt{1 + y^2}, \quad u = \sqrt{1 + v^2}$$

and

$$\frac{1}{2}\left(k + \frac{1}{k}\right) = \frac{1}{yv}.$$

As usual, the first two relations guarantee that *a* and *b* have determinant 1. The last one is more interesting; as explained in Note 6.2, it expresses the fact that the four Schottky circles make a tangent chain. It turns out that these exact relationships between the variables x, y, u, v and k guarantee that Tr *baBA* $= -2$. Thus *baBA* is parabolic, and it is possible to check directly[2] that, as one would expect, its fixed point is the tangency point of the circles C_a and C_b. You may like to try verifying that *a* and *b* pair the circles properly, but finish reading what we have to say about the arrangement of the four circles first!

[1] Grandma's recipe is on p. 229. The formulas look pretty horrendous, but it is a great method of setting up maps to make fractals.

[2] Truth to tell, direct computations are pretty hairy. In the last section of this chapter we shall describe some facts about calculating with traces which will make this easier to check.

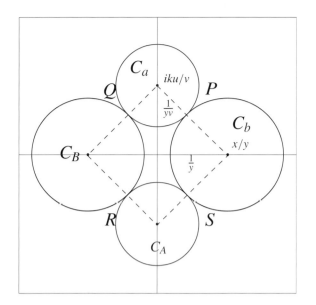

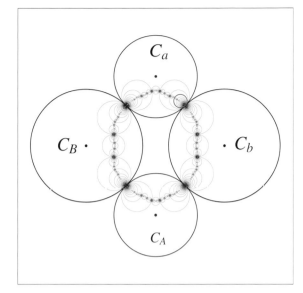

Figure 6.8. Four symmetrical tangent Schottky circles and their limit set. The central inner tile is bounded by the blue and red arcs.

These formulas provide us with a good first example of a **parameterized family** of groups. The entries in the generating matrices are given not as fixed complex numbers, but depend on certain carefully chosen variables or **parameters**. Once you have set up the basic programs, you can feed in any parameter values you please, producing not one but many different pictures at will. In this example, the parameters are x and u, from which the equations above allow us to determine (up to some choices of sign) the values of y, v and k.

As explained in Note 6.2, the circles C_B and C_b were chosen so as to be orthogonal to the unit circle. The circles C_A and C_a are orthogonal not to the unit circle, but to a concentric circle of a different radius k. To get the formula for a we first replaced the variable x in b by another variable u, then we rotated by $-\pi/2$, which we did by conjugating by $z \mapsto -iz$, and finally we conjugated by the scaling $z \mapsto kz$. The result is that a pairs the outside of the circle C_A centre $-iku/v$ radius k/v to the inside of the circle centre iku/v and the same radius.

If $k = 1$, all four circles are orthogonal to the unit circle. The most symmetrical example of all is if we set $k = 1$ and $x = u$. Solving the equations, we find that in this case $x = u = \sqrt{2}$ and $y = v = 1$. In this case, the four circles have their tangency points at $\pm(\cos 45° + i \sin 45°) = \pm e^{\pm i\pi/4}$. This is the rather special group which made Figure 6.5. The generators are

$$a = \begin{pmatrix} \sqrt{2} & i \\ -i & \sqrt{2} \end{pmatrix} \quad \text{and} \quad b = \begin{pmatrix} \sqrt{2} & 1 \\ 1 & \sqrt{2} \end{pmatrix}.$$

If $k > 1$, the transformation a stretches out: its fixed points move further apart and the Schottky circles C_A and C_a expand. If $k < 1$ on the other hand, the fixed points of a move together while the Schottky circles C_A and C_a contract. Everything is arranged so that the new circles still form a tangent chain. By a minor miracle, when we do this, the trace of the new commutator remains equal to -2, see Project 6.8.

Conjugating groups

There is an easy way to create new pictures from old. This is to conjugate the ones we already have. Remember that conjugation means looking at the same picture but from a different viewpoint, the one we get after Möbius transformation, that is, a coordinate change on the Riemann sphere. Conjugation will never produce anything fundamentally different because it is merely changing our point of view. In Chapter 3, we already studied the effect of conjugation on a single Möbius map. When you replace a transformation by a conjugate, you change the matrix but the dynamical behaviour remains the same.

There are two stages in conjugating by a map M. As we saw in Chapter 1, you replace every point z by the transformed point $\hat{z} = M(z)$, and you replace a transformation T by its conjugate $\hat{T} = MTM^{-1}$. There

---

Note 6.2: **How we arranged the circles**

We chose the centre of C_b to be x/y and its radius to be $1/y$ so that it would meet the unit circle at right angles. To arrange for two circles to meet at right angles at a point Q, we need to ensure that the lines from their two centres to Q are perpendicular. Suppose that C_b, whose centre is at $P = x/y$, meets the unit circle at Q. We work out $\overline{OP}^2 = (x/y)^2$ and $\overline{OQ}^2 + \overline{PQ}^2 = 1^2 + (1/y)^2$. So our condition $x^2 = y^2 + 1$ says exactly that $\overline{OP}^2 = \overline{OQ}^2 + \overline{PQ}^2$. Thus $\angle OPQ = 90°$, because Pythagoras' theorem holds in triangle OPQ.

The formula $\frac{1}{2}\left(k + \frac{1}{k}\right) = \frac{1}{yv}$ comes from requiring that the four Schottky circles be tangent. In other words, we want to find the relationship between the variables x, y, u, v and k which will ensure that C_a is tangent to C_b, C_b is tangent to C_A, and so on. By the

symmetry of the picture, it is clear that all we need to do is match up one pair, say for example C_a and C_b.

When two circles are tangent, the distance between their centres is exactly equal to the sum of their radii. Referring to Figure 6.8, the distance between the centre of C_a and the centre of C_b is $|iuk/v - x/y|$. Squaring and using that x, y, u, v, k are all real, we find $|iuk/v - x/y|^2 = u^2k^2/v^2 + x^2/y^2$. On the other hand, the sum of their radii is $k/v + 1/y$. Therefore, the condition for C_a and C_b to be tangent can be written as

$$u^2k^2/v^2 + x^2/y^2 = (k/v + 1/y)^2$$

which simplifies to

$$\frac{1}{2}\left(k + \frac{1}{k}\right) = \frac{1}{yv}.$$

is no particular reason to confine conjugation to a single transformation T. Suppose we try to conjugate a group formed by pairing four tangent circles. The image of each of the four circles in the initial configuration under the conjugating Möbius map M is a circle. Furthermore, since the four circles C_B, C_a, C_A, C_b touch each other in order forming a tangent chain, the same must be true of the four image circles $M(C_B)$, $M(C_a)$, $M(C_A)$, $M(C_b)$.

This changes coordinates, but how do we conjugate the group G? The answer is we have to replace each and every transformation g in G by the conjugated transformation $\hat{g} = MgM^{-1}$. The generators of the new group will be $\hat{a} = MaM^{-1}$ and $\hat{b} = MbM^{-1}$. The generator $\hat{a}$ pairs $M(C_A)$ to $M(C_a)$ because

$$\hat{a}(M(C_A)) = MaM^{-1}(M(C_A)) = Ma(C_A) = M(C_a).$$

Sometimes it may be preferable to write $M(C_A) = C_{\hat{A}}$, to indicate that $M(C_A)$ is the circle in the new chain which is paired by the new generator $\hat{a}$ to the new circle $M(C_a) = C_{\hat{a}}$. Then the equation would simply read

$$\hat{a}(C_{\hat{A}}) = C_{\hat{a}}.$$

As we prove formally in Note 6.3, the conjugated collection still forms a group. Usually for shorthand the **conjugated group** is written MGM^{-1}, which symbolizes the whole collection of transformations formed by taking each individual transformation g in G and replacing it by its conjugate MgM^{-1}.

Until you get used to it, the effects of conjugation can be rather surprising. Figure 6.9 shows four conjugate versions of the group shown in Figure 6.6. On the top left the group is shown on the sphere and in the

---

Note 6.3: **Conjugating a Möbius group**

Suppose G is a group of Möbius maps. We have to show that so is the set of conjugated maps $\hat{G} = MGM^{-1}$. Remember the defining properties of a group:

(1)   if S and T are in G then so is ST;
(2)   if S is in G then so is S^{-1}.

So now we check. Suppose that S and T are transformations in $\hat{G}$. By definition, this means that $S = MUM^{-1}$ and $T = MVM^{-1}$ for some members U

and V of G. To check property (1) we just compose: $ST = MUM^{-1}MVM^{-1} = MUVM^{-1}$. Since G is a group, UV is in G which means that $MUVM^{-1}$ is in $\hat{G} = MGM^{-1}$. To check property (2), suppose once again that $S = MUM^{-1}$ belongs to $\hat{G}$. Then $S^{-1} = (MUM^{-1})^{-1}$. Remembering the rule for inverting, we find $(M^{-1})^{-1}U^{-1}M^{-1} = MU^{-1}M^{-1}$. Since G is a group, U^{-1} belongs to U, so we have just proved that S^{-1} belongs to $\hat{G}$.

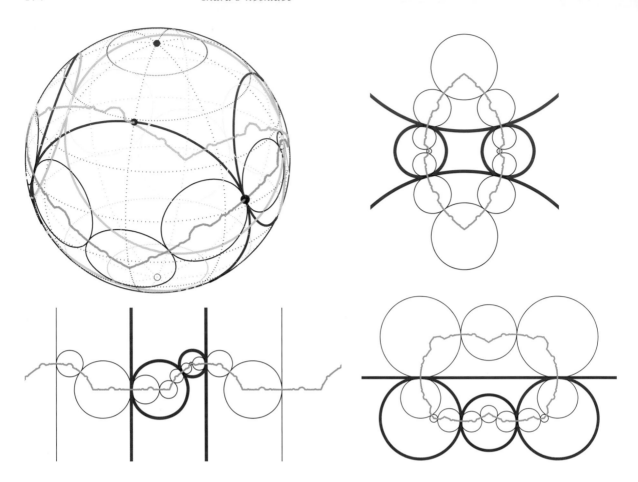

other three pictures it has been stereographically projected to the plane in three different ways. If you choose any pair of the three pictures on the plane, there is a Möbius map which conjugates between one picture and the other.

   On the top right we project using the top black dot on the sphere as the South Pole so that the region 'inside' the green limit set comes from the 'northern hemisphere' on the sphere; on the bottom left a tangency point of two red and blue Schottky circles was the North Pole and was therefore sent to ∞; and on the bottom right the black dot which is a non-tangency point of a red Schottky circle was the North Pole, sending the circle to a straight line. The region *inside* the limit set in the bottom right picture corresponds to the region *outside* in the picture above. One way to see this is to look carefully at which side of the limit set the spiky 'angles', most prominent on the sphere picture and in the top right. (Remember that stereographic projection and Möbius maps preserve angles but not size or shape!) The picture in Figure 6.6 is an inside-out version of the picture on the top right.

Figure 6.9. Conjugates of the circle chain group in Figure 6.6. On the top left, the group is shown on the sphere, the other three frames showing various projections onto the plane. For clarity, we omitted some of the black Schottky second-level circles on the back side of the sphere. The bottom right arrangement comes from Grandma's special recipe which we shall meet in Chapter 8.

Since conjugation preserves iteration and dynamics, all the objects we are interested in, like tilings and the limit set for a conjugated group MGM^{-1}, should be the images of the same object in the first group G under the conjugating map M. The limit set of the group G maps to the limit set of the group MGM^{-1}. You can see this in the pictures by tracing the fate of some of the Schottky circles and the green limit set through our four frames. In fact for each circle C in the Schottky array on the sphere, there is an equivalent circle $M(C)$ in each of the other frames.[1]

There is an important point about notation here. We have been describing our Schottky group in terms of its generating elements a and b. Sometimes we are thinking of a and b as specific matrices, for example $a = \begin{pmatrix} \sqrt{2} & i \\ -i & \sqrt{2} \end{pmatrix}$ and $b = \begin{pmatrix} \sqrt{2} & 1 \\ 1 & \sqrt{2} \end{pmatrix}$, while sometimes a and b are abstract symbols, if you like representing nodes on the word tree. When we conjugate a by a transformation M, for example anticlockwise rotation by $\pi/4$, we have to replace a with the new matrix

$$\begin{pmatrix} e^{i\pi/8} & 0 \\ 0 & e^{-i\pi/8} \end{pmatrix} \begin{pmatrix} \sqrt{2} & i \\ -i & \sqrt{2} \end{pmatrix} \begin{pmatrix} e^{-i\pi/8} & 0 \\ 0 & e^{i\pi/8} \end{pmatrix} = \begin{pmatrix} \sqrt{2} & ie^{i\pi/4} \\ -ie^{-i\pi/4} & \sqrt{2} \end{pmatrix}$$

giving us a new generator $\hat{a}$ which pairs the new circle $C_{\hat{A}}$ to the new circle $C_{\hat{a}}$. However, if we are just thinking in the abstract of the Schottky group with generators $\hat{a}$ and $\hat{b}$ which pair the circles $C_{\hat{A}}$ to $C_{\hat{a}}$ and $C_{\hat{B}}$ to $C_{\hat{b}}$, we might as usual have called the generators and the circles by their usual names a, A, b, B and C_a, C_A, C_b, C_B.[2]

Figure 6.10 shows six conjugated versions of the group with generators

$$a = \begin{pmatrix} \sqrt{2} & i \\ -i & \sqrt{2} \end{pmatrix} \quad \text{and} \quad b = \begin{pmatrix} \sqrt{2} & 1 \\ 1 & \sqrt{2} \end{pmatrix}$$

which we met in the last section. The arrangement of circles for the above generators is shown in frame (ii). The conjugates we have chosen reflect the ones in Figure 6.9. In each frame, we have consistently labelled the four circles C_a, C_A, C_b and C_B and we have labelled the tangency points of the four circles P for the tangency point of C_a and C_b, Q for the tangency point of C_a and C_B, and so on. Because of the difficulty of distinguishing insides from outsides, in each picture we have coloured the image of the *same half* of the region outside the four circles green. You can distinguish which is which by noting that in the green region, the points P, Q, R and S run *anticlockwise* round the boundary, while in the white region they run *clockwise*.

The top left frame (i) depicts the arrangement of circles on the Riemann sphere. Conceptually, this picture unifies all the other five. The four

---

[1] This explains the rule for conjugation: if the transformation g carries C to another circle $g(C)$, then MgM^{-1} carries $M(C)$ to $MgM^{-1}(M(C)) = MgM^{-1}M(C) = M(g(C))$.

[2] This shouldn't lead to much confusion. Conjugation preserves the group structure, which means that the new word $\hat{a}\hat{B}\hat{B}\hat{a}\hat{B}$ has exactly the same effect on the circle $C_{\hat{A}}$ as the word $aBBaB$ has on the circle C_A.

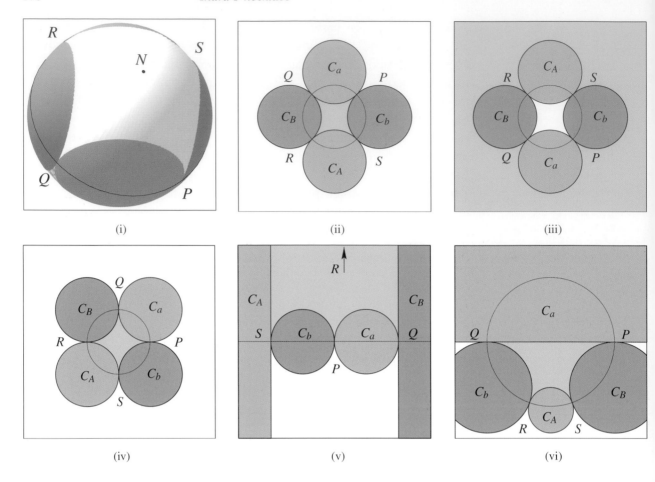

Schottky circles are arranged in a belt round the equator and you should imagine the two pairing transformations pushing them over one to the other across the North Pole. The region outside the circles and in the 'southern hemisphere', which you can just see peeping out underneath, is coloured green. To get frame (ii) you stereographically project to the plane from the sphere sitting on the plane in its usual position with the South Pole at the origin O.

To get frame (iii), we rotated the sphere through $180°$ interchanging the North and South Poles. In frames (ii) and (iii) the arrangement of the four circles looks the exactly the same, except that in one the green region is 'inside' and in the other it is 'outside' the four Schottky disks. Notice how the cyclic order of P, Q, R and S has been reversed. The conjugation which gets from (ii) to (iii) can be implemented in practice with the map $z \mapsto 1/z$. Since the original circles are orthogonal to the equator, in both (ii) and (iii) all four circles C_a, C_A, C_b and C_B are orthogonal to the unit circle.

Figure 6.10. Lots of conjugations. This picture shows different configurations of four tangent circles all obtained from each other by conjugation. You can find explicit formulas for the group generators and conjugations in Project 6.5.

To get frame (iv), you rotate round the North–South axis by $\pi/4$ and project in exactly the same way as you did to get frame (ii). Alternatively, conjugate the generators for the group in (ii) by the map $z \mapsto e^{-i\pi/4}z$. To get frame (v), rotate the sphere so the belt of tangent circles runs around a great circle joining the two poles, with R at the North Pole. As you can see, this means that two 'circles' C_A and C_B show up in the picture as parallel 'vertical' lines. This can be implemented in practice by conjugating (iv) by the inverse of the Cayley transform (see p. 87), which takes the unit circle to the upper half plane.

Finally, to get frame (vi), rotate the sphere so that the point on circle C_a midway between P and Q and on the edge of the green region is at the South Pole. This time, you will have to translate, rotate and scale the projected picture to come out with our arrangement in which P and Q are at ± 1. If instead we had put the midpoint on the edge of the white region at the South Pole, then the picture you see in frame (vi) would be turned inside out. You can find some more details about implementing these conjugations in Project 6.5.

The upshot of this set of examples is that the really interesting thing to do is study not groups of Möbius transformations, but groups of transformations *up to conjugation by Möbius maps*. In other words, for most purposes, two conjugate groups G and MGM^{-1} should be considered for all serious purposes 'the same'. This has a very practical aspect when making pictures. Often, by conjugating, you can produce a more pleasing picture because it sits more symmetrically (or just plain fits better) on the page. You should not be alarmed when we take this liberty. From now on, we shall feel free to show you conjugate images and leave you to pick out the essential features which make two pictures 'the same'.

Fuchsian subgroups and ghost circles

You probably noticed that the limit set in Figure 6.1 seems to be made up of lots of circular arcs, with rather sharp angles where they meet. Similar circular structures have been highlighted in Figure 6.11. They have a nice explanation, and there will be lots more circles like this from now on. Remember we said that a group whose limit set is a circle (or line) is called Fuchsian. Actually this term is also used to refer to a group whose limit set is just part of a circle – typically, fractal dust. Fuchsian groups like this can be made by starting with four disjoint Schottky circles all orthogonal to another circle, as in the left frame of Figure 6.12. We call them **Fuchsian Schottky groups**.

The explanation of all the circles in Figure 6.11 is that the group

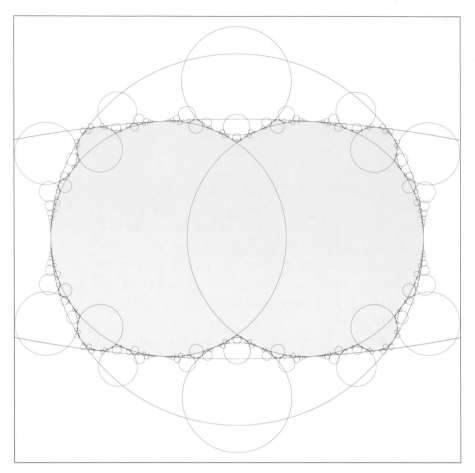

Figure 6.11. This limit set is a quasicircle which grazes many circles, touching each of them in infinitely many points. There are two families of these circles, one inside and one outside the limit set. They meet the limit set in 'ghost circles' made of fractal dust.

generated by *a* and *b* has lots of **subgroups**[1] which are Fuchsian, for example the subgroup generated by the transformation *a* and its conjugate *baB*. The limit set of this subgroup is both part of a circle (because it is Fuchsian), and part of the limit set of the whole group. The circle in question is the circumference of one of the large yellow disks in Figure 6.11, highlighted as one of the large glowing circles in the limit set in Figure 6.1. If we write points of the limit set as infinite words in *a* and *b*, then the limit points on this circle will be precisely all the words which can be written using only *a* and *baB*. The other large circle contains all the limit points of the group generated by *b* and its conjugate *abA*. One might call these circular limit sets 'ghost circles', because the limit points are spread thinly in fractal dust. All the other ghost circles correspond to subgroups which are conjugate to one of these two.

You can recognize a Fuchsian group because the traces of all its elements have to be real numbers. This is because the limit circle can always

[1] A subgroup of a group is just a collection of its members which happen to form a group in their own right. One way to specify a subgroup is just by naming its generators.

Immanuel Lazarus Fuchs, 1833–1902

Groups whose limit sets lie on a circle are called **Fuchsian** after Immanuel Lazarus Fuchs. Born near what is now Poznan in Poland, Fuchs was a leading member of the mathematical circle in Berlin. He systematically studied systems of differential equations in the complex plane, investigating how a solution would change when you made a circuit round a 'singularity' where a term in the equation took the value ∞ (as for example $1/z$ at the value $z = 0$). Fuchs realised that it might be simpler to study the *ratio* of two solutions, because when you make a circuit round a singular point, this ratio changes by a Möbius map. As you follow different paths round different singular points, the collection of all maps you get forms a group!

Shortly after the appearance of his paper at the beginning of May 1880, a rapid and lively correspondence ensued with a young Frenchman from the University of Caen, Henri Poincaré. Poincaré realised that the inverse of the ratio function was really a function on the surface associated to the group – the self-same type of surface that we have made by gluing up our tiles. This study led to his great insight into the connection between Möbius transformations and non-Euclidean geometry, of which more in the last chapter. Poincaré called the groups he discovered *'groupes fuchsiens'* in Fuchs' honour.

be conjugated into the real line, so any Fuchsian group can be conjugated into a subgroup of the upper half plane group which we met in Recipe I on p. 85. Traces don't change under conjugation, and the trace of any matrix in the upper half plane group is certainly real. In fact, as we shall see at the end of this chapter, it is possible to express the trace of any element in a group with two generators as a polynomial in $\mathrm{Tr}\,a$, $\mathrm{Tr}\,b$ and $\mathrm{Tr}\,ab$. It follows that if these three traces are real, so are all the rest. Any such group must be Fuchsian, as we explain in Project 6.6.

In our example, the subgroup with the two generators a and baB is Fuchsian because $\mathrm{Tr}\,a$ is real (a requirement of our recipe), $\mathrm{Tr}\,baB$ is also real (because conjugation preserves traces), and finally $\mathrm{Tr}((a)(baB))$ is also real because it is the parabolic commutator whose trace is -2. Similar situations will crop up repeatedly as we go along.

Sealing the gaps: following branches down the word tree

We learned in the last two chapters how to describe limit points as infinite paths in the word tree. What happens to this procedure when the circles kiss? In Figure 6.12, we compare the kissing Schottky group from Figure 6.5 with a 'nearby' true Schottky group with four nearly tangent but disjoint circles.

A crucial difference between the pictures is that there are certain pairs of infinite words which specify nests of circles which have distinct limit points on the left, but for which the corresponding nests of circles limit down on the *same* limit point on the right. In the left frame, where the circles are disjoint, there are big gaps in the limit set which include the gaps between 'adjacent' pairs of Schottky circles and more. On the right, these gaps have been closed and all the circles in what were two distinct nesting families become simultaneously tangent at one common point.

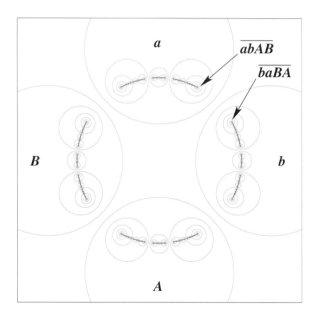
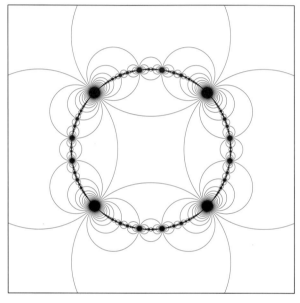

The detailed pattern in which limit points come together is best understood from Figure 6.13. Starting at the centre, imagine walking down the word tree first to a and then heading, from your viewpoint as you walk, always as far as possible to the right. After a, the next vertex you come to is ab, followed by abA and then $abAB$. The pattern begins to cycle and at the next level the label is $abABa$. Following the repeats, you find yourself headed out along the infinite word path $\overline{abAB}$, whose endpoint as we know is the attracting fixed point of $abAB$. In the right frame of Figure 6.12, this point is the top right tangency point of the four initial Schottky circles, the tangency point of C_a and C_b. On the left, it is the limit point inside C_a and closest to C_b.

Figure 6.12. Closing the gaps. On the left is a Schottky group with four nearly tangent circles and on the right the gaps between the Schottky circles have been closed. These groups were made by the θ-Schottky recipe in Project 4.2.

Suppose we did a different thing and headed first towards b, and then turned at each step as far as possible to the left. The pattern begins with $baBA$, after which it cycles giving the infinite word $\overline{baBA}$, which represents the attracting fixed point of $baBA$. Wait: the inverse of $baBA$ is just $abAB$! In the lefthand picture, $\overline{baBA}$ is the limit point inside C_b and closest to C_a, but in the righthand picture, the commutator has become parabolic so its two fixed points have coalesced into one. The two different word paths $\overline{abAB}$ and $\overline{baBA}$, which have different end points on the left, both end at one and the same limit point on the right.

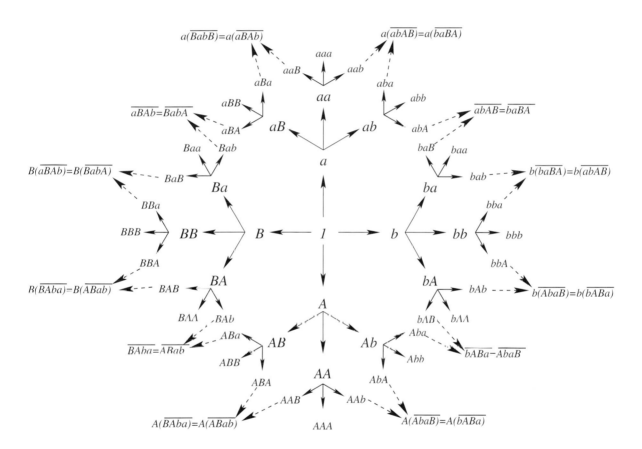

Figure 6.13. The gaps from the point of view of the tree of words. This figure is the same as Figure 4.5 except that we have added the merged infinite words where the Schottky circles now meet.

When the Schottky circles touch, the merging effect at the end of these word paths propagates all over the diagram. For example, consider the circles C_{bb} and C_{bA}. There are two paths, one on each side of this gap between these two circles, which look as if they should meet when the circles touch. You can see these paths emanating from the corresponding nodes in Figure 6.13. The first one starts at bb and always turns as far as possible to the right. This brings you down to the infinite word $bb\overline{ABa}$, which represents the attracting fixed point of $W = b(bABa)B$. The second path starts at bA, and then always turns as far as possible to the left. Its label is $\overline{bAbaB}$, so it limits on the attracting point of $b(AbaB)B$, which is of course nothing other than the repelling fixed point of $W^{-1} = b(bABa)B$. And why are these fixed points the same? It is just because $b(bABa)B = (b)(bABa)(b)^{-1}$, which is a conjugate of $bABa$. If $bABa$ is parabolic with just one merged fixed point, so is $b(bABa)B$!

This merging effect is exactly the same as what happens with decimals. If the initial Schottky circles touch, then the chains of nested circles at each level are mutually tangent and form a chain linking right around the unit circle. All the gaps are sealed, and the limit set becomes the whole circle. Since there are families of circles nesting down on every limit point, we have found that *every point of the unit circle is labelled by an infinite path on the word tree*. Most points have only one label, but there are ambiguities whenever the ends of two word paths merge. As we have just seen, this happens because the infinite word path $W\overline{ABab}$ now represents the same point as the infinite word $W\overline{BAba}$. Just as there are ambiguities in infinite decimal expansions, with $0.\overline{9}$ representing the same point as $1 = 1.\overline{0}$, we find that the same limit point may now be described by two different infinite words.

The fact that the limit set is contained in the unit circle in this example was not really important. All that matters is that the original four Schottky circles are tangent. This gives us an alternative viewpoint on the *necklace condition* in Box 18: the reason the commutator $abAB$ has to be parabolic is to close the gaps in the limit set. In terms of the word tree diagram in Figure 6.13, this merging can be visualized as bringing the infinite tips of the tree together into a schematic circle which manifests as a real circle or quasicircle whenever we have a concrete Fuchsian or quasifuchsian group.

Upgrading the program to draw curves

Now we really get down to the nitty-gritty: how can we adapt our program to draw quasicircles? The reason we made such a fuss over cyclic permutations of the four generators and the order in which the DFS algorithm explored the tree of words was precisely to ensure that when the limit set is a loop, the limit points pop out in the correct order as we follow along the curve. Like doing a child's puzzle, we can trace out the curve by 'joining the dots'. The dots are choices of limit points, numbered in the order they appear. All we do is connect them with line segments from one number to the next.

What this means in practice is that instead of plotting isolated points each time we reach a 'branch termination', we plot a short line segment joining the last limit point found to the next one calculated. More explicitly, when initializing, we want to declare an extra variable `oldpoint` as well as `newpoint` (see the `branch-termination` procedure on p. 148), and then write the procedure as follows:

```
branch_termination{
    circle = mobius_on_circle(word[lev-1],
    circ[tags[lev]]);
    if radius(circle) < epsilon then {
        newpoint =
          mobius_on_point(word[lev],fix[tags[lev]]);
        plot_line(oldpoint, newpoint);
        oldpoint = newpoint;
        RETURN(TRUE) }
    else RETURN(FALSE)
}
```

After plotting a line segment, the current `newpoint` is made into the `oldpoint`, ready for the next call to plot. Combined with the work we already did on the DFS algorithm, this is all we need! Provided the generators are correctly ordered, the routine will trace out the limit curve. However there are some important additional features which will considerably improve our program.

Do we need to calculate circles? Computationally, the most expensive part of the procedure is the calculation of `circle` based on the formulas described on p. 91. The circles are only used in the test to see if the forward searching should end; they are not used for plotting limit points. In the interests of speed, let's see if we can do away with that part of the program.

Our procedure has available an `oldpoint` and a `newpoint`. Since our goal is to roughly trace out a continuous curve, perhaps we should simply check that the distance between `oldpoint` and `newpoint`, that is, `abs(newpoint - oldpoint)`, is smaller than our threshold `epsilon`. That would reduce the code to the following:

```
branch_termination{
    newpoint =
      mobius_on_point(word[lev],fix[tags[lev]]);
    if abs(newpoint - oldpoint) < epsilon then {
        plot_line(oldpoint, newpoint);
```

```
    oldpoint = newpoint;
    RETURN(TRUE) }
  else RETURN(FALSE)
}
```

The danger of this change is that we have lost the information about the sizes of the Schottky disks. When the branch termination is tied to the disks, then ensuring they are all small provides absolute certainty that what we plot is a close approximation to the limit set. If, however, we simply measure the distance from one limit point to the next, we might accidentally jump from one point to another very close by, completely missing an extremely large and wiggly section of the limit set in between.

The problem is illustrated in the margin. As you can see, the distance between the two sides of Dr. Stickler's neck being small is no guarantee that all points along the stretch of curve between them – his head – will also be nearby! If the distance across his neck is less than our threshold `epsilon`, then our algorithm will jump straight across, and we will sadly miss his handsome profile. For obvious reasons, we call this the 'problem of necks'. We shall have to wrestle with the neck problem in various forms in later chapters.

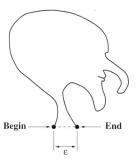

Figure 6.14. The problem of necks

Which point should we plot? Thus far in our algorithm, the limit points we have plotted have been images of the fixed points of our four generators a, A, b and B. But as we saw in Box 14, there are many other ways to plot the limit set. If we are not calculating the Schottky disks, it turns out that there is a better choice which fits extremely well with our DFS algorithm: plot the images of the fixed points of all the commutators.

The idea is intimately related to our discussion about closing the gaps in the word tree in the last section. Suppose our current word is *abbABB*. The limit point that our procedure calculates for `newpoint` is *abbABBā*. However assuming words with the same prefix are close, any other point with the same prefix should do just as well. Really, we are interested in the segment of limit set consisting of all points with prefix *abbABB*. The trick is to think about the 'first' and 'last' limit points of this segment. Imagine standing on a node of the word tree, and consider the three possible directions forward, moving deeper into the tree. The two extremes are always to turn right, towards the 'first point' or always to turn left, towards the 'last point'. Examination of Figure 6.13 reveals:

$$\text{First Point:} \quad = abbABB\ abABabAB\cdots \quad = abbABB\overline{abAB}$$
$$\text{Last Point:} \quad = abbABB\ AbaBAbaB\cdots \quad = abbABB\overline{AbaB}.$$

Just as when we headed towards extreme points of the 'gaps' in Figure 6.12, the repetends at the ends of the infinite words are commutators. Luckily, in kissing Schottky groups, the commutators are all parabolic and so have only one fixed point. This means that we can calculate the first and last limit points by applying the transformation *abbABB* to the fixed points of *abAB* and *AbaB*.

Now it really is our lucky day, because the discussion in the last section implies that the beginning point for the stretch of the limit set with a given prefix is the *same* as the ending point of the previous stretch – that is, our new beginning point is exactly the ending point which was computed during the most recent call of the DFS routine which resulted in a point being plotted. This works because the commutators are parabolic, or more precisely because the fixed points $\overline{abAB}$ and $\overline{baBA}$ are the same.

Here are some details about implementing this in our program. Based on the fixed point formulas in Note 3.3, is easy to write a subroutine `fix` for calculating the fixed point of a Möbius transformation. During the initialization and setup part of the algorithm, after our matrices *a*, *b*, *A* and *B* have been computed and stored in the array `gens`, we compute and store the fixed points of the commutators. Here are the appropriate commutator fixed points for a prefix that ends with generator `gens[1]` (which we have taken to be *a*):

```
begpt[1] = fix( gens[2]*gens[3]*gens[4]*gens[1] )
endpt[1] = fix( gens[4]*gens[3]*gens[2]*gens[1] )
```

For `begpt`, we filled out the tags with continual 'right' turns, while for `endpt` we used 'left' turns. We have to do this for each of the other generators as well, because which commutator we get depends on the last letter of the prefix, (although turning in the same direction always gives cyclic permutations of the same one).

Assuming we have calculated `begpt[i]` and `endpt[i]` for each of the generators 1 through 4, here is the new branch termination procedure:

```
branch_termination{
   newpoint = mobius_on_point( word[lev],
   endpt[tags[lev]] )
   if abs( newpoint - oldpoint ) < epsilon then {
      plot_line( oldpoint, newpoint );
      oldpoint = newpoint;
      RETURN(TRUE) }
   else RETURN(FALSE)
}
```

We used this branch termination criterion to plot almost all the limit sets pictures in this book. Actually if we hadn't used it, all our quasifuchsian limit sets would have ugly looking gaps. It turns out that, for the reasons explained in Note 6.1, it is very important to plot the parabolic limit points where infinite words have merged. We shall be going into this in more detail in the next chapter.

We do have to plot one beginning point to get started, usually $\overline{abAB}$. Beyond that, the algorithm doesn't make use of begpt. However the algorithm is also useful for plotting exotic Schottky dust limit sets, in which case it is essential to have access to the beginning points, since the two fixed points of each commutator no longer coincide.

Plotting stretches of the limit set. The final feature we shall describe is not so important as the other two, but can be extremely useful when we wish to 'zoom in' and plot only a small portion of the limit set. With the DFS algorithm, it is easy to specify starting and stopping points and only plot the stretch of limit set between. You can see an example if you skip ahead to Figure 8.17, in which we have plotted only the section of limit points corresponding to words beginning with the particular prefix ab^5. The first point plotted was $ab^5\overline{ABab}$ and the last one was $ab^5\overline{aBAb}$. The few modifications needed to do this occur only in the program initialization and in the procedure which checks whether or not to halt the enumeration of the word tree.

Suppose, for example, we want the beginning word *aaabA* and the ending word *aBBab*. (Check that the ending word is after the beginning word in the word tree ordering!) We input a beginning list of tags begtags which is an array like the list tags which keeps track of our current word. In this example, the values of begtags[i] would be 1, 1, 1, 2, 4, as i ranges from 1 to 5 (the starting level; call it beglev). Then in the initialization part of the program we would have a loop like

```
tags[1] = begtags[1]
word[1] = gens[begtags[1]]
for lev from 2 to beglev do {
   tags[lev] = begtags[lev]
   word[lev] = word[lev-1] * gens[begtags[lev]]
}
```

This establishes tags and word in agreement with the beginning word, and the program proceeds from that point. Remember that there are many infinite words starting with the same finite prefix; this algorithm starts the search at the *first* such one.

In a similar way, we can terminate the program at the *last* infinite word beginning with the prescribed ending prefix. Thus if the beginning and ending prefixes are the same, the program will draw all limit points corresponding to infinite words beginning with that prefix.

We suppose that a list of ending tags endtags[i] from i = 1 to endlev was established at the outset of the program. The test for termination should occur just after we stop going forward. There is no point in carrying out the backwards process, because this will not change tags or word, only lev. The following loop returns TRUE if the end has been reached, and FALSE otherwise.

```
if endlev = 0 and tags[1] != 2 then RETURN(FALSE)
else {
   for i from 1 to endlev do {
      if tag[i] != endtag[i] then RETURN(FALSE)
   }
}
for i from endlev+1 to lev do {
   if tag[i] != tag[i-1] - 1 mod 4 then RETURN(FALSE)
}
RETURN( TRUE )
```

This routine proceeds through a whole series of tests which must all be passed if the end has been reached. The first loop checks that the word matches the prescribed ending word. The second loop checks that only *left* turns occur after that point, meaning that the word is the *last word* (which passed the forward test) with that prefix.

There's a little bit of programming here: if the current level lev is *less than* endlev, then there is a problem of interpretation. However the current level will usually be vastly greater than the prescribed ending level, unless you typed in a very long ending word.

From pretzels to punctured tori

By now we have come to expect that whenever we have a tiling and a symmetry group, we can link geometry and topology with surfaces we called variously doughnuts, tyres and pretzels, made by gluing the sides of the tiles as dictated by the symmetries in the group. For example, Figure 4.8 showed how the swiss cheese tiles associated to Schottky groups made pretzels: in mathematical language **two-handled tori** or **surfaces of genus two**.

What happens to this construction when the four Schottky circles become tangent? In order to visualize this easily, we shall do the pretzel gluing in a different way: instead of pulling both pairs of circles *outside* the sphere as we did in Figure 4.8, we pull one pair *outside* but push the

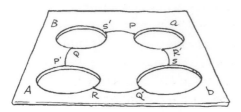

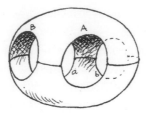

IN A PLANE WITH FOUR HOLES, WHAT BECOMES
OF THE ARCS R'→S, Q'→R, P'→Q, AND
S'→P WHEN WE GLUE A TO a AND B TO b?

ADD THE POINT AT INFINITY
TO MAKE THIS CLOSED
SURFACE.

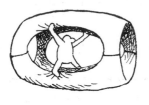

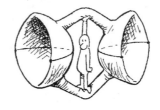

BRING A AND a TOGETHER.

GLUE A TO a.

SQUEEZE "TUBES" FOR
BETTER VISIBILITY.

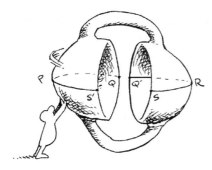

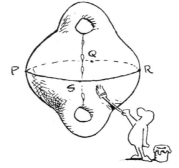

TWIST HEMISPHERES AROUND
TO BRING B TO b.

GLUE B TO b, MAKING A SURFACE
WITH TWO HANDLES. OUR ARCS
HAVE BECOME ITS "WAIST."

Figure 6.15. An alternative way of gluing the Schottky tile to make a pretzel. Note how the path given by the four arcs joining points marked P, P', Q, Q' etc., which separate the 'inside' of the basic Schottky tile from its 'outside', becomes the curve around the waist of the glued up pretzel. If you were to cut along the waist, the pretzel would fall apart into two halves, each of which would be a torus with a hole where the other half-surface used to be attached.

other *inside*, as shown in our new Figure 6.15. The advantage of doing this is that we can now trace the fate of a path which is set to collapse when the four circles kiss. The path is made from the following four segments:

- from S' on C_B to P on C_a;
- from $P' = A(P)$ on C_A to Q on C_B;
- from $Q' = b(Q)$ on C_b to R on C_A;
- from $R' = a(R)$ on C_a to S on C_b.

This looks like four disconnected segments, but when you glue up the surface, P and $P' = A(P)$ get matched together, as do Q and $Q' = b(Q)$, and so on until we come back around to the final point S which glues to $S' = B(S)$. If you follow through the steps in Figure 6.15, you will see exactly how the four pieces of path exactly glue together making a loop on the glued-up pretzel. The path corresponds to the commutator $aBAb$ because, if you transport the path segments around by the group in a picture like Figure 4.7, you get an *unbroken* path from S' to $a(P')$ to $aB(Q')$ to $aBA(R')$ to $aBAb(S')$.

As we bring the Schottky circles together, the points at the ends of the four arcs come together so the commutator loop on the pretzel gets shorter and shorter, squeezing the waistline ever tighter until finally, as the circles touch, the loop has been shortened to just a single point. The surface has been squeezed so much that, whoops, it pops! It is like pinching a balloon; after the pop, the surface is broken where the loop was and all that remains of the thin waist are two tiny pinpricks or **punctures**. Each of the two halves of the pinched off surface is a **torus with one puncture**, that is, a torus with one missing point on the surface, all that remains to show where the other half of the pretzel got popped off. Figure 6.16 shows directly how the tiles for a kissing Schottky group glue up. The official name for a group which gives you a pair of surfaces like this is a **once-punctured torus group**; it turns out that any quasifuchsian group has tiles which glue up in exactly the same way.

Trace matters!

As we saw on p. 168, the *necklace condition* for pairing opposite tangent Schottky circles to make a quasifuchsian group includes the vital requirement that $\mathrm{Tr}\, abAB = -2$. Checking this involves some pretty messy calculations! We promised above to reveal a shortcut which makes the checking somewhat easier: it turns out that the trace of the commutator $abAB$ can be expressed in terms of the three basic traces $\mathrm{Tr}\, a$, $\mathrm{Tr}\, b$ and $\mathrm{Tr}\, ab$ by the equation

$$\mathrm{Tr}\, abAB = (\mathrm{Tr}\, a)^2 + (\mathrm{Tr}\, b)^2 + (\mathrm{Tr}\, ab)^2 - \mathrm{Tr}\, a \cdot \mathrm{Tr}\, b \cdot \mathrm{Tr}\, ab - 2.$$

This formula is complicated but crucial: it was already an old friend for Klein in 1897.[1] If $\mathrm{Tr}\, abAB = -2$, it simplifies to a relation called the **Markov identity** after A.A. Markov who studied it extensively:

$$(\mathrm{Tr}\, a)^2 + (\mathrm{Tr}\, b)^2 + (\mathrm{Tr}\, ab)^2 = \mathrm{Tr}\, a \cdot \mathrm{Tr}\, b \cdot \mathrm{Tr}\, ab.$$

(Markov's interest is explained in Note 6.4.) We can regard the Markov identity as a quadratic equation for $\mathrm{Tr}\, ab$ in terms of $\mathrm{Tr}\, a$ and $\mathrm{Tr}\, b$. In other words, $\mathrm{Tr}\, ab$ must be one of the two roots of the quadratic

[1] *Vorlesungen über Die Theorie der Automorphen Funktionen*, 1897, Volume I, p. 337 ff. These relations are sometimes aslo known as **Fricke identities**. We are uncertain who discovered them first.

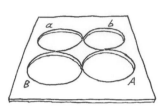

BEGIN WITH A PLANE MINUS
FOUR DISKS, EACH TANGENT TO
TWO OTHERS, IN THE ORDER
aBAb.

ADD THE POINT AT INFINITY
TO MAKE A CLOSED SURFACE.

STRETCH THE SURFACE
TO SEE ITS ESSENTIAL
SYMMETRY.

Figure 6.16. The gluing construction for a quasifuchsian group. We start from a chain of four kissing circles giving us a tile which comes in two pieces, each with four sides and four cusps. After gluing, we get two tori, each with a cuspidal 'puncture' where they are joined.

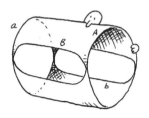

CONTINUE STRETCHING TO
PREPARE FOR FIRST GLUING.

PULL CIRCLES B AND b
TOGETHER...

AND GLUE THEM, FORMING
TWO TUBES.

BEND TUBES AROUND TO JOIN
CIRCLES a AND A.

GLUE TO FORM TWO PUNCTURED
TORI JOINED AT THEIR CUSPS.

equation

$$x^2 - \operatorname{Tr} a \ \operatorname{Tr} b \ x + (\operatorname{Tr} a)^2 + (\operatorname{Tr} b)^2 = 0.$$

What about the other root? Suppose we replace b by its inverse B. The Markov relation would now read

$$(\operatorname{Tr} a)^2 + (\operatorname{Tr} B)^2 + (\operatorname{Tr} aB)^2 = \operatorname{Tr} a \ \operatorname{Tr} B \ \operatorname{Tr} aB.$$

Since $\operatorname{Tr} b = \operatorname{Tr} B$, this shows that

$$(\operatorname{Tr} a)^2 + (\operatorname{Tr} b)^2 + (\operatorname{Tr} aB)^2 = \operatorname{Tr} a \ \operatorname{Tr} b \ \operatorname{Tr} aB.$$

In other words, $\mathrm{Tr}\,aB$ satisfies exactly the same quadratic equation as $\mathrm{Tr}\,ab$.

One way to verify these formulas is by 'brute force': plugging in 2×2 matrices for a and b and calculating all the products. Most mathematicians hate doing that – they want short elegant proofs. In the present case, a bag of formulas known as **trace identities** come to our aid, allowing us to calculate new traces from old. In fact, clever use of the trace identities shows that it is possible to express the trace of *any* word in the matrices a, b, A, B as a polynomial depending only on the three basic traces $\mathrm{Tr}\,a$, $\mathrm{Tr}\,b$ and $\mathrm{Tr}\,ab$, a point to which we shall return in Chapter 9.

The key is a basic result which we call the 'grandfather' of all trace identities, so elegant that we can't resist explaining. We have already seen that traces are invariant under conjugation: $\mathrm{Tr}\,MNM^{-1} = \mathrm{Tr}\,N$ for any two transformations M and N. We applied this when we cyclically permuted words and found, in particular, $\mathrm{Tr}(MN) = \mathrm{Tr}(NM)$. We get more identities if we invert. If $M = \begin{pmatrix} a & b \\ c & d \end{pmatrix}$ with $ad - bc = 1$, then $M^{-1} = \begin{pmatrix} d & -b \\ -c & a \end{pmatrix}$,

---

Note 6.4: **The Markov conjecture**

Markov's equation comes up in number theory where it is studied as the **Diophantine equation**

$$x^2 + y^2 + z^2 = xyz.$$

Saying an equation is Diophantine means we are concerned only with solutions with *integer* values for x, y and z, for example, $x = y = z = 3$. There is a beautiful connection between the integer solutions of this equation and the group generated by

$$a = \begin{pmatrix} 1 & 1 \\ 1 & 2 \end{pmatrix} \quad \text{and} \quad b = \begin{pmatrix} 1 & -1 \\ -1 & 2 \end{pmatrix}.$$

The $(3, 3, 3)$ solution comes from $\mathrm{Tr}\,a = \mathrm{Tr}\,b = \mathrm{Tr}\,ab = 3$. One can prove that if u, v are *any* pair of matrices generating this group, then their commutator $uvUV$ is conjugate to $abAB$ or its inverse $baBA$.

This means that $\mathrm{Tr}\,uvUV = -2$, so that

$$(\mathrm{Tr}\,u)^2 + (\mathrm{Tr}\,v)^2 + (\mathrm{Tr}\,uv)^2 = \mathrm{Tr}\,u\,\mathrm{Tr}\,v\,\mathrm{Tr}\,uv,$$

Since the trace of every element in this group is an integer, we get another integer solution $x = \mathrm{Tr}\,u$, $y = \mathrm{Tr}\,v$, $z = \mathrm{Tr}\,uv$! Markov proved that *every* integer solution to his equation comes up in this way.

Based on extensive calculation, he also made a famous conjecture. Suppose that p, q, r and p', q', r' are both integer solutions and suppose that $p \leq q \leq r$ and $p' \leq q' \leq r'$. Markov conjectured that if $r = r'$, then $p = p'$ and $q = q'$. Despite huge amounts of computation and many false attempts, whether or not Markov's conjecture is true remains unresolved to this day.

Box 19: Trace identities

The grandfather of trace formulas, from which all others follow, is

$$\text{Tr}\, MN + \text{Tr}\, M^{-1}N = \text{Tr}\, M \;\; \text{Tr}\, N.$$

Derived from this, we get

$$\text{Tr}\, abAB = (\text{Tr}\, a)^2 + (\text{Tr}\, b)^2 + (\text{Tr}\, ab)^2 - \text{Tr}\, a \cdot \text{Tr}\, b \cdot \text{Tr}\, ab - 2$$

which becomes the Markov identity

$$(\text{Tr}\, a)^2 + (\text{Tr}\, b)^2 + (\text{Tr}\, ab)^2 = \text{Tr}\, a \cdot \text{Tr}\, b \cdot \text{Tr}\, ab$$

if $\text{Tr}\, abAB = -2$.

---

showing that

$$M + M^{-1} = \begin{pmatrix} a + d & 0 \\ 0 & a + d \end{pmatrix} = (a + d)I.$$

Multiply both sides by a second matrix N and we obtain

$$MN + M^{-1}N = (\text{Tr}\, M)\, N.$$

Taking traces of both sides yields the grandfather identity:

$$\text{Tr}\, MN + \text{Tr}\, M^{-1}N = \text{Tr}\, M \;\; \text{Tr}\, N.$$

   The proof of the formula for the trace of the commutator is just a question of applying the grandfather identity to a series of pairs of words in a and b. How it works is explained in Note 6.5.

Note 6.5: Proof of the commutator relation

Here are the applications of the grandfather identity needed to prove the Markov identity. On the left we have listed a pair of words (T, U) and then we have applied the grandfather equation in the form $\text{Tr}(TU) + \text{Tr}(T^{-1}U) = \text{Tr}\, T \;\; \text{Tr}\, U$. Using the observations we made earlier about inverses, conjugates, and cyclic rearrangements, we know that $\text{Tr}\, A = \text{Tr}\, a$, $\text{Tr}\, B = \text{Tr}\, b$, $\text{Tr}\, bAB = \text{Tr}\, a$, $\text{Tr}\, AB = \text{Tr}\, ab$, and $\text{Tr}\, I = 2$. Juggling all these identities gives the proof.

(a, b)	$\text{Tr}(ab) + \text{Tr}(Ab) = \text{Tr}(a)\,\text{Tr}(b)$
(a, bAB)	$\text{Tr}(abAB) + \text{Tr}(AbAB) = \text{Tr}(a)\,\text{Tr}(bAB)$
(Ab, AB)	$\text{Tr}(AbAB) + \text{Tr}(B^2) = \text{Tr}(Ab)\,\text{Tr}(AB)$
(B, B)	$\text{Tr}(B^2) + \text{Tr}\, I = (\text{Tr}\, B)^2$

Projects

6.1: When the trace of a commutator equals 2

Suppose S and T are two Möbius maps. If we change the sign of S, by multiplying all its entries by -1, then we also must change the sign of its inverse S^{-1}. Check that changing the sign of either S or T produces *no change* in the commutator $STS^{-1}T^{-1}$.

Now here's a fact about Möbius maps: if $\operatorname{Tr} STS^{-1}T^{-1} = 2$, then S and T must share a common fixed point P, that is, $S(P) = P$ and $T(P) = P$. Verify that this is *impossible* if S and T pair four disjoint circles according to the Schottky rules.

Here's how you can check the fact about fixed points. Suppose $S(z) = kz$ and $T(z) = (az + b)/cz + d$. With a large piece of paper, verify that

$$\operatorname{Tr}(STS^{-1}T^{-1}) - 2 = -bc(k - 1)^2/k.$$

Hence $\operatorname{Tr}(STS^{-1}T^{-1}) = 2$ implies either $b = 0$ or $c = 0$ or $k = 1$. Show that in all three cases, S and T share a fixed point. Now in this calculation, S had a special form. But most Möbius maps S can be put in this form by conjugation. Use this to show the result holds for any loxodromic or elliptic S. Parabolic S's can be conjugated to the form $S(z) = z + a$. Work out a simpler fact about the trace of the commutator for such S's. Altogether, this covers everything.

6.2: The four tangency points lie on a common circle

An important use of conjugation is that sometimes you can see much better in one setup something which is hidden in another. It is a remarkable geometrical fact that if four circles form a tangent chain, then their tangency points are themselves on a circle. To prove this, conjugate by a Möbius map in such a way that one of the four tangency points is sent to ∞, so that two of the circles become parallel straight lines. Now try filling in the other two circles and find a simple argument to show that the three remaining tangency points must lie on a straight line.

6.3: Finding four tangent circles

Here is a way of creating a chain of four tangent circles, using Project 6.2. Start from the four tangency points P, Q, R and S, which can be any four points arranged in order around your favourite circle C. Draw any circle you wish through P and Q. This will be the circle C_a. Now draw another circle through P and S and tangent to C_a. This will be the circle C_b. The circle C_B is drawn in a similar way. Here comes the crunch: use some geometry to convince yourself there is a circle through R and S which is tangent to both C_B and C_b. Be careful: in general you can't find a circle through two given points with two given tangent directions! There has to be something special about the points and tangents, and if P, Q, R and S weren't on a circle, it just wouldn't work.

6.4: Finding four tangent circles in Apollonius' way

There is another quite different way of creating a chain of four tangent circles, using some very pretty geometry going back to the Greek geometer Apollonius.

Call the circles C_a, C_b, C_A, C_B and denote their centres and radii by c_a, c_b, c_A, c_B and r_a, r_b, r_A, r_B respectively.

(1) Verify that C_a and C_b are tangent if $|c_a - c_b| = r_a + r_b$, and similarly for the other 3 pairs.

(2) Hence show that the choice of any 3 points c_a, c_b, c_A and any positive number r_b between $(|c_a - c_b| + |c_b - c_A| - |c_a - c_A|)/2$ and the lesser of $|c_a - c_b|$ and $|c_A - c_b|$, gives a chain of 3 tangent circles.

(3) The tricky part is choosing the fourth circle C_B tangent to both c_a and c_A. (Apollonius' problem was to find *all* circles tangent to a given pair.) Write the requirements for c_B and r_B using (1) and eliminate r_B to show that

$$|c_B - c_a| = (r_a - r_A) + |c_B - c_A|.$$

In other words, c_B is a point such that the *difference* between its distances to the two known points c_a and c_A equals a fixed constant. Aha: this is the 'old-fashioned' way of defining a hyperbola! Try checking it is a hyperbola algebraically by using complex arithmetic and squaring the formula above, squaring again, and so on.

(4) We have found that the centres of all possible circles C_B lie on a hyperbola passing through the special point c_B given by the formula in (3). This hyperbola must pass through c_b, because C_b is also a circle tangent to C_a and C_A. You may remember that hyperbolas come in two pieces. One part comes from the centres of the circles tangent to C_a and C_A from the outside, and the other part comes from the centres of the circles tangent to C_a and C_A which enclose them! Looking at the figure, tell which branch is which.

6.5: Lots of conjugations

This project refers to Figure 6.10. We want explicit formulas for the group generators. The generators for frame (ii) given in the text are a special case of the formulas on p. 170 with $x = u = \sqrt{2}$. Try getting the generators for frames (iii) and (iv) by conjugating these matrices by $z \mapsto 1/z$ and $z \mapsto e^{-i\pi/4}z$ respectively.

To get frame (v), conjugate the group in frame (iv) by the inverse of the Cayley transform K, see p. 87. Check that the four initial circles become circles orthogonal to the real axis with tangency points at $-1, 0, 1$ and ∞. The result is

$$a = \frac{1}{\sqrt{2}}\begin{pmatrix} 3 & -1 \\ -1 & 1 \end{pmatrix} \quad \text{and} \quad b = \frac{1}{\sqrt{2}}\begin{pmatrix} 3 & 1 \\ 1 & 1 \end{pmatrix}.$$

Alternatively, observe that a is a loxodromic transformation which carries 1 to ∞, 0 to -1, and with trace $2\sqrt{2}$. Why does this determine a?

In frame (vi) the generators are

$$a = \begin{pmatrix} \sqrt{2} & i \\ -i & \sqrt{2} \end{pmatrix} \quad \text{and} \quad b = \begin{pmatrix} \sqrt{2} - i & \sqrt{2} \\ \sqrt{2} & \sqrt{2} + i \end{pmatrix}.$$

(This is Grandma's special recipe on p. 229.) Verify that $R = -(1 + 2\sqrt{2}i)/3$ and $S = (1 - 2\sqrt{2}i)/3$ and then check that $b(R) = S$.

Figure 6.17. The fourth circle must have its centre on the orange hyperbola.

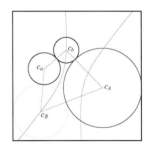

6.6: Three real traces imply Fuchsian

We want to prove that if the traces of the three matrices a, b and ab are all real and $\text{Tr } abAB \neq 2$ then the group they generate is conjugate to a subgroup of the upper half plane group $SL(2, \mathbb{R})$. Although this is true in general, it is a little easier if we concentrate on the case in which a and b are hyperbolic, so that neither $\text{Tr } a$ nor $\text{Tr } b$ is between -2 and 2.

The idea is to keep conjugating to make a and b as simple as possible. First put a in the standard form $z \mapsto k^2 z$ where k is real. Now conjugate again by a rotation keeping a fixed, in such a way that one of the fixed points of b becomes real. Then let $b = \begin{pmatrix} u & v \\ w & z \end{pmatrix}$ where $uz - vw = 1$ and use the assumption that $\text{Tr } b$ and $\text{Tr } ab$ are real to show that u and z must be real. Use that one of the fixed points $(u - z \pm \sqrt{(\text{Tr } b)^2 - 4})/2w$ is real to show that w and v are real too.

6.7: Closing the gaps in the θ-Schottky groups

One can use the θ-Schottky groups we met in Project 4.2 to illustrate how the four Schottky circles come together to form a tangent chain. If θ is near 0 the circles are small and far apart. As θ increases, then, as pictured in Figure 6.12, they gradually expand until we reach the value $\theta = \pi/4$ when all four touch. Check that at this value, the generators of the θ-Schottky group are exactly the same as the generators for our special family of quasifuchsian groups for the parameter values $x = u = \sqrt{2}$.

Now verify that

$$abAB = \begin{pmatrix} 4i - 1 & 2\sqrt{2}(1 - i) \\ 2\sqrt{2}(1 + i) & -4i - 1 \end{pmatrix}.$$

(The calculations are simplified if you remember that $abAB$ must be of the form $z \mapsto (uz + v)/(\bar{v}z + \bar{u})$ with $u^2 - v^2 = 1$.) Check directly that $\text{Tr } baBA = -2$ and there is only one fixed point at $(i + 1)/\sqrt{2} = e^{i\pi/4}$.

To find the fixed points of the other three commutators $bABa$, $ABab$ and $BabA$, you don't have to compute everything all over again. For example, our cyclic permutation trick gives $bABa = A(abAB)a$, showing that the fixed point of $bABa$ must be at $A(e^{i\pi/4}) = e^{-i\pi/4}$. Find the fixed points of the other commutators and check you have got them in the right place by comparing with the pattern in Figure 6.13.

6.8: The Markov equation and our special family

It involves a bit of slightly tricky algebra to check that the traces of the generators of our special family of groups on p. 170 satisfy the Markov identity. We already know that $\text{Tr } a = 2u$ and $\text{Tr } b = 2x$. Check by directly multiplying out that $\text{Tr } ab = 2ux + ivy(k - 1/k)$.

Now try substituting these values directly in the Markov identity. To see it works, you will need to use the relations $x^2 - y^2 = 1$ and $u^2 - v^2 = 1$ and it may help to notice that squaring both sides of the relation $1/2(k + 1/k) = 1/yv$ leads to $1 - y^2 v^2 = \frac{1}{4}(k - 1/k)^2$.

CHAPTER 7

The glowing gasket

Four circles to the kissing come,
The smaller are the benter.
The bend is just the inverse of
The distance from the centre.
Though their intrigue left Euclid dumb
There's now no need for rule of thumb.
Since zero bend's a dead straight line
And concave bends have minus sign,
The sum of the squares of all four bends
Is half the square of their sum.

The Kiss Precise, Sir Frederick Soddy[1]

The lacy web in Figure 7.1 is called the **Apollonian gasket**. Usually, it is constructed by a simple geometric procedure, dating back to those most famous of geometers, the ancient Greeks. We shall start by explaining the traditional construction, but as we shall disclose shortly, the gasket also represents another remarkable way in which the Schottky dust can congeal. The pictures you see here were actually all drawn using a refinement[2] of the DFS algorithm for tangent Schottky circles.

The starting point of the traditional construction is a chain of three non-overlapping disks, each tangent to both of the others. A region between three tangent disks is a 'triangle' with circular arcs for sides. This shape is often called an **ideal triangle**: the sides are tangent at each of the three vertices so the angle between them is zero degrees.[3] The gasket is activated by the fact that in the middle of each ideal triangle there is always a unique 'inscribed disk' or **incircle**, tangent to the three outer circles. It is really better to think of the gasket as a construction on the sphere. Insides and outsides don't matter any more, so we may as well start with *any* three mutually tangent circles. You can see lots of disks and incircles in Figure 7.2.

In the figure, we show two initial configurations of three tangent blue disks. When you take out the three blue disks you are left with two red ideal triangles. Each red ideal triangle has a yellow incircle. See how each yellow incircle divides the red triangle into three more triangles.

[2]The refinement has to do with ensuring that circles appear with the desired visual accuracy.

[3]That would have made high school geometry a lot simpler! Perhaps that's why triangles like this are called 'ideal'.

[1] Reproduced from Nature, **137**, 1936.

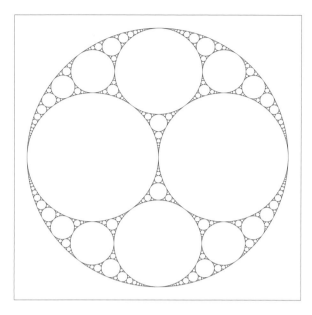

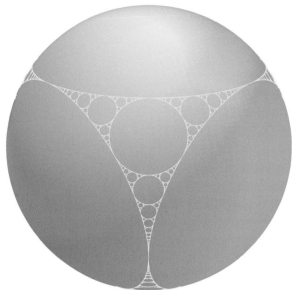

Figure 7.1. The Apollonian gasket. The lacy web in the two pictures is the same; on the left, it is drawn in the complex plane and, on the right, on the sphere. As you might imagine, many people have tried calculating the Hausdorff dimension of the gasket. Curt McMullen has found the most accurate value, which is estimated as about 1.305688.

[1] We are grateful to Ms. Alexandra M. Wright and Ms. Julie M. Wright for calling our attention to this reference.

[2] The gasket is also sometimes called the **Apollonian packing**, since we have packed in as many tangent circles as we possibly can. You can find lots more information on the web.

For repetitive people (a necessary quality in this subject, you might say), it is only natural to draw the incircles in these new triangles, resulting, of course, in even more triangles of the same kind. The bottom frame shows this subdivision carried out twice more, with green and then even smaller purple disks. In *The Cat in the Hat Comes Back*,[1] the cat takes off his hat to reveal Little Cat *A*, who then removes his hat and releases Little Cat *B*, who then uncovers Little Cat *C*, and so on. Now imagine there are not one but three new cats inside each cat's hat. That gives a good impression of the explosive proliferation of these tiny ideal triangles. Carry out this process to infinity, and *Voom*, the Apollonian Gasket appears.[2]

The Apollonian gasket is indeed very pretty, but the reason for introducing it here is that, remarkably, it is also the limit set of a Schottky group made by pairing tangent circles. Exactly the same intricate mathematical object can be created by completely different means! You can see better how this works in the beautiful glowing version in Figure 7.3. The solid red circles in this picture are the initial Schottky circles in a very special configuration which we will look at closely in the next section. The glowing yellow limit set can be recognized as the same as the Apollonian gasket of Figure 7.1. The picture was made by pairing four tangent circles arranged in the configuration shown in Figure 7.4. The four circles are tangent not only in a chain; there are also extra tangencies between C_a and C_A, and between C_b and C_B.

As you iterate the pairing transformations a and b, the extra tangency proliferates, with the effect that inside each image disk D in Figure 7.3

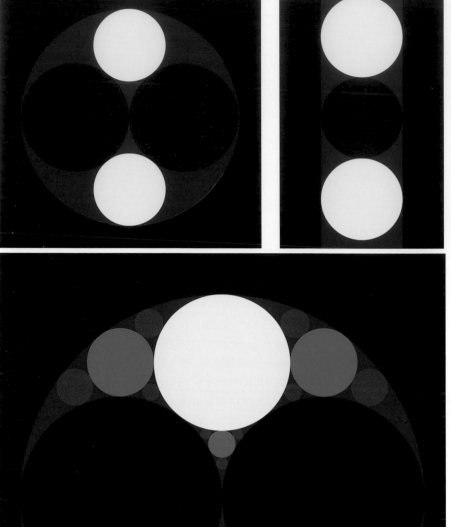

Figure 7.2. Activating the gasket. On the top left, the three largest disks which activate the gasket are shown in blue. One of the three disks appears as the outer blue region which frames the picture, because it contains the point ∞, the North Pole on the Riemann sphere. On the top right, two of the disks are tangent at ∞, so that you see them in the picture as parallel blue strips. Removing the blue circles leaves two red ideal triangles. Inside each ideal triangle is an incircle, coloured yellow. Below, we see more levels of incircles.

you see three further Schottky disks tangent to D and each of the other two. In our version, the circles have been coloured depending on their level, starting with red at the first or lowest level, gradually changing to yellow, green and then blue. The small yellow and blue circles pile up, highlighting the limit set with a mysterious glow.

In this chapter, we shall be exploring various features of the gasket. Notwithstanding the extra tangency, it turns out that each limit point is still associated with exactly one or two infinite words in the generators a, b, A and B. You will be able to make your own version of the glowing gasket by running our DFS algorithm for the group generated by the

Figure 7.3. *(Overleaf.)* **The glowing gasket.** This picture was created by applying the generating transformations a, A, b, B to the chain of four solid red disks bounding the black ideal triangles. At higher levels, the image circles tone from red to orange to yellow, through green to blue, finally cycling back to red. Don't let the picture fool you – the red circles are *not* the circles which activate the gasket in the traditional construction. The ones which appear in the traditional construction are the 'dual' circles with the yellow glow.

Apollonius, circa 250–200 BC

Apollonius, known to his contemporaries as the Great Geometer, lived in Perga, now part of Turkey. One of the giants of Greek mathematics, he was famed for his 8 volume treatise *Conics* which studied ellipses, hyperbolas and parabolas as sections of a cone by a plane at various angles. His writings swiftly became standard texts in the ancient world. Many are now lost and we know them only through mention in other commentaries, among them works on regular solids, irrational numbers, and approximations to π. Ptolemy credits Apollonius with the theory of epicycles on which he based his theory of planetary motion.

One of Apollonius' lost works is a book called *Tangencies*, reported to provide methods of constructing circles tangent to various other combinations of lines and circles, for example finding a circle tangent to two given lines and another circle. You can think of the problem of finding the incircle of an ideal triangle in this way. The most difficult problem, that of constructing the two circles tangent to three other given disjoint circles, was probably not solved in ancient times, however Sir Isaac Newton wrote down a proof. According to Pappus, *Tangencies* gave a formula for the radius of the incircle to an ideal triangle in terms of the radii of the circles which bound its three sides. Be that as it may, exactly such a formula was described by Descartes in 1643, and a version was known in eighteenth century Japan. In fact this formula seems to have been rediscovered many times, most recently by Sir Frederick Soddy, in whose honour the incircles are sometimes known as **Soddy circles**. Awarded the Nobel prize in 1921, for the discovery of isotopes, Soddy had a natural interest in how to pack spherical atoms of differing size.

Soddy was so taken with the formula that he published it in the unusual form of a poem, which appeared in the journal *Nature* in 1936. The central part is contained in the middle verse quoted at the head of this chapter. For those who feel more comfortable with symbols, suppose the radii of the chain of three circles are a, b and c, and that the incircle has radius d. Soddy's formula is:

$$\left(\frac{1}{a} + \frac{1}{b} + \frac{1}{c} + \frac{1}{d}\right)^2 = 2\left(\frac{1}{a^2} + \frac{1}{b^2} + \frac{1}{c^2} + \frac{1}{d^2}\right).$$

transformations a and b. The algorithm draws this complicated lacework as a continuous curve, which is hard to imagine until you see it in progress on a computer screen. The curve snakes its way through the gasket, apparently leaving one region for quite a while until finally weaving its way

back. Animation is the true reward of successfully implementing the program we have been learning to build.

Generating the gasket

The configuration of tangent circles which produced the gasket is shown in the right frame of Figure 7.4. The picture has been arranged so that C_a goes through ∞, hence it appears in the figure as a straight line.

In addition, C_A and C_a are tangent at 0 and C_B and C_b are tangent at $-i$. You can see how this picture is made by creating extra tangencies among a kissing chain of four circles by comparing with the nearby arrangement of four circles in the left hand frame.

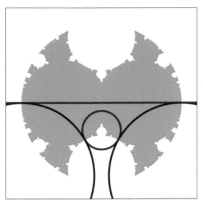
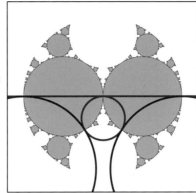
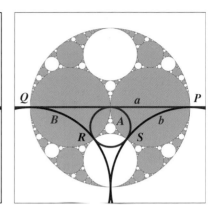

Figure 7.4. The right frame shows the starting configuration of tangent Schottky circles which produces the glowing gasket. The red circles C_a and C_A are paired by the transformation a and the blue circles C_b and C_B by b. Using notation from the last chapter, the tangency points P, Q, R and S are at $1, -1, -0.2 - 0.4i$ and $0.2 - 0.4i$ respectively. On the left is a nearby Schottky configuration of circles which are not quite tangent and a and b are loxodromic. This is similar to the configuration shown in frame (vi) of Figure 6.10. The centre frame is an intermediate stage where a is parabolic but b is not.

The generating matrices for the gasket are quite simple:

$$a = \begin{pmatrix} 1 & 0 \\ -2i & 1 \end{pmatrix} \quad \text{and} \quad b = \begin{pmatrix} 1-i & 1 \\ 1 & 1+i \end{pmatrix}.$$

We shall have more to say about how we arrived at these particular formulas later on. Note that $\text{Tr}\, a = \text{Tr}\, b = 2$, so a and b are parabolic. Looking at the arrangement of Schottky circles in Figure 7.4, you see the fixed point of a is 0, the tangency point of the circles C_a and C_A. In Figure 7.3, you can see two chains of tangent circles nesting down on 0 from above and below. The same phenomenon occurs at $-i$, the tangency point of C_b and C_B and the fixed point of b. Notwithstanding extra tangencies, the generators a and b still pair opposite circles in the initial tangent chain C_a, C_b, C_A and C_B. This means that for nesting circles we still need the commutator condition $\text{Tr}\, abAB = -2$, which is not hard to check.

We have been speaking as if there is only one Apollonian gasket, but could we not get different gaskets by starting with different tangent chains? Not really, because it turns out that any chain of three tangent circles can be conjugated to any other three. As you can work out in Project 7.1, this stems from the fact that there is always a Möbius map carrying any three points to any other three. Since the gasket is activated by its initial ideal triangle, and since the procedure at each step consists in adding incircles, a Möbius map which conjugates one ideal triangle to another carries the whole gasket along in its wake. This explains why it

makes sense to talk about *the* Apollonian gasket, because up to conjugation by Möbius maps there is really only one.

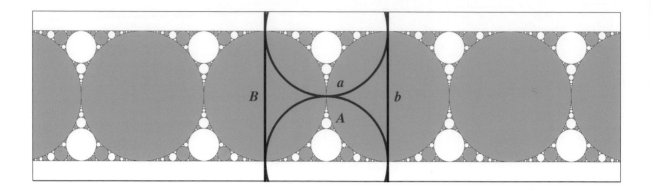

Another famous version of the gasket can be seen in Figure 7.5. To get
this we conjugated so that C_b and C_b are tangent at ∞, so that they are
vertical lines. The map a now has a fixed point at i. The generators for
this group are:

$$a = \begin{pmatrix} 2 & -i \\ -i & 0 \end{pmatrix} \quad \text{and} \quad b = \begin{pmatrix} 1 & 2 \\ 0 & 1 \end{pmatrix}.$$

**Figure 7.5. The strip
gasket.** This shows the
gasket as it appears when
we conjugate so that the
extra tangency point of C_b
and C_B is at ∞. Any
parabolic with a fixed point
at ∞ is a Euclidean
translation, in this case
$b(z) = z + 2$, which explains
the translational symmetry
along the infinite strip.

Pinching tiles

Figure 7.6 is a wonderful picture of what happened when we introduced
Dr. Stickler to Apollonius! It is a pretty intricate arrangement, so let's take
a bit of time understanding what has happened to the tiles. To get a grasp
on the situation, look back at the three pictures in Figure 7.4, and watch
the progression across the three frames. On the left the limit set is a loop
or quasicircle, so the ordinary set – what is left when you take away the
limit set – has two parts, a pink inside and a white outside. In the central picture, the pink part has collapsed into a myriad of tangent disks,
and the red Schottky circles C_a and C_A touch at 0. On the right, the gasket group, the 'horns' of the pink region have also come together, causing
the white outside to fracture into disks as well. Notice how the memory of
which was inside and which was outside still persists, because what were
the 'inside' disks are pink while the 'outside' ones are white.

In each picture, the initial Schottky circles are blue and red. Watch them
to follow the fate of the tiles. On the left, as usual for a kissing Schottky
group, they surround the central inner four sided tile. If we transported this
tile around by the group, we would see a tessellation of the pink region
similar to the one in Figure 6.6. (There is also an outer tile, the region

outside the four Schottky circles, which as usual you can see more clearly by imagining it on the Riemann sphere.) The inner and outer parts of the ordinary set are invariant under the group, so if you apply any transformation of the group to any tile in the pink region 'inside' the limit set, you get another tile which is also 'inside'.

In the central picture, where a has become parabolic, the inner tile has been pinched into two halves. Each half-tile is an ideal triangle, with two red sides and one blue. You should think of this pair of triangles as one composite two-part tile. Moved around by the group, the composite tile will cover all the pink circles. There is an outer tile in this picture too, which (on the Riemann sphere) remains in one piece.

On the right, in the gasket group, both a and b have been pinched so that now C_b and C_B also touch, at $-i$. Now there are *four* basic half-tiles. The two pink ones will produce a tiling of the pink circles and the white ones will make a tiling of the white circles. In the glowing gasket picture, these four tiles are black. The upper two 'horizontal' ideal triangles are the remnants of the inner Schottky tile, while the lower 'vertical' triangle is a remnant of the outer one. If you look carefully, you can just see its twin peeping out in the bottom centre of the page.

Now we can go back to the picture of Dr. Stickler meeting Apollonius. The party is taking place in the remnants of the 'pink' circles. If you compare with the half-tiles in Figure 7.4, something rather odd has happened to Dr. Stickler – when the original tile split in two, his head ended up in the green half-tile and his feet in the blue one. Fortunately, there is a transformation of the group (namely B) which carries the blue Stickler to the green one, moving the blue half-tile containing the blue feet to the yellow half-tile containing the green feet. Had we not pointed out his difficulties you might not even have noticed that anything was wrong. After gluing the yellow half-tile to the green half-tile, the relieved (but still slightly greenish) Dr. Stickler stands reunited in a new and complete tile whose images under the group map to all the Sticklers in the picture.

And pinching surfaces

What happened to the tiles in the last section, has, of course, also an interpretation in terms of surfaces. Looking back to the picture on p. 190 which showed how tiles were glued up in a kissing Schottky group, we can work out what happens when we bring the four circles together to make the

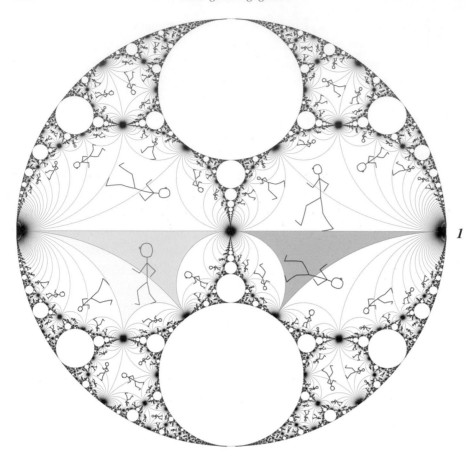

Figure 7.6. Dr. Stickler meets Apollonius. Placing Dr. Stickler in the Apollonian gasket, we let the group of symmetries carry him around. He appears exactly in those disks which were pink in Figure 7.4. If we had started him off in a white one, his images would fill the white disks instead. The symmetry a is parabolic and on both sides of its fixed point 0, circles of Sticklers are streaming out and then circling back in. A startling feature is the circle of Sticklers streaming out from and into 1. Every alternate Stickler is standing on his head! The upright Sticklers are just powers of $abAB$ (which fixes 1) applied to the Stickler standing on the right hand horizontal axis, while the upside down ones are the images of this same Stickler under $(abAB)^n A$.

gasket. It takes a bit of stretching and squeezing to do this, which we have illustrated in Figure 7.7.

The result, shown in the last panel, is our old friend the pretzel with *three* circles pinched to points or cusps: the waist as in the last chapter and, in addition, loops around the top and bottom tori. Both top and bottom are now 'spheres' with three cusps or punctures each. One pair of cusps on each sphere are joined together like 'horns', and these two 'horned spheres' are themselves joined together at the last two cusps.

The gasket group is called **doubly cusped** because we have pinched two extra loops, a and b. It is also sometimes called **maximally cusped**, because, after all this squeezing, there are no more curves left to pinch. In Chapter 9, we shall see that you can make many variants of the gasket group by imposing more complicated relationships between the curves we choose to pinch on the top and bottom halves of the pinched pretzel.

Figure 7.7. Pinching curves. How gluing up the gasket configuration of tangent circles leads to a pair of **triply-punctured spheres**. The *a* and *b* curves we have to shrink are are marked *L* and *M*. Instead of pulling the upper and lower partially glued-up cylinders together right away, as we did in Figure 6.16, it now takes some effort first to twist them relative to each other in such a way that when we glue-up, the dotted loops are in their proper position ready to be shrunk.

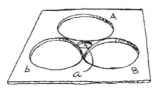

BEGIN WITH A PLANE WITH FOUR HOLES, EACH OF WHICH JUST TOUCHES THREE OTHERS. WHAT HAPPENS WHEN WE GLUE A TO a (THE SMALL ONE) AND B TO b?

FIRST PULL a AWAY FROM A AND B AWAY FROM b ALONG THE ARCS L AND M. NOW WE SEE A DISTORTED BUT FAMILIAR PICTURE: ABab.

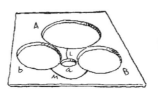

ADD THE POINT AT INFINITY TO WRAP UP THE PLANE.

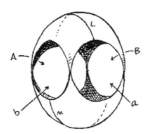

STRETCH TO MAKE THIS FAMILIAR SURFACE. NOTE ARCS L AND M.

NOW SOMETHING STRANGE: PULL TOGETHER THE TOP HALVES OF A AND a AND THE BOTTOM HALVES OF B AND b.

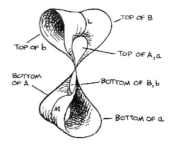

GLUE TOP OF A TO TOP OF a, BOTTOM OF B TO BOTTOM OF b.

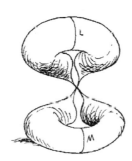

NOW JOIN THE REST OF A TO a AND B TO b.

FINALLY, SHORTEN THE ARCS L AND M...

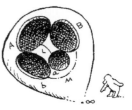

UNTIL EACH ARC SHRINKS TO A POINT, MAKING TWO TRIPLY-PUNCTURED SPHERES.

Tiling the inner disks

Figure 7.6 is made up of lots of disks full of Dr. Sticklers, each tiled by ideal triangles shown in grey. These disks are the remnants of the pink region in Figure 7.4. For most of the rest of this chapter, we shall be occupied with the tiling of just one of these disks. The same tiling fills out the insides of each of the glowing circles in Figure 7.3. The group of symmetries which goes with this very special disk tessellation is called the

modular group[1] and has been the well-spring of a huge body of mathematics.

Since the tiling in each disk is the same, we may as well focus on the large disk through -1 and 0, shown in yellow in Figure 7.8. To understand how these ideal triangle tiles cover the yellow disk we need to find the subgroup of all the transformations in the gasket group which map the inside of this disk to itself. This subgroup (which is of course also a group in its own right), or any of its conjugates, is what we call the modular group. The basic tile is made up of *two* ideal triangles, the ones coloured green and yellow in Figure 7.6. The two triangles together form one of our familiar four-sided pinched-off tiles with four circular arc sides. Moved around by the modular group, they tile the whole yellow disk.

[1] Strictly speaking, it is the 'level 2 congruence subgroup of the full modular group', but since it is the main group of interest to us, we have simplified the terminology. The 'full modular group' will be described in the next section.

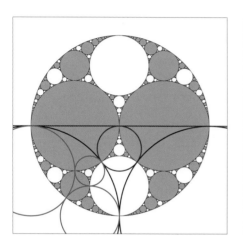
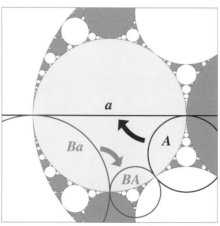

Figure 7.8. The set-up of tangent circles for the modular subgroup. The left frame is just to get oriented. The red circles C_A and C_a, together with the level 2 green circles $B(C_a) = C_{Ba}$ and $B(C_A) = C_{BA}$ form the boundary of the new four-sided tile. The arrows show how a pairs C_A to C_a and BAb pairs C_{Ba} to C_{BA}. The boundary of the yellow disk is the limit set of the modular group generated by a and BAb.

We worked out the labels of the boundary circles C_a, C_A, C_{BA} and C_{Ba} in Figure 7.8 of the four-sided tile by going to part of the level-two Schottky chain for the gasket group. (You may find it easiest to check the arrangement in a picture like the left frame of Figure 7.4 without all the extra gasket tangencies first.) Notice the four tangency points of these circles are all on the boundary of the yellow disk. As you can see, the four circles form a new chain of tangent circles. As usual, a pairs C_A to C_a. In addition, BAb pairs C_{Ba} to C_{BA} because:

$$BAb(C_{Ba}) = BAb(B(C_a)) = BA(C_a) = B(C_A) = C_{BA}.$$

Inside the gasket group we have found another mini-chain of four tangent circles, together with a pair of transformations which match them together in pairs!

This construction shows that the modular group is a new kind of 'necklace group', made by disregarding all the rest of the gasket and looking only at the disks produced by acting with *a* and *BAb* on the four circles which bound the new tile. The new group is generated by the transformations *a* and *BAb*. Indeed in Figure 7.3, you can actually pick out chains of image disks nicely shrinking down onto the glowing limit circle through −1 and 0. The only difference from the kissing Schottky groups we met in the last chapter is that the two generators pair not *opposite* circles but *adjacent* ones. As we shall explain in more detail on p. 213 ff., the image circles shrink because *a*, *BAb* and their product *aBAb* are all parabolic.

The same pattern of pairing circles is repeated all over the gasket. Every pink disk is the image of the yellow one under some element in the gasket group, which conjugates our modular group to another 'modular group' acting in the new disk. The white disks are different from the pink ones, because you can never get from pink to white using transformations in the gasket group. However you can still find a chain of four tangent circles matched in the same pattern, as described in Project 7.4.

Note 7.1: **Uniqueness of the modular group**

Suppose that U, V and UV are all parabolic (and therefore not the identity!) and that the fixed point of U is z_U and the fixed point of V is z_V. We are trying to conjugate U and V to the generators of the modular group. We have seen that we can find a Möbius transformation M that maps z_U to 0 and z_V to ∞. Conjugating our original transformations U and V by M arranges that $MUM^{-1}(0) = 0$ and $MVM^{-1}(\infty) = \infty$, and still the two transformations MUM^{-1} and MVM^{-1} are parabolic. Since we can simultaneously conjugate them in this way, we may just as well assume the original transformations U and V have fixed points 0 and ∞, respectively.

A parabolic transformation that fixes ∞ is always conjugate to any other, up to a minus sign. (See Chapter 3.) Let's arrange by conjugation and possibly multiplying by −1 that V corresponds to the matrix $\begin{pmatrix} 1 & 2 \\ 0 & 1 \end{pmatrix}$. Now all that's left is U. Since $U(0) = 0$, after again possibly multiplying by −1, we

can conclude that the matrix of U is

$$\begin{pmatrix} 1 & 0 \\ x & 1 \end{pmatrix}$$

for some number x.

That brings us to the last hypothesis that UV is parabolic. Let's multiply this out:

$$\begin{pmatrix} 1 & 0 \\ x & 1 \end{pmatrix}\begin{pmatrix} 1 & 2 \\ 0 & 1 \end{pmatrix} = \begin{pmatrix} 1 & 2 \\ x & 1+2x \end{pmatrix}.$$

The trace of UV under these assumptions is $2 + 2x$. This is ± 2 for precisely two values of x, namely, $x = -2$ and $x = 0$. In the latter case, U is the identity, which we are definitely excluding. That means $x = -2$, and we have shown that U and V are simultaneously conjugate to

$$\begin{pmatrix} 1 & 0 \\ -2 & 1 \end{pmatrix} \text{ and } \begin{pmatrix} 1 & 2 \\ 0 & 1 \end{pmatrix}.$$

(We may have to multiply one or both matrices by −1 to arrange that they both have trace 2.)

You might well imagine that we should be set to repeat everything we did in the last chapter. By taking four tangent circles and pairing them in this new pattern we should presumably get a whole new lot of quasifuchsian groups. Not so! It turns out that the rigours imposed by specifying that the two generators and their product are all parabolic actually 'freeze' the group. Without any mention of circle chains, we prove in Note 7.1 the remarkable fact that all groups made with pairing conditions like this are, up to conjugation, 'the same'. What this means in more detail is this. Suppose that U and V are *any* two parabolic Möbius transformations with the property that UV is also parabolic, and such that the fixed points Fix U and Fix V of U and V are not the same.[1] Then there is always a conjugating map M such that:

$$MUM^{-1} = \begin{pmatrix} 1 & 0 \\ -2 & 1 \end{pmatrix}, \quad MVM^{-1} = \begin{pmatrix} 1 & 2 \\ 0 & 1 \end{pmatrix}.$$

This explains why there are so many circles in the gasket group, and why you get an identical tiling pattern in each one.

The modular group of arithmetic

The result just discussed shows that the modular group is conjugate to a very famous group of great importance in number theory.[2] It is made by arranging the four Schottky circles with their tangency points at $-1, 0, 1$ and ∞. You can see these, coloured red and green, in the left frame

[1] We need the condition about distinct fixed points because otherwise the choice $U = \begin{pmatrix} 1 & 2 \\ 0 & 1 \end{pmatrix}, V = \begin{pmatrix} 1 & 4 \\ 0 & 1 \end{pmatrix}$ would provide a counterexample.

[2] Number theory is the study of properties of whole numbers, for example, making predictions about which are prime. A crowning glory of 20th century number theory was Andrew Wiles' proof of *Fermat's last theorem*, which deeply involves the modular group.

Figure 7.9. The modular tessellation of the upper half plane. The left frame shows the tiling or tessellation of the upper half plane by ideal triangles belonging to the smaller modular group coming from pairing tangent circles, while the right frame shows the richer tessellation associated to the full modular group with its added symmetries. Each tile on the left is subdivided into twelve tiles on the right.

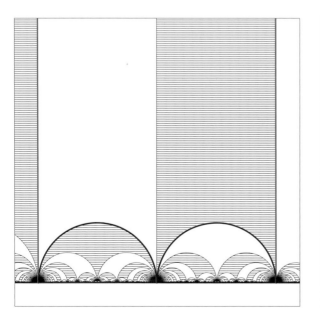
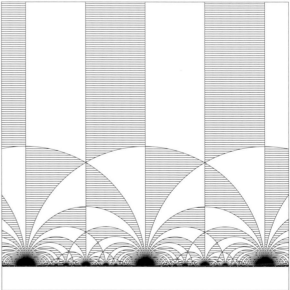

**Carl Friedrich Gauss, 1777–1830
and the modular group**

Rudimentary versions of Figure 7.9 were drawn by the king of mathematicians, Carl Friedrich Gauss. There are several sketches in his notebooks, the first ever pictures of tilings whose symmetries are Möbius maps. Unfortunately, Gauss only left unsystematic fragments published posthumously. Some, interestingly, were annotated by Klein and Fricke.

 Gauss was interested in how many ways you could express a number N by a **quadratic form**, that is an expression like $ax^2 + 2bxy + cy^2$, where the coefficients a, b and c and the variables x and y are all integers. To take a simple example, 34 can be represented by the form $x^2 + y^2$, because $34 = 3^2 + 5^2$. Möbius maps come in because it is not hard to see that if you replace x and y by $px + ry$ and $qx + sy$, where $\begin{pmatrix} p & r \\ q & s \end{pmatrix}$ is a matrix with integer entries and determinant 1, you get a new quadratic form (whose coefficients depend on a, b, c, p, q, r and s) which can also be used to express N. Gauss realised he only need study a, b and c up to the effect of conjugation by a map in the modular group, which explains why he was sketching the modular tiling! We return to the word modular in the last chapter.

of Figure 7.9. Since one of the tangency points is the point at infinity, two of the circles show up as green vertical lines. These green lines are paired by $b = \begin{pmatrix} 1 & 2 \\ 0 & 1 \end{pmatrix}$, while the two red circles tangent at 0 are paired by $a = \begin{pmatrix} 1 & 0 \\ -2 & 1 \end{pmatrix}$. Notice how a and b match adjacent circles in the chain in exactly the pattern of the red and green arrows in Figure 7.8. In fact, as you can easily calculate, ab is the *parabolic* transformation $\begin{pmatrix} 1 & 2 \\ -2 & -3 \end{pmatrix}$.

 It is no coincidence that the entries of these three matrices are integers. The right frame of Figure 7.9 is a more complicated picture which shows all the symmetries of the tiling on the left. Each ideal triangle has been subdivided into three hatched and three unhatched sub-triangles. (The sub-triangles are not quite ideal, because only one of their angles is 0.) The group of symmetries of this more complicated tiling is, from the point of view of Möbius maps, the simplest group of all: just the set of all 2×2

Figure 7.10. The modular tiling from our ancestral home. The tessellation generated by the full modular group, conjugated over to the unit disk. This beautiful rendition is Figure 35 in Vol. 1 of *Vorlesungen über elliptischen Modulfunctionen* by Klein and Fricke.

matrices $\begin{pmatrix} p & q \\ r & s \end{pmatrix}$ with integer entries p, q, r and s and determinant $ps - qr$ equal to 1. To distinguish from the group of the left frame, we sometimes call it the **full** modular group. The matrices in the (smaller) modular group of the left picture are just those matrices with integer entries for which q and r are even and p and s are odd.[1]

There is a very beautiful connection between the modular tessellation and fractions: *the points where the ideal triangles meet the real axis are exactly the rational numbers*. Although it is something of a digression here, we want to take the time to explain the pattern, which turns out to be indispensible when we come to map making in Chapter 9.

Figure 7.11 shows the first few levels in the modular tessellation. The basic tile is the four-sided region, called an ideal quadrilateral, which was bounded by the coloured lines in Figure 7.9. Images of this ideal quadrilateral are shown bounded by solid arcs. The dotted arcs divide them into the two ideal triangles which we saw, half hatched and half white, on the

[1] Think of this as saying the matrix has a little bit in common with the identity: it is described technically by saying that the matrix is 'congruent' to the identity modulo 2, hence the name 'level 2 congruence subgroup'. The full modular group is often symbolized as $SL(2, \mathbb{Z})$.

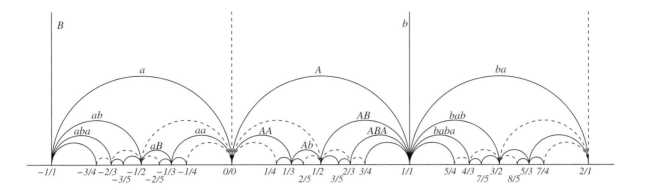

Figure 7.11. The largest ideal triangle tiles in the modular tessellation and their vertices. The word immediately above (i. e. just outside) a Schottky circle is its label. For example, from left to right the level 2 circles shown are $C_{ab}, C_{aB}, C_{aa}, C_{AA}, C_{Ab}, C_{AB}$ and C_{ba}. For more discussion of the labelling, see the next section.

left in Figure 7.9. As the group acts on the basic tile, we get more and more smaller and smaller tiles nesting down to the real axis. The vertices of all these tiles meet the real axis in points which are all fractions. Several things can be read off from a careful examination of this intricate pattern:

- All vertices of the tiles are rational numbers p/q.
- If r/s and p/q are two vertices of the same tile, then $ps - rq = \pm 1$.
- If $r/s < p/q$ are the outer two vertices of a tile, then the third vertex between them is $(r + p)/(s + q)$.

Check this out! For instance, between $2/3$ and $1/2$, we get $(2 + 1)/(3 + 2)$, that is $3/5$.

It's easy to see why this happens. As we have seen, a typical matrix in the modular group will look like $M = \begin{pmatrix} p & q \\ r & s \end{pmatrix}$ where p, q, r and s are integers and $ps - rq = 1$. If M acts on the vertices of the initial triangle with vertices $0 = 0/1$, $1 = 1/1$ and $\infty = 1/0$, then we get the new triangle with vertices $M(0) = r/s, M(\infty) = p/q$ and $M(1) = (p + r)/(q + s)$. Assuming all four entries are positive, we have $r/s < (p + r)/(q + s) < p/q$ (you can see this by multiplying out). This is just what we found in Figure 7.11. If p, q, r, s are not all positive, there are half a dozen other cases in which the order of the points $M(0), M(1)$ and $M(\infty)$ is different but we get the same result. The same thing happens if we start from the other triangle $M(-1), M(0), M(\infty)$.

Any two fractions r/s and p/q such that $ps - qr = \pm 1$ are called **neighbours**. Thus any two vertices of an ideal triangle in the modular tessellation are neighbours. If p/q is a fraction, then, as we explain in Project 7.5, the process of finding its neighbours is essentially Euclid's two thousand year old algorithm for finding the highest common factor of two numbers, surely one of the most useful and clever algorithms of all time. The rule for finding the 'next' point $\dfrac{p+r}{q+s}$ between two neighbours is every student's dream of what addition of fractions should be. This simple form of fraction 'addition' is sometimes called **Farey addition**[1], which one might want to symbolise with a funny symbol like:

$$\frac{p}{q} \oplus \frac{r}{s} = \frac{p+r}{q+s}.$$

Farey addition gives a neat way of organising the rational numbers. Instead of the usual way of arranging them in increasing order (which is difficult, because you never know which one should come 'next'), fractions can be described by a sequence of left or right moves, reflecting the choice at each stage of whether we choose the new pair of neighbours to the right, or the pair of neighbours to the left.

For positive fractions, the starting point are the two fractions 0/1 and

[1] John Farey was a British geologist, who in 1816 published some observations concerning this strange form of addition. We shall meet the actual series Farey was studying in Chapter 9.

Note 7.2: Continued fractions

Expressions like

$$\frac{3}{10} = \cfrac{1}{3 + \cfrac{1}{3}}$$

and

$$\frac{2}{19} = \cfrac{1}{2 + \cfrac{1}{9}}$$

are called **continued fractions**. It turns out that every fraction p/q can be written in a similar way. Assuming p/q is between 0 and 1, then you can always write it in the form

$$\frac{p}{q} = \cfrac{1}{a + \cfrac{1}{b + \cfrac{1}{c + \cfrac{1}{\dots}}}}$$

where a, b, c and so on are positive integers. Actually you can do the same even for an *irrational* number. The difference is that if the original number is irrational, then the terms a, b, c and so on continue without end. The sequence $a, b, c, \dots$ always

describes the number of left–right turns in the Farey algorithm we described in the text.

If we successively 'reduce' a continued fraction by decreasing its final entry by 1, down to the value 1, and then shrinking its length, we recover the Farey process for homing in on the fraction. For example, for the fraction 3/5, we get:

$$\frac{3}{5} = \cfrac{1}{2 + \cfrac{1}{1 + \cfrac{1}{2}}}$$

$$\frac{2}{5} = \cfrac{1}{2 + \cfrac{1}{1 + \cfrac{1}{1}}} = \cfrac{1}{2 + \cfrac{1}{2}}$$

$$\frac{1}{3} = \cfrac{1}{2 + \cfrac{1}{1}}$$

$$\frac{1}{2} = \frac{1}{2}.$$

The process of turning a fraction into a continued fraction is very closely related to Euclid's algorithm. We give some hints in Project 7.5.

<aside>
[1] We realize your high school teachers might be rolling over in their graves if they knew we were discussing the fraction 1/0, but an important threshold in anyone's mathematical education is to learn not to let your teachers stifle your creativity.
</aside>

1/0,[1] which we can regard as special honorary neighbours because they are connected by a side of our initial triangle, the vertical imaginary axis. Farey addition gives the in-between fraction $0/1 \oplus 1/0 = 1/1$.

Now we have a choice: go to the 'left' and look in the interval between 0 and 1, or go to the 'right' and look in the interval between 1 and ∞. Suppose we are aiming for the fraction 3/5. Then we turn to the left and apply the Farey addition $0/1 \oplus 1/1 = 1/2$. At the next stage, we choose the right interval and Farey add to get $1/2 \oplus 1/1 = 2/3$. Finally, we choose the left interval and Farey add $1/2 \oplus 2/3 = 3/5$. An exactly similar procedure could be applied to home in on any fraction p/q. Our choice of left–right turns is a driving map: 3/5 is given by the instructions 'left, right, left'. This arrangement of fractions and sequence of right-left moves is closely related to a way of writing fractions as what are called **continued fractions**, explained in Note 7.2.

The pairing pattern of the modular group

The modular group is a new kind of 'necklace group'. It is still made by pairing four tangent circles, and the only difference from the kissing Schottky groups we met in the last chapter is that the generators pair not *opposite* circles but *adjacent* ones. Whenever we have an arrangement of paired tangent circles like this, something like the *necklace condition* on p. 168 must still be true, but because we are pairing the circles in a different pattern, we can expect that different elements must be parabolic to cause the image circles to shrink.

With the notation of the figure beside Box 20, we have $a(P) = R$ and $b(R) = P$, so that the four tangency points of the circles are $S = \text{Fix}(a)$, $Q = \text{Fix}(b)$, $P = \text{Fix}(ba)$ and $R = \text{Fix}(ab)$. By similar reasoning to that in Chapter 6, in order for the image circles near S and Q to shrink, the generators a and b must be parabolic. Moreover ba must also be parabolic, to make the circles shrink at P. Notice that ab and ba are conjugate (since $b(ab)B = ba$) so saying that ab or ba must be parabolic is really one and the same thing. The wonderful thing is, that as we proved in Note 7.1, all groups with these three elements parabolic are automatically conjugate. This is so important to us that we summarize it in Box 20.

Because the pattern of pairing circles is different, so is the arrangement in which the labelled circles are laid down in the plane. The Schottky circles in Figure 7.11 are labelled according to our usual rules, so for example, C_{ba} still means the image of circle C_a under the map b. However if you look carefully, you will see that the order of the circles along the line is not the same as our original order round the boundary of the word tree

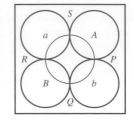

Box 20: The modular necklace

A modular necklace is a tangent chain of four circles in which *adjacent* disks are paired by two transformations *a* and *b* as in the figure in the margin. The transformations *a*, *b* and *ab* are all parabolic and $S = \text{Fix}(a)$, $Q = \text{Fix}(b)$, $a(P) = R$, $b(R) = P$ so that $P = \text{Fix}(ba)$ and $R = \text{Fix}(ab)$. The group generated by *a* and *b* is always conjugate to the 'standard' modular group generated by

$$\begin{pmatrix} 1 & 0 \\ -2 & 1 \end{pmatrix} \quad \text{and} \quad \begin{pmatrix} 1 & 2 \\ 0 & 1 \end{pmatrix}.$$

The four points *P*, *Q*, *R* and *S* always lie on a circle (or line) which is the limit set of the group. The limit circle is perpendicular to all circles in the chain. Both inner and outer tiles have their sides matched in the same way and the surfaces made by gluing up these tiles are each spheres with three punctures or cusps.

on p. 104. The labels can be read off in their correct order from the revised version in Figure 7.12. (To see this you will have to twiddle the diagram around so the arrows from the vertex you are interested in are pointing 'down' rather than 'up'.) There is a subtle difference from our original word tree, because there the cyclic order round a vertex was a, B, A, b while now it is a, A, b, B. The ramifications of this seemingly minor change propagate down the tree.

Because some elements in the modular group are parabolic, the infinite endpoints of certain paths down the word tree merge. For example, the path which starts at *a* and heads always as far as possible to the left (from the viewpoint of a person walking down the branch) is $aaa\cdots$, ending at the attracting fixed point $\text{Fix}^+ a$. On the other hand, starting from *A* and always turning right gives the path $AAA\cdots$, ending at $\text{Fix}^+ A$. Since *a* is parabolic, the end points of these two paths are the same. In a similar way, other coincidences of endpoints are caused by the merging of the fixed points of *b* and *ab*. You should compare the details with Figure 6.13.

A similar phenomenon is repeated at all levels. For example, the extreme left and extreme right paths starting at aB end at $aB\bar{a}$ = $\text{Fix}^+ aBabA$, and $aB\overline{A}$ = $\text{Fix}^+ aBAbA$. Notice that $(aBAbA)^{-1} = aBabA$ and that $aBabA$ must be parabolic because it can be written $(aB)a(aB)^{-1}$, so the two endpoints merge. Every tangency point of the many circles in Figure 7.9 comes about because of a similar conjugacy to one of the three basic parabolics *a,b* or *ab*.

Figure 7.12. The tree of words rearranged in the pattern of the modular group.

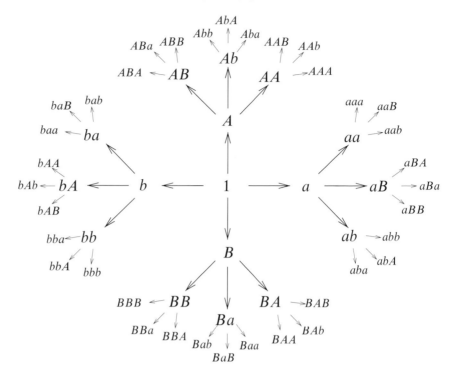

Let's mention one final difference from the kissing Schottky groups in Chapter 6. The chain of four initial circles still divides the Riemann sphere into two four-sided tiles. However, because the pairing is different, so is the result of gluing up the tiles. Dr. Stickler is puzzling this out in Figure 7.13. In contrast to the pair of once-punctured tori we got in Figure 6.16, the outcome is now a pair of triply-punctured spheres.

The problem of gaps

We end this chapter by returning to the gasket and programming, with the confession that we cheated slightly to make pictures like Figures 7.1 and 7.4. Figure 7.14 shows the same picture plotted with our current algorithm. It contains a slight but thoroughly annoying imperfection – if you look closely, you will see that at many places where we allege the limit set is 'pinched', it does not actually quite meet itself, but contains what are in reality quite large gaps.

Gaps and other imperfections in scientific pictures are a common nuisance, but they sometimes have greater significance. Mandelbrot recounted that his first detailed pictures of what later came to be known as the **Mandelbrot set** seemed to be plagued by specks of dirt. He and his assistant made a complete inspection of their program, computer system and

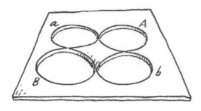

BEGIN WITH A PLANE MINUS FOUR
DISKS, EACH TANGENT TO TWO
OTHERS, IN THE ORDER aBbA.

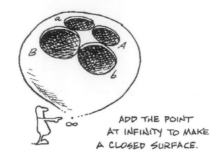

ADD THE POINT
AT INFINITY TO MAKE
A CLOSED SURFACE.

Figure 7.13. The gluing construction for the modular group. We start from a chain of four kissing circles giving a tile with two pieces, each a rectangle with four cusps. After gluing according to the modular recipe, we get two spheres, each with three cuspidal 'punctures' where they are joined.

STRETCH THE SURFACE TO
SEE ITS ESSENTIAL
SYMMETRY.

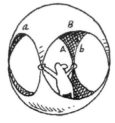

NOTE THAT TWO OPPOSITE
CUSPS WILL BE GLUED
TOGETHER.

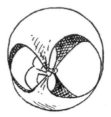

JOIN THEM.

FINALLY, PULL THE ORDINARY,
NON-CUSP POINTS OF THE
CORRESPONDING CIRCLES
TOGETHER.

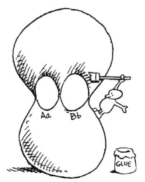

GLUE CIRCLES TO FORM TWO
TRIPLY-PUNCTURED SPHERES
JOINED AT THEIR CUSPS.

printer, only to find that the specks were indeed correct. Investigating the specks led to the discovery of the vast and complicated spider web of filaments connecting all the various parts of that stunningly beautiful and now famous icon of the fractal world.[1]

What is happening in the present case is that our algorithm develops tremendous inertia as it approaches the fixed points of parabolic or nearly parabolic words. It's as if it starts to run in slow motion, simply not having the energy to go all the way out to the end. For this particular picture,

[1] See p. 291 for a picture.

Figure 7.14. The gaps in the gasket.

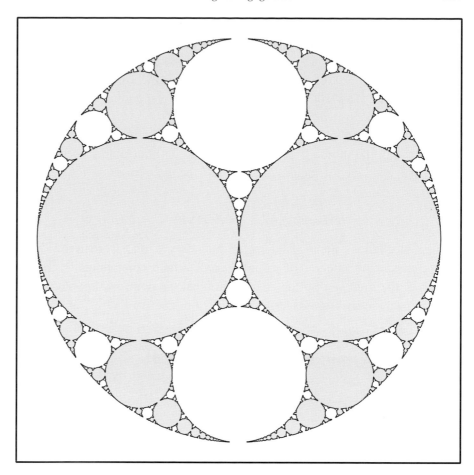

[1] Here as in other pictures the unit is set by our choice of generating matrices for the group. In this case the centre of the picture is O and the outer circle has radius 1.

the plotting threshold was a reasonably small 0.005[1] and the word length crept up to more than 400 as the plot approached the gaps. Still the visible gaps are substantially larger than 0.005.

We could make better plots if we modified the algorithm to take account of the fact that the words a and b are parabolic. The basic idea is that we should plot not only points with repetends which are cyclic permutations of the basic commutator $abAB$ (itself parabolic), but in addition those whose repetends are any of a, A, b or B. Pictures like Figure 7.4, plotted with the new algorithm, are the measure of our success.

Here is a more detailed look at the modified algorithm. We start by telling the program that the three parabolic elements a, b and $abAB$ are 'special'. For each generator, the program then determines the repetends it must consider when enumerating the infinite words. They are listed in Note 7.3.

This chart is used in the following way. Suppose we are considering a word w which ends with the tag gens[1] = a, for example $w = BABaa$. We are trying to plot the part of the limit set corresponding to words with prefix w, so we want to look at those places where this segment of the limit curve unexpectedly gets stretched out. This section of limit set, the w-section as we may call it, consists of points corresponding to all the infinite words between $w\overline{bABa}$ and $w\overline{BAba}$.

Because a is parabolic, the fixed points $\overline{a}$ and $\overline{A}$ are the same, and therefore, although the infinite words are different, the limit points corresponding to $w\overline{a}$ and $w\overline{A}$ are equal. Notice that the infinite word $w\overline{A}$ reduces slightly to

$$w\overline{A} = BABaaAAAAA\cdots = BABAAA\cdots = BAB\overline{A}.$$

These cancellations mean that, on the boundary of the word tree, $BAB\overline{A}$ is far outside the interval of infinite words which correspond to the w-section of the limit set. So the chances are that the point $w\overline{a}$ is going to be stretched way away from its expected position, and we had better check it

---

Note 7.3: **Which repetends should be considered?**

Here is the list of repetends the program must consider when enumerating infinite words:

$$\text{gens[1]} = a \;\Rightarrow\; \begin{cases} \text{repet[1,1]} = bABa \\ \text{repet[1,2]} = a \\ \text{repet[1,3]} = BAba \end{cases}$$

$$\text{gens[2]} = b \;\Rightarrow\; \begin{cases} \text{repet[2,1]} = ABab \\ \text{repet[2,2]} = b \\ \text{repet[2,3]} = aBAb \end{cases}$$

$$\text{gens[3]} = A \;\Rightarrow\; \begin{cases} \text{repet[3,1]} = BabA \\ \text{repet[3,2]} = A \\ \text{repet[3,3]} = baBA \end{cases}$$

$$\text{gens[4]} = B \;\Rightarrow\; \begin{cases} \text{repet[4,1]} = abAB \\ \text{repet[4,2]} = B \\ \text{repet[4,3]} = AbaB \end{cases}$$

There are some obvious omissions. For example, the generator b is also parabolic, so that $w\overline{b}$ is the same point as $w\overline{B}$. Should we perhaps check out $w\overline{b}$ as well? Well, since w ends in a, neither of the infinite words $w\overline{b}$ and $w\overline{B}$ collapse at all, so both $w\overline{b}$ and $w\overline{B}$ are in the small stretch of infinite words which correspond to the wiggly w-section. There seems to be no special need to bother with these points, in fact looking at plots we learned by experience that they don't stick particularly far out at all. In other words, if w ends with a, then there is no need to look at repetends b or B. Similar reasoning leads to the pattern in the chart. The repetends which are important to consider after a word w which ends in the tag c (one of a, b, A or B) are precisely those which themselves end in c. This is because the inverse of the repetend will begin with C (letters in the inverse word, remember, appear in reverse order) so that there are cancellations in wC. This means that the point $w(\overline{C})$ is not in the part of the boundary of the word tree corresponding to infinite words with the prefix w. You can imagine that its presence pulls the point $w(\overline{c})$ (which *is* in the w-section) towards it, causing distortions which our program needs to check out.

out when we are making our plot. We explain in Note 7.3 why it is only worth checking out the special repetends in the list and not, for example, words ending in *a* followed by the repetend *b*.

Figure 7.15. Slow motion gasket. This is a piece of the Apollonian gasket plotted with epsilon=0.1. The width and height of the frame are 0.4. We have started numbering the limit points at $1 = \overline{bABa}$ in the upper left corner. One has to run through quite a few limit points after number $20 = b\overline{A}$, before we return to that same point labelled by the different infinite word $b\overline{a}$ at number 128. The limit points through which the curve passes twice have repetends *a*, *A*, *b* or *B*.

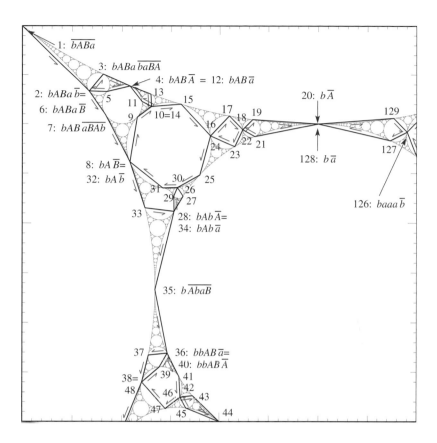

At the beginning of the program, our routine computes the attractive fixed point fp[i,j] of each of the repetends repet[i,j] listed in the chart. The only parts of the running algorithm which need to be changed are the termination and plotting subroutines. Suppose we are at the finite word word ending in the generator gens[i]. Then we compute the three points

$$z_j = \text{word(fp[i,j])}$$

for $j = 1, 2, 3$. Given the plotting gap epsilon, we require both $|z_2 - z_1| \le$ epsilon *and* $|z_3 - z_2| \le$ epsilon in order to terminate the forward enumeration along that particular branch of the word tree. If the criterion is met, we draw a line segment from z_1 to z_2, and a line segment from z_2 to z_3. That is all the modification we need.

Figure 7.15 shows a slow motion execution of this special word algorithm for the gasket. As above, a, b and the commutator have been listed as special words. A piece of the gasket has been plotted at the rather large gap size of `epsilon=0.1`. As you can see, all the contact points between far away parts of the limit set curve are completely filled in and the algorithm produces limit sets with no visible gaps. You will see better how the special word algorithm works if you follow through the description in Note 7.4.

Many fractals are created through a process of iteration which is not terribly sensitive to the order of execution of the program. Seeing limit sets like the gasket drawn as curves is too wonderful to leave to such a procedure. When testing your program it is essential to scrutinize with excruciating care the order of infinite words and corresponding limit points. For lack of space (it's already a crowded picture), not all the limit points in Figure 7.15 have been labelled. If you really wish to test your

---

Note 7.4: **Tracing the gasket**

To understand the workings of our special word algorithm better, you may like to try following along segments of the curve in Figure 7.15 carefully as we describe here.

Let's focus on the segment of limit set between points 1 and 35. This is the segment corresponding to all words with prefix bA. As you can see by referring to our chart of repetends, $1 = bA\overline{BabA}$ and $35 = bA\overline{baBA}$. (In the figure, these points are written slightly differently, simply because we chose a different point starting point for cycling the repetend. So for example $bA\overline{BabA}$ is exactly the same sequence as $\overline{bABa}$.) As you can see by following the numbers closely, the segment of limit set between points $1 = bA\overline{BabA}$ and $35 = bA\overline{baBA}$ passes through the bad fixed point $20 = bA\overline{A}$. At this scale the distortion is not too bad – in comparison to the distance between 1 and 35, the point 20 is not too far away. Things get worse when you go to the next level and study the subsegment corresponding to infinite words with prefix bAA. The initial point of this section of limit set is $15 = bAA\overline{BabA}$ and the final point is

$25 = bAA\overline{baBA}$. The outlier is still 20, which can also be written as $bAA\overline{A}$. See how 20 is pulled quite far out in comparison to the distance between 15 and 25. This is happening because in fact $20 = b\overline{A}$ is the same point as $128 = b\overline{a}$, which is in quite a different section of the path round the boundary of the word tree. Our algorithm would only stop at this prefix if the distances from 15 to 20 *and* from 20 to 25 were both less than the cut-off value `epsilon`. At the next stage, look at the interval between $17 = bAAA\overline{BabA}$ and $23 = bAAA\overline{baBA}$. The outlier is still $20 = bAAA\overline{A}$. Now the disproportion between the distances from 17 to 23, as opposed to the distances from 17 to 20 and 20 to 23, is really getting large. As you go further and further into the spike, the disproportion gets ever worse, so the finer you want to plot your pictures the more important it is to check out these outliers. If we weren't using the special words algorithm, the plot would terminate far too soon, chopping off the piece of limit set which sticks out into the spike.

understanding of the algorithm, you should try to fill in some missing labels, which can be done given the dictionary ordering and the desired special words.

Projects

7.1: Uniqueness of ideal triangles

How many ideal triangles can you draw with given vertices P, Q and R? If the vertices are 0, 1 and ∞, show that the sides must be the two 'vertical' lines through 0 and 1 and the circle with centre $1/2$, radius $1/2$. Thus there are exactly two ideal triangles with these vertices, but only in the 'upper' triangle are the vertices 0, 1 and ∞ in anticlockwise order round the edge. Sometimes this is called the standard ideal triangle. Find its incircle.

Suppose now you have another ideal triangle with vertices P, Q and R in anticlockwise order. Project 3.2 showed there is exactly one Möbius map M which carries 0, 1 and ∞ to P, Q, R in that order. What properties of Möbius maps show that M carries the standard ideal triangle into the new one? If there were another ideal triangle with the same vertices in the same order, what would happen when you applied M^{-1}? Why does any ideal triangle have exactly one incircle?

7.2: An instance of Soddy's formula

Calculate the radius of the incircle C_A in the right frame of Figure 7.4 using Soddy's formula, and check your answer with a more conventional computation using Pythagoras' theorem. Be careful: one circle has infinite radius! Use your result to show that the points S and R in the figure are at $\pm 0.2 - 0.4i$, and then check the generators pair the tangency points properly.

7.3: The Ford circles

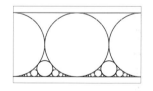

There is a remarkable pattern for the radii of the circles which touch the real axis in Figure 7.5, commonly credited to Lester Ford. The pattern is easiest to analyze if we shift and scale so that the largest pink circles are tangent to the real axis at integer points, so they have diameter 1, and the strip is 1 unit high. Then there is exactly one circle (pink or white) tangent at every rational point p/q. Find the pattern that describes which rationals p/q are tangent to a white circle, and which to a pink one.

The formula for the radius of the circle tangent at p/q is quite simple. It depends only on q. For example, at each integer point $n/1$ the denominator is 1 and the radius is $1/2$. At each half-integer, $\pm 1/2$, $\pm 3/2$, etc., the denominator is 2 and the radius is $1/8$. Those are all the hints we'll give.[1]

[1] For more information, see H. Rademacher, *Higher Mathematics from an Elementary Point of View.*

7.4: Another modular group in the gasket group

In the text (see p. 213), we focussed on the modular group acting in what were the pink disks in Figure 7.4. (The tiling of these disks is shown in Figure 7.6.) The modular group acts in the white disks too. In the right frame of Figure 7.4, look at the lower blue circles C_B and C_b. Their images $a(C_B)$ and $a(C_b)$ are also tangent circles, actually the reflections of the first two in the real axis. These four circles are tangent and cut the white exterior of the unit disk into four quarters. Twist yourself inside out to see that in this white exterior we also have the gluing pattern for the modular group: four tangent circles with neighbouring circles being paired. Show that the two tangency points correspond to the parabolic elements b and abA, while the other two are the products $abAB$ and $BabA$, which are parabolic with fixed points at 1 and -1 respectively. The limit set for this subgroup is the unit circle.

7.5: Farey fractions and Euclid's algorithm

To show that every rational number p/q is a vertex of the modular tiling in Figure 7.9 we need to find two other integers r and s such that $ps - qr = \pm 1$. Why will this do what we want?

The procedure for finding r and s like this is exactly the famous **Euclidean algorithm** for finding the highest common factor of two numbers. In this case, we may as well suppose that p/q is in its lowest terms so the highest common factor is 1. Let's suppose $p > q > 0$. The algorithm says we shall be able to find integers $a_1, r_1, a_2, r_2, \ldots$ and so on such that:

$$p = a_1 q + r_1, \quad 0 < r_1 < q,$$
$$q = a_2 r_1 + r_2, \quad 0 < r_2 < r_1,$$
$$r_1 = a_3 r_2 + r_3, \quad 0 < r_3 < r_2,$$
$$\vdots$$
$$r_{n-1} = a_{n+1} r_n + 1.$$

Using these equations you can work backwards to find r and s. For example, if $p = 14$, $q = 3$ then $14 = 4 \cdot 3 + 2$, $3 = 2 \cdot 1 + 1$ and so $14 - 4 \cdot 3 = 2 = 3 - 1$ giving $14 - 5 \cdot 3 = -1$. Now try to spot the connection with continued fractions:

$$\frac{14}{3} = 4 + \frac{2}{3} = 4 + \frac{1}{\frac{3}{2}} = 4 + \cfrac{1}{1 + \cfrac{1}{2}}.$$

7.6: The modular group and odd–even fractions

Show that if $\begin{pmatrix} a & b \\ c & d \end{pmatrix}$ is in the (small) modular group then a and d are odd and b and c are even. Why is a fraction in the orbit of 0 under the (small) modular group exactly when it is 'even/odd'? What can you say about odd/odd and odd/even? (Notice that in any triple of Farey neighbours you get one fraction of each kind.)

7.7: The modular group or not?

It is interesting to compare the group which generated Figure 6.5 in the last chapter with the modular group. The group in Chapter 6 has generators

$$a = \begin{pmatrix} \sqrt{2} & i \\ -i & \sqrt{2} \end{pmatrix} \quad \text{and} \quad b = \begin{pmatrix} \sqrt{2} & 1 \\ 1 & \sqrt{2} \end{pmatrix}.$$

The conjugated picture of this group in frame (v) of Figure 6.10 looks exactly the same as the basic tile for the modular group in Figure 7.9. Are these two groups the same up to conjugation or not?

7.8: A special Schottky group

In Project 4.1 we introduced a family of Schottky groups depending on 2 real numbers s and t which mapped the real axis to itself. These groups were designed so that the order of the Schottky circles along the real axis was the same as for the modular group. In fact, we organised things so that when $s = 0$ and $t = 1$, we get the modular group itself! Just as we did on p. 179 ff. in the last chapter, we can use this family to see explicitly what happens as we bring the four disjoint circles together until they touch.

Check that if $s = 0$, then both generators are parabolic. Which of the four Schottky circles touch? Show that if $s = 0$ then Tr $ab = 2 - 4/t^2$ and the distance between its fixed points is $4\sqrt{1 - t^2}/t^2$. Hence verify directly that as t tends to 1, ab also becomes parabolic with trace -2 and fixed point at -1, the tangency point of C_a and C_B. Which group element corresponds to the tangency point of C_A and C_b? Why does it become parabolic at the same time as ab?

7.9: Nesting circles which don't shrink!

It is not quite so easy to find an explicit group whose generators a and b pair opposite Schottky circles in a tangent chain in such a way that $abAB$ is not parabolic. If instead we use circles paired according to the modular group pattern, it is much simpler. The non-parabolic element will be the product ab.

We do it with a slight alteration to the modular group which is easiest to describe in terms of the vertices of the red and green ideal quadrilateral in Figure 7.9. Fix the vertices at $0, -1$ and ∞ but move the vertex 1 to a point r on the real axis slightly to its right (so $r > 1$). Check that

$$b = \begin{pmatrix} 1 & 1+r \\ 0 & 1 \end{pmatrix} \quad \text{and} \quad a = \begin{pmatrix} 1 & 0 \\ -1 - 1/r & 1 \end{pmatrix}$$

pair the new chain of tangent circles, matching tangency points correctly.

Verify that a and b are parabolic with fixed points at ∞ and 0. Now calculate ab and show it is hyperbolic whenever $r > 1$.[1] Where are its fixed points? If you have a suitable program up and running, you may wish to draw some pictures to check that the diameters of the nested circles do not shrink properly near -1 and r.

[1] The best way to prove this is to use the well-known inequality $r + 1/r > 2$ which holds for any $r > 1$. You can test this by examples or prove it using calculus or algebra. You may even recognise it as a disguised form of a well known inequality between geometric and algebraic means!

CHAPTER 8

Playing with parameters

'I could spin a web if I tried,' said Wilbur, boasting. 'I've just never tried.'
'Let's see you do it,' said Charlotte...
'OK,' replied Wilbur. 'You coach me and I'll spin one. It must be a lot of fun to spin a web. How do I start?'[1]

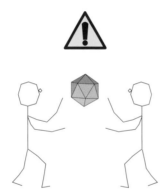

Caution: mathematicians at play

As any mathematician who has revealed his (or her) occupation to a neighbour on a plane flight has discovered, most people associate mathematics with something akin to the more agonizing forms of medieval torture. It seems indeed unlikely that mathematics would be done at all, were it not that a few people discover the *play* that lies at its heart. Most published mathematics appears long after the play is done, cloaked in lengthy technicalities which obscure the original fun. The book in hand is unfortunately scarcely an exception. Never mind; after a fairly detailed introduction to the art of creating tilings and fractal limit sets out of two very carefully chosen Möbius maps, we are finally set to embark on some serious mathematical play. The greatest rewards will be reaped by those who invest the

---

[1] From *Charlotte's Web* by E.B. White, Hamish Hamilton, 1952. © J. White 1952. Reproduced by permission of Penguin Books Ltd.

224

time to set up their own programs and join us charting mathematical territory which is still only partially explored.

All the limit sets we have constructed thus far began from a special arrangement of four circles, the Schottky circles, grouped into two pairs. For each pair, we found a Möbius map which moved the inside of one circle to the outside of the other. Our initial tile was the region outside these four circles. By iterating, we produced a tiling which covered the plane minus the limit set, near which the tiles shrank to minute size. Depending on how we chose the initial Schottky circles, the limit set was either fractal dust, a very crinkled fractal loop we called a quasicircle or, in certain very special cases, a true circle.

The problem with this approach is that it is just too time-consuming to set up the circles and maps which pair them. Free-spirited play shouldn't be ruined by too much preparation. Why not throw the Schottky circles away, take *any* pair of 2×2 matrices for our generators a and b, run our limit point plotting program, and see what we get?

Hold on though – how exactly will this work? The shrinking disks were so reassuring, and the limit set was so comfortably nestled within them, that it is hard to see why we won't get chaos in their absence. No matter, the worst that is likely to happen is that the hard disk crashes, so why not give it a try? Luckily, on p. 182 ff. we already upgraded the DFS code to remove the calculation of Schottky disks from the branch termination procedure. All we need do is take the plunge and run the very same algorithm for any pair of transformations a and b.

The reward is the glorious Figure 8.1! See intricate dance of spirals of two loxodromic transformations. This is a quasifuchsian group very different from the circle groups we met in Chapter 6. As usual, once a certain feature appears, the Mobius transformations in the group transport it around. Theoretical knowledge is one thing, but it wasn't until we got our programs up and running that the first ever pictures of exploding spirals brought the reality home. The authors, and later the participants in 1980 Thurston Theory Conference at Bowdoin College, could not suppress their awe at the eerie glowing image of the limit curve snaking its way across an old Tektronix terminal.[1]

[1] At 300 baud; those were the days!

There's one question here you may be asking: how did we choose a and b? The answer is, we built a machine. When engineers design a new sports car, do you really think they first test it by creeping down the driveway carefully at 5 miles per hour, then 10 miles per hour, and so on? Of course not; they push it madly through its paces to see how it drives. In order to carry out our explorations, we needed an easy-to-use program so we could quickly test out all sorts of possible matrices. In the next section, we shall explain a recipe which allows us to easily make as many

Figure 8.1. Mating snails?
The limit set of a group generated by two maps *a* and *b* with complex conjugate traces $t_a = 1.87 + 0.1i$ and $t_b = 1.87 - 0.1i$. This group is quasifuchsian because its limit set is a continuous loop which never crosses or meets itself. Curves like this are called Jordan curves: the celebrated *Jordan Curve Theorem* innocently states that every Jordan curve divides the plane into two parts, an 'inside' (gray) and an 'outside' (white). The proof of this seemingly obvious result is not easy, and this picture gives some idea of just how complicated a Jordan curve can be. We had to devise a handcrafted algorithm to colour the inside.

variants of pictures like this as we please. The recipe depends on just two complex numbers, the traces of a and b. These will be our parameters; just feed them in and let the program fly.

Actually you may get even greater satisfaction by varying the parameters continuously and watching the limit set writhing in response. For that, you will have to write another small program which plots a sequence of frames of limit sets whose parameters are just slightly changed step-by-step. Several people have done this, but stills are the best we can offer in a book. Peter Liepa has made some rather beautiful pictures of evolving limit sets which you can find under his name on the Vimeo website. For the Macintosh, we recommend a program by Masaaki Wada called *OPTi*, available at `delta.math.sci.osaka-u.ac.jp/OPTi/index.html`.

Grandma's recipe

To make pictures, we need two Möbius maps a and b, given by matrices

$$a = \begin{pmatrix} a_1 & a_2 \\ a_3 & a_4 \end{pmatrix} \quad \text{and} \quad b = \begin{pmatrix} b_1 & b_2 \\ b_3 & b_4 \end{pmatrix}.$$

Numerical inputs to a device or program are often called **parameters**. On the face of it, two matrices means eight complex numbers which means sixteen real numbers: that's quite a few! To build our 'easy-to-use' program, we need reduce the parameters to a minimum. We can get the number down to six by assuming that each matrix has determinant 1. We can further reduce the number by remembering that the interesting thing is to study groups up to conjugation. In practice this means that after we have studied one particular group G, we no longer need study any of the conjugate groups hGh^{-1} for any conjugating Möbius map h (apart of course from the fun of getting a quite different 'view' of the limit set). A definite choice among all the conjugate groups hGh^{-1} is called a **normalization** for G.

Choosing a particular normalization allows you to eliminate three further parameters, because there is always exactly one Möbius map which carries any three points to any other three. This means that you can pre-specify the position of three points; for example, you might specify that the attracting fixed points of a, A and b are 0, 1, and ∞ respectively. This is exactly what we did on p. 207 when we proved that, up to conjugation, the modular group is unique. The upshot is that up to conjugacy, we should be able to reduce the number of complex parameters necessary to describe a two-generator group from eight to just three. The question is, which three? From our experience in the last two chapters, a good guess might be the three traces $\text{Tr}\,a, \text{Tr}\,b$ and $\text{Tr}\,ab$. These numbers don't change when you conjugate, moreover we have already seen in some special cases that, up to conjugation, they completely determine the group.

Our upgraded algorithm is going to work by moving systematically round the boundary of the word tree, plotting limit points in order when it detects they are close. This means that for the program to work reasonably efficiently, it will be best if the limit set is still, at least roughly speaking, a connected loop. In the situation of pairing opposite Schottky circles this happens provided all four basic commutators are parabolic with traces equal -2. As we saw on p. 189, we can arrange this by choosing $\text{Tr}\,ab$ to satisfy the Markov identity

$$(\text{Tr}\,a)^2 + (\text{Tr}\,b)^2 + (\text{Tr}\,ab)^2 = \text{Tr}\,a\;\text{Tr}\,b\;\text{Tr}\,ab.$$

So for most of this chapter, we shall insist that $\text{Tr}\,abAB = -2$, or equivalently that our three parameters t_a, t_b and t_{ab} satisfy the Markov equation. Give or take some trouble with square roots, this reduces our parameter count from 3 to 2, namely t_a and t_b. We shall give the name **parabolic commutator groups** to those groups in which $\text{Tr}\,abAB = -2$. They are also sometimes known as **once-punctured torus groups**, because of the topological picture on p. 190.

In Box 21 we have revealed Grandma's treasured family recipe for the specially normalized two-parameter family which we used for most of our own explorations.[1] The matrix entries of the two generators a and b are written down entirely in terms of the parameters t_a and t_b, which you can set equal to any two complex numbers you care to choose. As you can see, the recipe is designed so that these numbers are the traces of a and b. Among all possible normalizations and hence many different possible recipes she might equally well have tried, Grandma selected this one mixed with some special spices to make the pictures come out really nice. If you put in real values for t_a and t_b, you get the group which pairs Schottky circles arranged in the pattern in frame (vi) on p. 176. The same formula gave us the generators for the Apollonian gasket on p. 201. There are some hints on how to verify that just knowing t_a and t_b really does fix the group in Project 8.4.

Gosh, it's so easy; why is there any need to explain? As you can see, in the second step of her recipe Grandma arranged that t_{ab} satisfies the Markov identity, thereby ensuring that $\text{Tr}\,abAB = -2$. We had better check that multiplying a and b gives the formula we have written down for ab, and that the determinants of a and b are both 1. You may wish to resort to your favourite symbolic algebra program, or, for the traditionalists, we recommend beginning with a good pile of blank scratch paper, copying

[1] As you will immediately appreciate if you try to play with Grandma's formula, you will see that Grandma had some considerable degree of mathematical talent. In fact, not only did the senior author's aunt attend Girton College in 1916 (see p. 36), his grandmother Edith Read was awarded a double first in mathematics at Newnham in 1892, only two years after Philippa Fawcett had been placed above the Senior Wrangler in the Mathematical Tripos, creating so much stir that a national petition was drawn up that women be awarded Cambridge degrees. It gained 5000 signatures and only narrowly missed being passed by the Senate. Women were finally awarded degrees by Cambridge in 1948. Like many other women of her generation, Edith Read took her degree *ad eundem* from Trinity College, Dublin.

Box 21: **Grandma's special parabolic commutator groups**

(1)   Choose any complex numbers t_a and t_b.

(2)   Choose one of the solutions x of the quadratic equation

$$x^2 - t_a t_b x + t_a^2 + t_b^2 = 0$$

and set $t_{ab} = x$.

(3)   Compute

$$z_0 = \frac{(t_{ab} - 2)t_b}{t_b t_{ab} - 2t_a + 2i\, t_{ab}}.$$

(4)   Compute the generator matrices:

$$a = \begin{pmatrix} \dfrac{t_a}{2} & \dfrac{t_a t_{ab} - 2t_b + 4i}{(2t_{ab} + 4)z_0} \\ \dfrac{(t_a t_{ab} - 2t_b - 4i)\, z_0}{2t_{ab} - 4} & \dfrac{t_a}{2} \end{pmatrix}$$

$$b = \begin{pmatrix} \dfrac{t_b - 2i}{2} & \dfrac{t_b}{2} \\ \dfrac{t_b}{2} & \dfrac{t_b + 2i}{2} \end{pmatrix}.$$

(5)   It's worth noting that the product ab is also quite simple:

$$ab = \begin{pmatrix} \dfrac{t_{ab}}{2} & \dfrac{t_{ab} - 2}{2z_0} \\ \dfrac{(t_{ab} + 2)z_0}{2} & \dfrac{t_{ab}}{2} \end{pmatrix}.$$

One could compute b and ab first, and then find a by multiplying ab on the right by the inverse of b, that is, $a = (ab)B$.

---

the two matrices a and b carefully, and multiplying them out very slowly indeed.[1]

All the groups made using Grandma's recipe have a rather beautiful symmetry, which Grandma felt was a very flavourful ingredient in her groups. You may notice that the diagonal entries in both a and ab are the same. This has the consequence, immediately noticeable in all our pictures, that the limit set of any group made using her recipe is symmetrical under the $180°$ rotation about the origin O. How this works is explained in Note 8.1.

Lastly, what about that mysterious number z_0 in the off-diagonal entries of a and ab? Grandma could just have left it out of her recipes altogether,

[1] If it doesn't come out right, do it two more times: the right answer usually wins two out of three.

and then z_0 would have been none other than the fixed point of the commutator $abAB$. By conjugating by a map that moves z_0 to 1, Grandma has added a little extra style to her pictures. To get the hang of her recipe, you may like to work through Projects 8.1 and 8.2.

Let's play (gently at first)

To ensure all is working smoothly and to gain familiarity with what to expect from Grandma's recipe, we are going start our play rather gently with groups in which the traces of the generators a and b are both real.

Figure 8.2 shows the outcome of our first experiment. We made it by running the program many times, keeping the two traces t_a and t_b equal and real-valued, sliding down from the initial value 3 to the final value 2. These groups can be made by pairing tangent circles, and we have shown the Schottky circles. In the first three frames, the limit set is just the unit circle and the group is Fuchsian. The arcs rotate as the traces decrease until they reach the symmetrical position in frame (iii). You may recognize this picture – it is exactly the group in frame (vi) on p. 176. As we move past the symmetric position, something dramatic happens. The limit set crinkles up and the group has become quasifuchsian. As we keep moving, the lowermost limit points (these are actually the fixed points $\overline{b}, \overline{B}$) become corners with evermore pronounced angles, until finally they come together like a crab's pincers, chopping the region enclosed by the limit set

---

Note 8.1: **Grandma's symmetry**

Suppose M is a matrix whose diagonal entries are equal, in other words with the special form $M = \begin{pmatrix} r & s \\ t & r \end{pmatrix}$. Such transformations have a special symmetry, encoded in the equation

$$-M(-z) = -\frac{r(-z)+s}{t(-z)+r} = \frac{rz-s}{-tz+r} = M^{-1}(z).$$

To interpret this equation, write j for the 180° rotation $z \mapsto -z$. Using the relation $j^{-1} = j$, this equation says that $jMj^{-1} = M^{-1}$. In other words, conjugating the transformation M by 180° rotation about O carries M to M^{-1}.

In Grandma's recipe, both a and ab have this property, which means $jaj = A$ and $jabj = BA$. Since j is its own inverse, these imply $jAj = a$ and $jBAj = ab$. By combining these relations, we can show that any word in a, b, A and B is conjugated by j into some other word in a, b, A and B. That is, conjugation by j does not change the group G generated by a and b, nor does it change the complete collection of infinite words in the generators. That is enough to tell us that the limit set of G is unchanged by the transformation j. As an example, consider applying j to the infinite word $abaBA \cdots$:

$$j \, abaBA \cdots = jabj \, jaj \, jBAj \cdots = BA \, A \, ab \cdots.$$

Figure 8.2. Varying parameters: from $t_a = t_b = 3$ to $t_a = t_b = 2$. We wrote down the generators for the group in frame (iii) in Project 6.5, and frame (vi) is the Apollonian gasket.

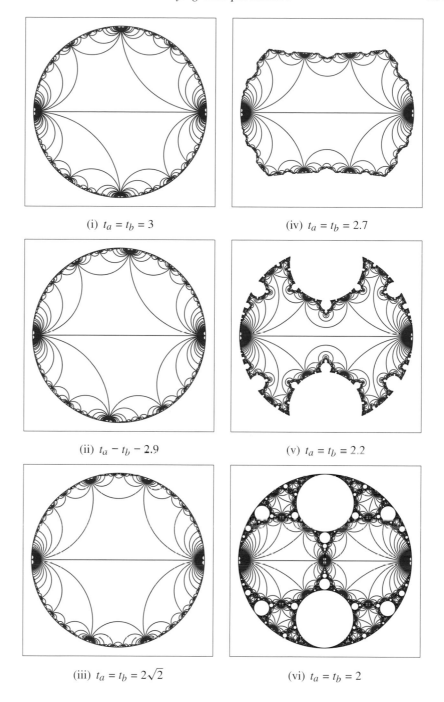

(i) $t_a = t_b = 3$

(iv) $t_a = t_b = 2.7$

(ii) $t_a = t_b = 2.9$

(v) $t_a = t_b = 2.2$

(iii) $t_a = t_b = 2\sqrt{2}$

(vi) $t_a = t_b = 2$

into a myriad of tiny disks. This last frame should look familiar too – we have arrived at our old friend the Apollonian gasket from Chapter 7!

What happens if we decrease the traces just a little bit further and try $t_a = t_b = 1.9$? *Warning, warning, danger, danger!* The Schottky circles will start to overlap, and it becomes not at all clear what to expect. You can see what we are worried about in Figure 8.3. In this picture, we chose a pretty large cut-off value `epsilon = 0.1` in comparison to the frame size, 2.2 by 2.2. In contrast to the previous pictures, you can actually see the line segments drawn by the DFS program. Some of them are actually much larger than `epsilon`. That is because no matter how far we go down the branch, limit points which are supposed to be neighbours never get truly close. The branch is terminated only by the built-in maximum depth `lev_max`. It is lucky we built in this safeguard; otherwise our program would be stuck running a never ending loop. The truth is, Figure 8.3 should be a solid black square.

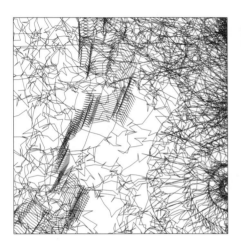 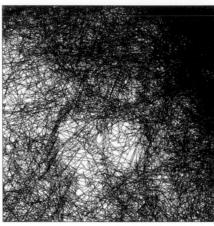

Figure 8.3. The program hits chaos. The left side is the result of running the DFS algorithm for just a short length of time; on the right we had slightly more patience. The picture resulting from running the program forever (that is, without the safety cut-off `lev_max`) is slightly less interesting. Groups whose limit set look like this are called **non-discrete**.

The groups in Figure 8.2 are actually conjugates of the groups made from our original circle pairing recipe on p. 170 in Chapter 6. To see this, let $x = u = t_a/2$. The connection between the two constructions is shown in Figure 8.4 which was drawn using Grandma's recipe for the quasifuchsian group with $t_a = t_b = 2.2$. Our original recipe gave roughly equal weight to each of the four generators a, A, b and B. By contrast, Grandma's recipe emphasizes symmetry relative to the alternative generators a and ab.[1] The word tree comes out distorted so that words beginning with a occupy half the picture, the other half being divided among words beginning with the other three letters A, b and B.

The basic tile in Figure 8.4 is the one which has one side of each colour. The red side is part of the circle C_a, the blue side of C_b, and so on. Its vertices are the fixed points of the four commutators $abAB$, $bABa$, $ABab$ and $BabA$. As you can see, the red part is exactly half the limit set. The other half can be obtained by reflecting through the origin, using the map $j : z \mapsto -z$. (This trick was in part the original motivation for Grandma's

[1] Instead of iterating the transformations a and b, we could just as well iterate the pair a and ab. This is because $b = A(ab)$, so that any word which can be expressed in terms of a and b can equally well be written in terms of a and ab. This is what we mean by saying that a and ab generate the same group as a and b. By contrast the pair of words a and $abAB$ cannot be used as generators, because we will never get to the word b.

Figure 8.4. The anatomy of the limit set for the group specified by Grandma's recipe with $t_a = t_b = 2.2$. The different colours show the pieces of the limit set that begin with different one letter prefixes: red for those beginning with a, green for A, blue for b, and yellow for B. We have marked certain limit points by their infinite words, for example $\bar{a}$ marks the attracting fixed point of a. Since it is represented by the infinite word $aaaa\cdots$, it appears in the red section. This section also contains the fixed point of aba^{-1} whose infinite word is $a\bar{b} = abbbb\cdots$, and aBa^{-1} with infinite word $a\bar{B} = aBBBB\cdots$. Notice the various fixed points which seem to be coming together as the traces t_a and t_b get near 2.

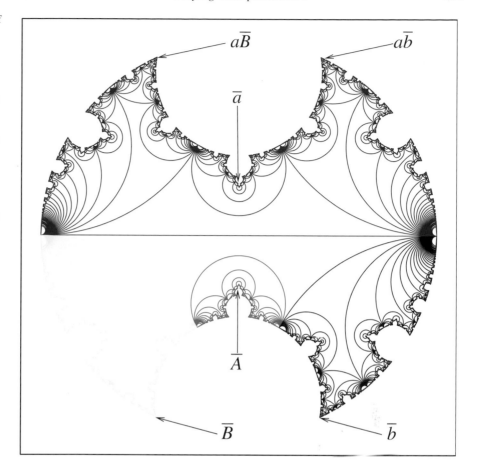

normalization: we had only to plot a quarter of the limit points, and by reflecting got the rest of the picture for free.)

The fun begins: traces go complex

Our play has been kept artificially gentle by restricting to examples in which t_a and t_b are both real. Such groups always come equipped with a chain of four tangent Schottky circles, so plotting their limit sets is really nothing new. The real fun starts when t_a and t_b go properly complex. Figure 8.5 shows the results of gingerly testing the waters.[1] What were the Schottky circles appear to have gone haywire, but at least the limit sets are still loops.

The curling and twisting you can see in these limit sets is caused by tiny spiralling motions of loxodromic transformations. Transformations with non-real traces are always loxodromic, so as soon as we make t_a and t_b complex, we can expect curling to occur. The amount of curling of a

[1] In our choice of parameters we have two complex and therefore four real degrees of freedom, a lot of possibilities to pick. After a great many hours playing (and more of real mathematics), the authors now have the benefit of hindsight; however, in the early days, we really had very little idea what values would give interesting results. We were reduced to making probes very like the ones we shall run through here.

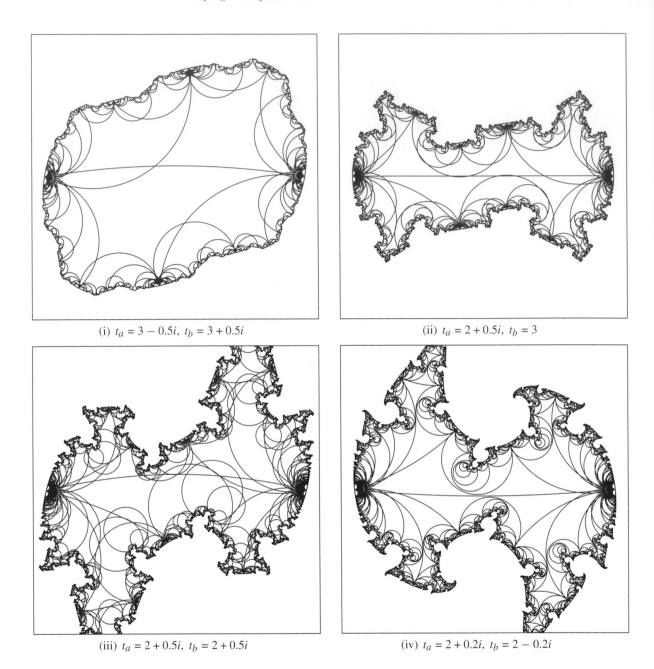

(i) $t_a = 3 - 0.5i$, $t_b = 3 + 0.5i$ (ii) $t_a = 2 + 0.5i$, $t_b = 3$

(iii) $t_a = 2 + 0.5i$, $t_b = 2 + 0.5i$ (iv) $t_a = 2 + 0.2i$, $t_b = 2 - 0.2i$

transformation depends not so much on the imaginary part of its trace as on the tightness of the spiral motion near its fixed points. Referring back to Chapter 3, you will see the spiral is tightly coiled if the multiplier is near 1, so the trace is near 2, and if in addition the imaginary part is comparatively small. You can see this in evidence in the substantial curling in

Figure 8.5. Testing some complex values of the traces. For comparison we have made the viewing window the same in all four frames.

Figure 8.6. Dr. Stickler blown about inside the limit set of a quasifuchsian group with $t_a = t_b = 1.91 + 0.05i$. You can see a full view of this limit set in Figure 8.1. We made this picture by implementing a tiling type plot with the prostrate Dr. Stickler as the initial seed.

the last two frames in Figure 8.5, where t_a and t_b are near 2 and only slightly complex. There are no Schottky circles, but on the 'inside', we have drawn red circular arcs meeting at the fixed points of the four basic commutators. (We made the first arc perpendicular to the direction of the parabolic at the fixed point.) You can tell they are no longer pieces of Schottky circles, because Schottky circles never intersect. You might wonder if the same group might be constructed as a Schottky group starting

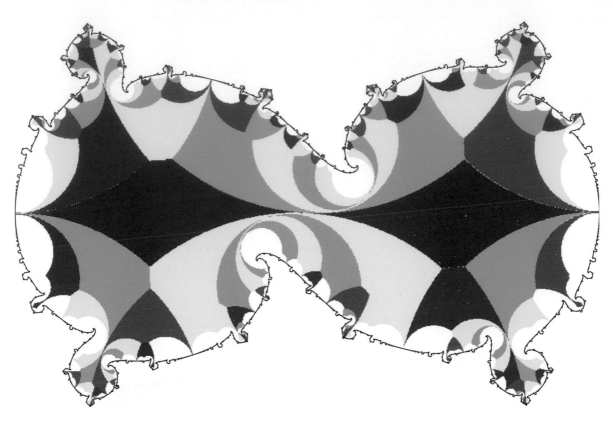

from a different set of circles, and with different generators doing the pairings. Whether or not this is possible, we don't know.

As you can see, all these groups are still quasifuchsian, meaning that the limit set is a connected curve which doesn't cross itself and which divides the plane into an 'inside' and 'outside'. These are the first groups whose limit sets we genuinely could not have drawn using our old Schottky circle algorithm. It would be nice to explore the region inside the limit set, but since there are no Schottky circles to work with, it can become a very tricky problem to find suitable tiles. Undeterred, we blew up a small part of a nice limit set with $t_a = t_b = 1.91 + 0.05i$ and set Dr. Stickler lying flat on his back in red on the righthand side. You can see him in Figure 8.6 spinning around, carried into every nook and cranny, so that there is exactly one Dr. Stickler for every word in the group.

In Figure 8.7 we have actually found a tiling for a quasifuchsian group with $t_a = 2 + 0.1i$, $t_b = 3$. Notice that the tiles are no longer four- but six-sided. As ever, the different tiles are carried onto each other by the transformations in the group. They get exceedingly skinny in the middle: if we

Figure 8.7. All the tiles are obtained from the red one by applying words in the group. What is the red tile? In Figure 8.6, note that the red Dr. Stickler is larger than all his copies. We could try putting a miniscule Dr. Stickler anywhere and see if he is larger than his copies. The places where this happens are the points of the red tile.

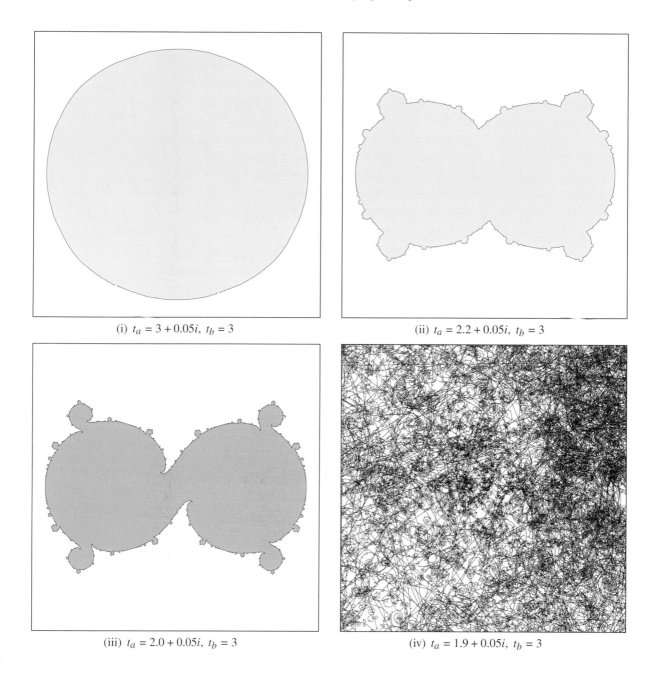

(i) $t_a = 3 + 0.05i$, $t_b = 3$

(ii) $t_a = 2.2 + 0.05i$, $t_b = 3$

(iii) $t_a = 2.0 + 0.05i$, $t_b = 3$

(iv) $t_a = 1.9 + 0.05i$, $t_b = 3$

Figure 8.8. Our first probe. In this sequence, $t_b = 3$ and $t_a = x + 0.05i$ with x varying from 3 down to 1.9.

varied our parameters just a little bit each of these tiles would fall apart into two halves. We found this particular tiling by a completely different method explained briefly in the caption.

The curling in the last two frames of Figure 8.5 piqued our interest, hinting at directions which might be interesting to explore. Just how much curling is possible? The interest seems to centre on traces near 2, but ever

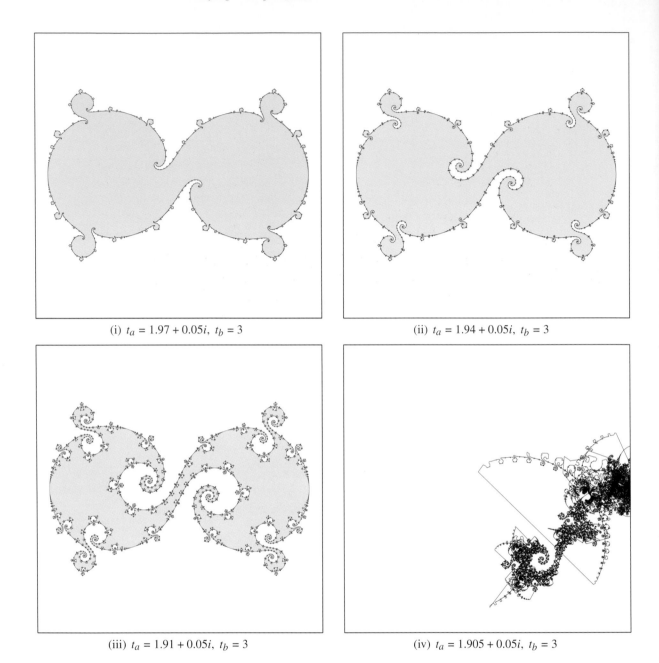

(i) $t_a = 1.97 + 0.05i$, $t_b = 3$

(ii) $t_a = 1.94 + 0.05i$, $t_b = 3$

(iii) $t_a = 1.91 + 0.05i$, $t_b = 3$

(iv) $t_a = 1.905 + 0.05i$, $t_b = 3$

so slightly complex. To investigate, we shall run an experiment in which we fix the trace t_b safely equal 3, and then let t_a run through values $x + 0.05i$, where x is a real number which starts at 3 and slowly decreases to some dangerous transitional value, as yet unknown. Figure 8.8 shows the preliminary results.

The first frame is just a slightly wobbly circle. The second and third

Figure 8.9. In this refined probe, $t_b = 3$ and $t_a = x + 0.05i$ with x varying from 1.97 down to 1.905. In (iv), the program was terminated prematurely, when the erratic nature of the plot became clear.

frames show some bumps forming, with the first hints of spiralling in the third frame at $x = 2$. From the chaotic fourth frame we deduce that somewhere between $x = 2.0$ and $x = 1.9$, we stepped over the boundary. The live version of the last frame is more interesting: the DFS algorithm frantically criss-crosses the picture trying to draw a solid black square.

To locate the transition point more exactly, in Figure 8.9 we decrease x by finer increments from 2.0 to 1.9. From 1.97 to 1.94 to 1.91, you can see the bumps on the limit set developing into pronounced and ever more tightly whirling spirals. We know from our earlier probe that the boundary lies above $x = 1.90$; when we ever-so-carefully stepped to 1.905, the DFS plot tried to fake sanity for a while, until its turbulent behaviour at last manifested and we terminated the program, allowing us at least to see some of the spirals that are still evident. We have pinned down the transition to madness somewhere between $x = 1.91$ and 1.905.

Transition to madness

We have now three times bumped into places where our limit set plot has gone wild. What is going on? As we warned at the outset, there is no reason to expect that our limit set plot, bereft of the Schottky circles, will produce anything reasonable at all. The greater miracle, perhaps, is that it ever does!

What is happening here is this. If you multiply a large number of matrices together, then you would expect that the resulting matrix product would automatically get 'large'. In the groups which produce reasonable limit sets this is certainly the case. What is going wrong when the limit set goes haywire is that a word in the group which is a long matrix product of many generators unexpectedly turns out to be actually very 'small'. The matrices cancel in a mysterious way, so the manner in which this particular group element moves points around in the plane is not at all what you might expect. Of course, not all entries can be near 0 because the determinant of any matrix in our group is always to equal to 1. So what we mean by saying that a matrix is 'small' is that it is very near the identity matrix I. From this point of view, I is the 'smallest' matrix and we measure how large a matrix is by measuring the distance of its entries from those of $\begin{pmatrix} 1 & 0 \\ 0 & 1 \end{pmatrix}$. Groups for which large products stay away from I are called **discrete**.

You can look for discreteness in our plots by seeing how close $M(z)$ can get to z, for any point z in the plane. In the groups we have studied so far, you can always find tiles which cover all of the ordinary set, that is, all parts of the plane not occupied by the limit set itself. If you sit at a point z in the middle of one tile, no points in any other tile can be too close, because no point can be nearer to you than the edges of your tile. But if M was very near to I, then $M(z)$ would be very near to $I(z) = z$. This shows that very nasty cancellations can never occur as long as there are some 'limit set free' regions in the plane which can be covered by tiles.

It turns out that there is yet another layer of complication because there are groups for which *there is no ordinary set at all but which are still discrete*. We will look at these in Chapter 10 where we shall be meeting some amazing pictures of groups which are discrete in the strict sense that they contain no matrices too close to I, but for which the limit set plot *fills out the whole plane*, in a magically organised and yet extremely complicated way. For now, though, looking at whether or not the computer plot goes wild and seems to be filling up the page will not lead you far wrong. If you can find a tiling, then it shows you visually that the group is discrete, and if there is no tiling, you had better watch out!

You can see that the whole question of the transition from discrete to non-discrete groups is pretty delicate by comparing frame (iii) of Figure 8.9 with with frame (i) of Figure 8.14. The values for t_b are the same and the values $t_a = 1.91 + 0.05i$ and $t_a = 2$ only differ by about a tenth, but this small difference in the input produces dramatic change. We are warned: small steps near these transition points may change the limit set by giant leaps.

Moreover, there is really no reason to suppose that as we decrease x along this path there is only one crossing into non-discrete groups. We might stay with non-discrete groups for a bit, then cross back into the discrete region, only to cross out again shortly thereafter. For some choices of probe this is certainly what happens, as we shall shortly see.

We have been speaking as if all the discrete groups occupied one region, and the non-discrete ones another. Unfortunately, things are a bit more complicated. Remember from Chapter 2 what happened when we plotted the orbits of a rotation $z \mapsto e^{i\theta}z$. If θ is a rational multiple of π (or, equivalently, p/q degrees for some integers p, q), then the orbit returns to its beginning after a finite number of steps. But if not, the orbit keeps spinning around until it fills up a whole circle (see Figure 2.4).

Such trouble can be avoided by a very careful choice of traces, for example we could set $t_a = t_b = 1.90211303259032$, a close approximation

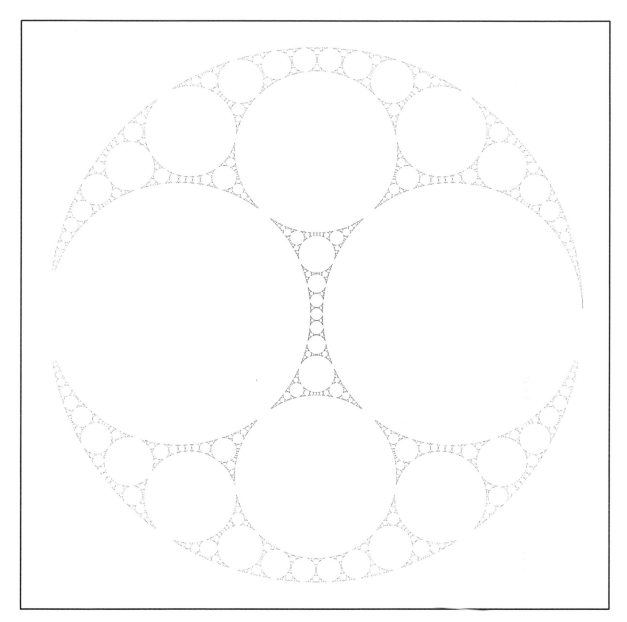

Figure 8.10. The DFS algorithm with the values $t_a = t_b = 2\cos\frac{\pi}{10}$. A lone sailor in a sea of non-discreteness, this group is actually discrete! To get the picture we ran the DFS algorithm with these values, setting it to plot only limit points and not, as it did in our other pictures, the line segments in between.

to $2\cos\frac{\pi}{10}$. This makes both generators elliptic, rotating about their fixed points through the angle $360°/10$ so that the transformations a^{10} and b^{10} are both *equal* to the identity matrix I. This group is certainly discrete. If it weren't, we should get a solid black square. The resulting picture is shown in Figure 8.10

If you try to run your program with these values, be warned that the equation $a^{10} = b^{10} = 1$ causes tremendous problems because the plot constantly retraces parts of the limit set. In addition, the termination procedure no longer works, and you will have to interrupt your program rudely to make it stop. How to adapt our algorithm to work in situations like this is a complicated problem which we shall only be able to touch on Chapter 11.

In the next chapter, we shall investigate the boundary between discrete and non-discrete much more thoroughly. Initially, though, one should be content to stumble around pretty blindly, accepting that on occasion we shall bump into non-discrete groups.

Spirals

The spiralling limit set in frame (iii) on p. 238 was so beautiful it deserves a closer look. We have blown it up in Figure 8.11, decorated with flow lines indicating the movement of the generators between their fixed points.

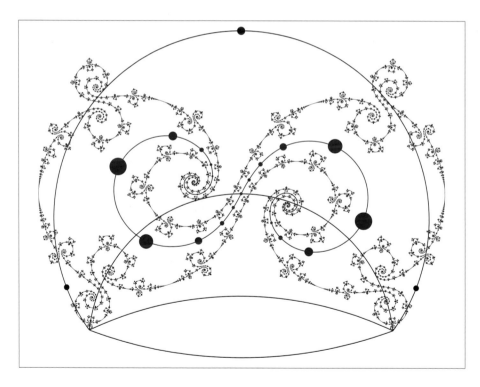

Figure 8.11. Spirals. This is the limit set with $t_a = 1.91 + 0.05i$ and $t_b = 3$. The red spiral is a flow line for the highly loxodromic generator a and the red dots show the orbit of 0. The blue flow lines show the movement of the strictly hyperbolic generator b. Because the multiplier of b is 2.618, the blue dots disappear very rapidly at its two fixed points. You can track how various spirals in the limit set are moved to other spirals by a and b. All spirals ultimately come from the ones centred at the fixed points of a, acted on by some word in the group.

Now let's try another experiment: since $t_a = 1.91 + 0.05i$ worked so well, why not try it for t_b as well? Will the a and b generators put out tendrils that curl away from each other? Perhaps they will collide making

Figure 8.12. Spirals on spirals. In this picture both generators are nearly parabolic with $t_a - t_b = 1.91 + 0.05i$. The loxodromic flow lines for a, b and abA are coloured red, blue, and green respectively.

the group non-discrete. Not so; another spin of our program produces Figure 8.12 in which the spirals have truly bloomed.

Extending our explorations beyond the realm of Schottky groups is beginning to pay off. Examining Figure 8.12, it is pretty easy to see that it is impossible to find a chain of Schottky circles matched by the generators a and b. In fact, comparing the tightly interwoven spirals of Figure 8.12 with any of the frames of Figure 8.2, makes it almost inconceivable that you could find a chain of kissing Schottky circles for *any*

Note 8.2: **Tight spirals**

The prominent spirals in Figure 8.12 are controlled by the multiplier of the generator a, the unique number k such that a is conjugate to the standard transformation $T(z) = kz$. Referring back to Note 3.5, we see that

$$k = \frac{t_a^2 - 2 \pm t_a \sqrt{t_a^2 - 4}}{2}$$

with the sign chosen so that $|k| > 1$. For $t_a = 1.91 + 0.05i$, this yields

$$k = 0.95566 + 0.68692i$$

which has absolute value about 1.17 and argument about 35.7°. Since the multiplier is so close to 1, powers of k do not increase in modulus especially rapidly, while the argument is large enough to cause a very noticeable rotation. The phenomenon occurs when a trace gets very close to ± 2 but with non-zero imaginary part. We call such elements nearly parabolic. They cause tight spirals and create far more distortion in the limit set than transformations which are genuinely parabolic.

Figure 8.13. Tighter spirals correspond to smaller imaginary part. The parameters for this picture are $t_a = 1.95 + 0.02i$ and $t_b = 3$.

pair of generators of this group.[1] We shall discover some rather remarkable substitutes for the Schottky circles in the course of the next chapter.

During our original computations we made many probes like the ones above. It is worth trying a few more sequences: say traces $t_b = 3$ and $t_a = x + 0.02i$ with x as varying downwards from 3. The spirals here are much tighter, due to the smaller imaginary part, while the transition between discrete and non-discrete occurs at a slightly larger value of x. Figure 8.13 shows the tight spiral when $t_a = 1.95 + 0.02i$. Decreasing x to 1.94, we get a non-discrete group.

Enlarging the imaginary part of t_a on the other hand, the spirals slightly unfurl. Figure 8.1, at the beginning of the chapter, was a particularly nice shape at the values $t_a = 1.87 + 0.1i$, $t_b = 1.87 - 0.1i$. We chose one trace to have negative imaginary part; try to interpret the effect of this in the picture by comparing with Figure 8.11.

[1] Just saying it looks impossible won't satisfy your typical mathematician; it is quite a challenge to furnish a rigorous proof.

Figure 8.14. The result of varying traces t_a and t_b separately. As explained in Project 8.7, these two groups are actually conjugate, though this is pretty hard to visualize! The conjugating transformation maps the light grey region in the lefthand picture to the light grey region on the right.

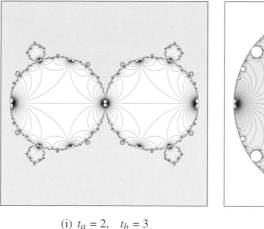 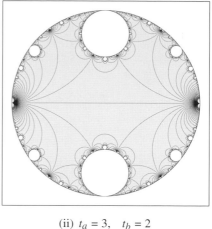

(i) $t_a = 2, \quad t_b = 3$ (ii) $t_a = 3, \quad t_b = 2$

Hitting the edge

Figure 8.14 shows two pictures in which one of the two traces is 2 and the other is 3. In other words, one of the two generators a and b is parabolic but the other is not. Both pictures are rather like the gasket picture in frame (vi) of Figure 8.2, but on the left only the circles C_a and C_A have come together with an extra point of tangency, while on the right the tangency is between the circles C_b and C_B.[1] This may be easier to see if you compare the left frame of Figure 8.14 to Figure 8.4. See how the fixed points of a have come together pinching off the lefthand part of the picture from the right. If, on the other hand, we fix t_a and send t_b to 2, the upper and lower pincers come together resulting in the righthand frame of Figure 8.14.

[1] On the right we have only shown the tiling of the inner region; to see the tangency of the b circles you have to draw the Schottky circles in the outer white region.

The myriad small circles in these pictures appear for exactly the same reason as they did in the last chapter. If for example a is parabolic, then so is bAB, and so also is $abAB$. Thus the two elements a, bAB generate a subgroup conjugate to the modular group, which as we know means we expect to see circles in the limit set. Well, here they are!

Groups like these in which one element is parabolic are called **cusp** groups, because they can be explained in terms of pinching points on surfaces to cusps. Some groups, like the ones in our pictures here, have one 'extra' parabolic element (in this case b) and so are called **single cusps**. Some, like the gasket, have *two* 'extra' parabolics, in which case they are called **double cusps**. In the single cusp groups in Figure 8.14, the curve traced out by our DFS algorithm touches itself at 'horns' where the newly formed set of white circles close up. Nevertheless, the other half of the ordinary set has not collapsed. Single cusp groups are especially nice to study because choosing one element to be parabolic, for example b, there

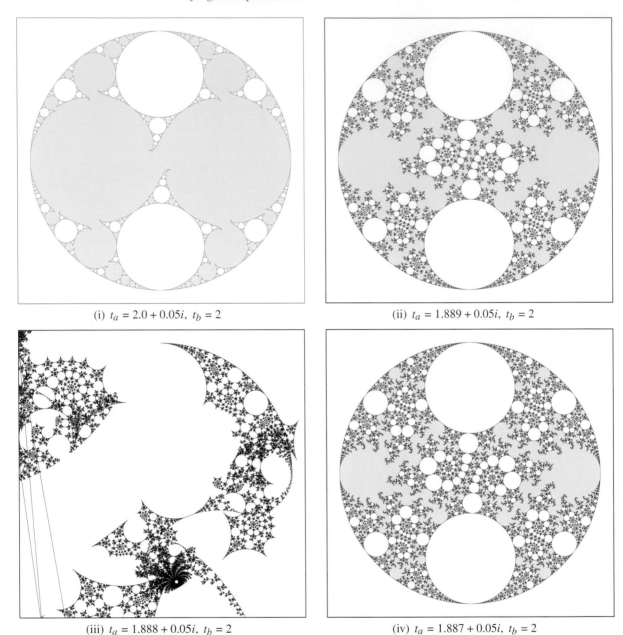

(i) $t_a = 2.0 + 0.05i$, $t_b = 2$

(ii) $t_a = 1.889 + 0.05i$, $t_b = 2$

(iii) $t_a = 1.888 + 0.05i$, $t_b = 2$

(iv) $t_a = 1.887 + 0.05i$, $t_b = 2$

Figure 8.15. A charlatan in our midst! Shrinking $\mathrm{Re}\, t_a$, the group in frame (iii) pretends to be discrete, but its histrionics are soon evident. But the group with a slightly *lower* $x = 1.887$ is a genuine free discrete group. To get so much fine detail, we had to choose the extremely small stepping size `epsilon= 0.0003`.

is still quite a lot of freedom to vary these groups, which we can do by changing the parameter t_a.

Figure 8.15 shows a whole sequence of groups in which we do just this. Since b is parabolic, the whole limit set is tidily packed up inside the circle of radius 1. This time, the group contains a Fuchsian subgroup generated by b and abA. Like all other features of the limit set, the images of this outer white circle are transported all around the picture,

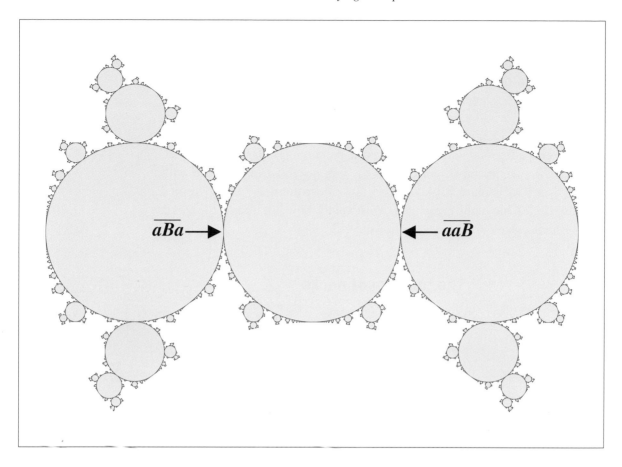

$\overline{aBa}\longrightarrow$ $\longleftarrow \overline{aaB}$

Figure 8.16. We calculated this cusp from scratch by finding parameter values for which the word *aaB* has trace 2. Comparing with the left frame of Figure 8.14, our new pinching process has divided this central part of the limit set not into two tangent disks, but three. This picture was plotted using the special word algorithm described on p. 260.

attractively arranged inside the outer white circle in elegant yin-yang spirals whose meaning, while perhaps not mystical, is certainly quite profound. We shall come back to these spiralling chains of circles in the next chapter.

Figure 8.15 was actually made to illustrate a sobering discovery we made early on in our explorations by probes. Our previous efforts may give the impression that the transition between quasifuchsian and non-discrete can be found by smoothly moving along lines. Not so! The probe finely homes in on the boundary between discrete and non-discrete groups. After fixing $t_b = 2$, we set $t_a = x + 0.05i$. The probe explored what happens as x decreases from 3. The transition point appears to be a bit less than 1.90, the lower limit we discovered in the last section when $t_b = 3$. To locate it, we found we needed to move by very small steps indeed. At $t_a = 1.889 + 0.05i$ the group is certainly discrete, as it is at $t_a = 1.887 + 0.05i$. But what has happened in between? The group for $t_a = 1.888 + 0.05i$ pretends to be discrete for a while, but you can soon see that something drastic has gone wrong. The sequence shows that there is a barrier of non-discrete groups placed between $\mathrm{Re}\, t_a = 1.889$ and $\mathrm{Re}\, t_a = 1.887$. We seem to be observing an island of cusp groups separated from the mainland by a strait of non-discreteness.

In all the cusp groups we have drawn thus far, the extra parabolic elements are one or other of the generators a and b. But other elements can become 'accidentally parabolic' in a similar way. Figure 8.16 shows a singly cusped group in which the extra parabolic is the element aaB. In this particular example, we choose $t_b = 3$, so you can compare with the left frame of Figure 8.14, and see how the pattern of newly formed pinched circles has subtly changed. We explain how we found the value for t_a which makes aaB parabolic in Note 8.3.

The problem of necks

All right, *we confess!* If you have been tinkering with code as you have been reading, you may have noticed that your plots don't quite have the

---

Note 8.3: Calculating the aaB cusp

Suppose we fix $t_b = 3$. Here is a neat way to find a point where aaB is parabolic, that is, where $t_{aaB} = 2$. There are the right number of variables to find an exact solution to our problem, because the three basic parameters t_a, t_b and t_{aB} are related by the Markov equation

$$t_{aB}^2 + t_a^2 + t_b^2 = t_a t_b t_{aB}.$$

According to the theory on p. 189 ff., we should be able to express t_{aaB} in terms of the traces t_a, t_b and t_{aB}. The trick is to use the grandfather trace identity $t_{uv} + t_{uV} = t_u\, t_v$. With $u = a$ and $v = aB$, this becomes

$$t_{aaB} = t_a\, t_{aB} - t_{abA} = t_a\, t_{aB} - t_b.$$

(We made the last step by remembering that trace is unchanged by conjugation, and noting that abA is a conjugate of b.)

Setting $t_b = 3$ and solving the Markov equation

with the quadratic formula gives the two solutions

$$t_{aB} = \frac{3t_a \pm \sqrt{9t_a^2 - 4t_a^2 - 36}}{2}.$$

Choose one of these solutions and substitute into the expression for t_{aaB}. If we set this expression equal to ± 2, we have an equation for t_a. One solution is

$$t_a = 2 - i, \qquad t_{aB} = 2 + i.$$

(We aren't going to tell you how we got this, at least not yet. Computers do have their uses; thank goodness they came along in our lifetimes.) Anyway, we can verify that

$$t_{aaB} = t_a t_{aB} - t_b = (2 - i)(2 + i) - 3 = 2$$

as promised.

There are other solutions, and which one to choose is a very interesting question. It is worth experimenting with the various solutions to see what our limit set program will produce. Some explosions may occur.

same detail as ours. Yes, we did cheat slightly: we did make our pictures using the DFS algorithm, but with a souped-up version designed to overcome the neck problem so graphically suggested in Figure 6.14. The DFS algorithm, remember, works by marking chosen neighbouring limit points along the curve and terminating whenever they are closer than a preset value `epsilon`. If two neighbouring marked points happened to be less than `epsilon` apart, the program would join them with a straight line, omitting any features of the limit set between them.

Figure 8.17. The problem of necks. The picture shows the segment of limit set corresponding to words beginning with the prefix ab^5, so the two limit points at the bottom have infinite words that agree up to level 6. The trace parameters for this limit set are $t_a = t_b = 1.91 + 0.05i$. See Figure 6.14 to admire the family resemblance to Dr. Stickler.

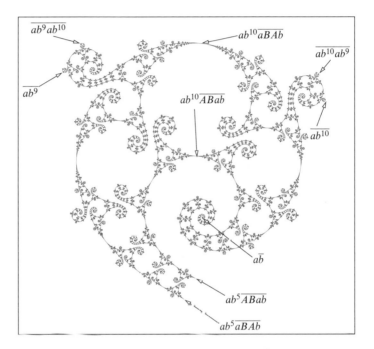

The problem is illustrated by Figure 8.17, which shows the part of the limit set for the values $t_a = t_b = 1.91 + .05i$ between $ab^5\overline{ABab}$ and $ab^5\overline{aBAb}$, (where we have abbreviated the string *bbbbb* by b^5 in the usual way). These two words agree up to the sixth digit, and are the first and last infinite words beginning with the prefix ab^5 in the anticlockwise order round the boundary of the word tree. A computation which plotted words only up to level 6 (covering $4 \times 3^5 = 972$ different words) would jump across the 'neck' at this point, entirely missing the rather large spiral in between. The root of the problem is that the centre point of the spiral $a\overline{b}$ is the fixed point of *abA* which is very nearly parabolic. The difficulty can be very serious: in proportion to the size of the spiral, the width of the neck can be as small as you please.

As we discovered in the course of many experiments, the problem centres on the presence of elements which we call **nearly parabolic**. A nearly
parabolic element is one which is loxodromic, but whose trace is almost
equal to ±2. In Chapter 6 we already noted the problem of 'clumping',
caused by the fact that near to parabolic fixed points, nested Schottky disks
shrink down much more slowly than they do elsewhere. Nearly parabolic
loxodromic elements exacerbate this difficulty, thereby creating 'the problem of necks.' The trouble is that two limit points which correspond to
infinite words which agree up to a very high level may actually be quite
far separated in the plane.

To truly understand this problem as it affects the DFS algorithm, we
need to watch the algorithm in 'slow-motion' as it draws the limit points
of the group. We can do this by choosing our threshold `epsilon` reasonably large so that we can easily see the line segments that are drawn.
As our test case, we will use the gently curling Figure 8.1 whose vertical scale ranges from -2.4 to 2.4. That plot has over 700,000 points,
with an `epsilon` of 0.001. We will 'zoom' in on the upper half and
enlarge `epsilon` to 0.4. Then only a handful of points are plotted. The
result with each limit point labelled and numbered in order of appearance
is shown in the top frame of Figure 8.18.

The first thing that becomes glaringly obvious from Figure 8.18 is that
there is no spiral! The threshold is just too large, and the algorithm chops
off the heads of the spirals. The neck of the largest spiral is from point
number 10 to point number 11 in the top frame of Figure 8.18. Denoting point number 10 as z_{10} and so on, these two points have infinite
words

$$z_{10} \longrightarrow aB\overline{BaBAb} = aB\overline{BabA}, \qquad z_{11} \longrightarrow aB\overline{BAba} = aB\overline{ABab}.$$

We get the two different representations because the commutator $abAB$ is
parabolic, so the infinite words $\overline{aBAb}$ and $\overline{BabA}$ both represent the fixed
point of $aBAb$. The representations to pay attention to here are $z_{10} = aB\overline{BabA}$ and $z_{11} = aB\overline{BAba}$. They show that z_{10} and z_{11} are the first and
last points of the set of limit points beginning with the prefix aBB.

In the lower left frame of Figure 8.18, the threshold is lowered to the
point where the algorithm just squeezes past the neck and, *whoosh*, a large
spiral appears. The figure shows 80 limit points plotted between $aBB\overline{abAB}$
and $aBB\overline{AbaB}$. Following along the spiral, we reach another point where
the neck narrows to less than the threshold. This occurs between points
numbered 55 and 56, corresponding to the words

$$z_{55} \longrightarrow aB^{10}\overline{abAB}, \qquad z_{56} \longrightarrow aB^{10}\overline{AbaB}.$$

Figure 8.18. At a very large threshold (`epsilon=0.4`), we can see the order of plotting of the limit points in the top frame. All the limit points plotted in that frame have a repetend equal to a commutator. The limit points are numbered in the order that the algorithm computes them. At that threshold, the large faded spiral is cut off. The lower left frame shows the missing spiral plotted with `epsilon=0.15`, which is small enough for the algorithm to be forced to 'squeeze through the neck.' The lower right frame shows the same region with the souped-up 'special words' algorithm with `epsilon=0.4`.

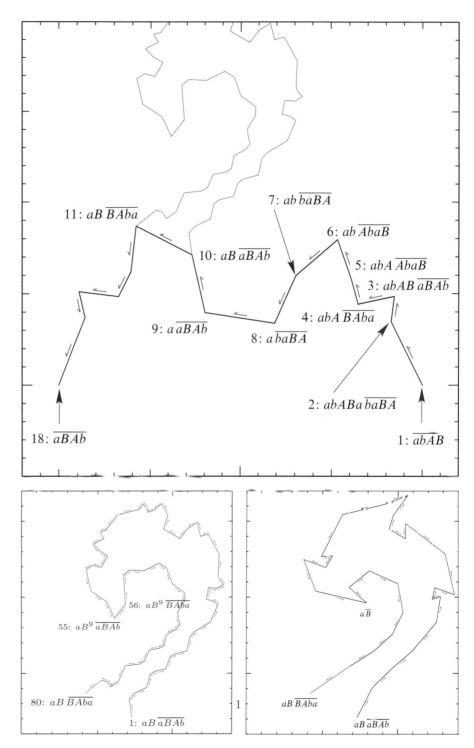

The missing piece of the spiral is the section of the limit set corresponding to infinite words with prefix aB^{10}. The 'centre' of the spiral is the infinite word $a\overline{B}$. Our problem is that the algorithm does not actually plot limit points whose infinite words end in B's.

Since we are still young enough to enjoy the thrill of an infinite spiral, let's see if we can't modify our algorithm to force it to travel all the way to the centre $a\overline{B}$. The reason this word is so troublesome is because B is a nearly parabolic element of our group, and its loxodromic flow is pompously grandiose. The answer is to simply instruct the program to be sure to plot the infinite words which have periodic ending equal to the special word B, because these points are at the centre of the spirals. That is exactly the same answer as the one we arrived at for the 'problem of gaps' in plotting the limit set of the Apollonian gasket. If you were anticipating another bout of tinkering with the program, you can relax. All the modifications were carried out in Chapter 7. All we have to do now is to inform the program that it must keep track of the 'nearly parabolic' words a and b, in the same way that it had to keep track of them when they were honestly parabolic. For this reason, we call this the **special words algorithm**, as it relies on our identification of certain special repetends to monitor.

Let's return to our discussion of Figure 8.17. Since our prefix $w = aB^{10}$ ends with B, the first and last limit points along the segment of the limit set in question are $w\overline{abAB}$ and $w\overline{AbaB}$. In between is the problem point $w\overline{B}$. Previously, when we got to the prefix w, we looked only at points $w\overline{abAB}$ and $w\overline{AbaB}$, plotting the line segment joining them if their distance was less than epsilon. Now, we inject the intermediate point $w\overline{B}$ and we only give the command to plot if *both* line segments from $w\overline{abAB}$ to $w\overline{B}$ and from $w\overline{B}$ to $w\overline{AbaB}$ are less than epsilon in length.

The proof this works is in the picture! The lower right frame of Figure 8.18 shows exactly the same limit set, except that the threshold has been raised back to the level epsilon=0.4 of the first frame. The generator B has now been identified as a special word. The result is that the spiral is not chopped off, and we get the same basic outline as that on the lower left having plotted far fewer points: 31, to be precise.

The new repetend B is our **special word**. Of course, one could work with several special words at once. In most of the pictures in this chapter the generators are nearly parabolic and the generators a, A, b and B were all designated as special words. On p. 260, we shall see how the routine works when aaB is parabolic, so it and its cyclic permutations become the special words.

Fractal dimension and quasicircles

[1]There are limit sets which can be drawn by our curve-tracing algorithm whose fractal dimension is actually 2. We shall meet some of these in Chapter 10.

[2]A similar problem was actually investigated by the British scientist Lewis Fry Richardson, in the course of investigating his hypothesis that the number of wars between two countries might depend on the length of their common frontier. When he came to look into this, he found that the recorded lengths of frontiers varied dramatically depending on who was making the measurements. His obscure 1951 paper *The problem of contiguity*, Appendix to *Statistics of Deadly Quarrels*, (reprinted in *The Collected Papers of Lewis Fry Richardson* Cambridge University Press, 1993) was rediscovered by Mandelbrot, thus becoming seminal for the whole modern subject of fractals.

The fractal or Hausdorff dimension which we met in Chapter 5 links very nicely to our algorithm for stepping around the limit set. Remember that the Hausdorff dimension of an object is a numerical measure of its complexity. It interpolates our usual idea of dimension (zero for a point, one for a line, two for a surface and so on) so as to give an idea of how convoluted and crinkled or fragmented the object is at fine scales. The wild quasicircles we have been drawing will have fractal dimension somewhere between one (corresponding to a Fuchsian group whose limit set is a line or a circle) and two (corresponding to the whole plane).[1]

To understand how fractal dimension relates to our algorithm, let's begin by thinking about the way you measure the length of a reasonably 'normal' curve. Move along the curve, marking points at roughly distance ϵ apart. Count the total number of points, say $N(\epsilon)$, and multiply by ϵ. All being well, the result $N(\epsilon) \times \epsilon$ should give a reasonable approximation to the curve's length L. You would expect the approximation to improve the smaller you chose the step size ϵ. The line segment plot which approximates the curve improves and $N(\epsilon)$ should roughly equal L/ϵ. This method of calculating the length of a curve is a foundational principle of integral calculus. If the curve in question is a circle, you can do the exact calculation as explained in Note 8.4.

Now let's try applying this idea to a real life example. Suppose you want to measure the length of the Maine coastline.[2] You survey the scene by placing markers at equally spaced intervals round the coast, say 10 miles apart. To get the length of coast you multiply the number of posts by 10. To get a more accurate value, place marker posts every mile and

---

Note 8.4: **The circumference of a circle**

In the case of a circle, one can exactly calculate the number of segments needed to go round a circle when the step size is ϵ. Of course, the segments will probably not fit exactly round the circle, so the final one will have to be a bit too short or too long. In the long run, because we may take ϵ very small, discrepancies like this will not matter, so we may as well assume we always choose a step size which exactly fits.

If the radius of the circle is R, we can work back-

wards and find by some simple trigonometry that the length of the side of an N-sided polygon inscribed in the circle is $2R \sin \pi/N$. Thus if the step size is ϵ, then by setting $\epsilon = 2R \sin \pi/N$ we find that $N(\epsilon)$ is roughly $\pi/\arcsin(\epsilon/2R)$.

As N gets large, π/N gets small, so that using the approximation $\sin x \sim x$ for small x, the value of ϵ is roughly $2R\pi/N$. The circumference of the circle is (exactly in this very special case!) $2R\pi/N \times N = 2R\pi$.

repeat the process. At the next stage, assuming you have the interest and resources, you should place your posts 0.1 miles apart, multiply the number of posts by 0.1, and so on. According to the above method, the length of coastline should be well approximated by adding up the number of points and multiplying by epsilon, the distance between posts. The smaller epsilon, the story goes, the more accurate the answer will be. The trouble is, that this idea fails to take account of the small scale intricacies of the coast. As anyone knows who has tried it, walking from A to B round the Maine coast takes a lot longer than going from inlet to inlet by road. As you decrease your step size ϵ (or epsilon) from 10 miles to 1 mile to 0.1 miles and so on, then, to take account of the complicated ins and out of coastline at the scale you are measuring, the distance you will have to travel will dramatically increase. In fact, as we explain in Note 8.5, if the coastline has Hausdorff dimension d, then the number $N(\epsilon)$ of steps you have to take is proportional, not to $1/\epsilon$ as it would be for a 'smooth' curve, but to $(1/\epsilon)^d$. This means that if you plot $\log N(\epsilon)$ against $\log \epsilon$ you get a line of slope d. Richardson's paper mentioned above, contains plots just like this, where you can see for example that the highly indented British coast has Hausdorff dimension roughly 1.25, as opposed to the almost linear value 1.02 for South Africa.

Just the same idea applies if we want to find the Hausdorff dimension of

---

Note 8.5: **Approximating the fractal dimension**

According to the definition in Note 5.7, to find the fractal dimension of a set X, we have to cover the set by $N(\epsilon)$ disks of diameter ϵ and then find a value d such that $N(\epsilon) \sim C/\epsilon^d$ when ϵ is small. Of course, we don't know in advance the value of the constant of proportionality C. This doesn't matter, because taking logs of both sides, we find $\log N(\epsilon) = \log C - d \log \epsilon$. Since ϵ tends to 0, $\log \epsilon$ tends to $-\infty$, $\log(1/\epsilon) = -\log \epsilon$ tends to $+\infty$ and so

$$-\log N(\epsilon)/\log \epsilon \to d.$$

Here is a list of the number of line segments (i.e. disks!) plotted for the limit set of Figure 8.1 as a function of the step size ϵ. The entries in the last column give successive approximations to the fractal dimension.

ϵ	$N(\epsilon)$	$-\log N(\epsilon)/\log \epsilon$
0.1	297	2.47
0.01	10099	2.00
0.001	344699	1.85
0.0001	11766613	1.77

a quasicircle. Our movement round the limit set joining points if they are at distance ϵ apart is exactly like the surveyor measuring the Maine coast. As we have seen, a relatively coarse step size means we miss a lot of delicate detail and are in danger of cutting off many spiralling inlets as we go. As we reduce the step size ϵ, new necks which we had completely missed at a larger value of ϵ will suddenly appear. As we decrease ϵ the new protuberances will cause a significant increase in $N(\epsilon)$, raising the dimension d well above 1.

The result of a detailed investigation for the curve in Figure 8.1 is shown in the table in Note 8.5. Remember that we made our plots by moving along the limit set curve plotting certain limit points in order. The step size between points was governed by our cut-off level `epsilon` (alias ϵ), so we are stepping our way around the limit set by steps of at most ϵ. It is possible that the next step may be very much smaller than ϵ, in which case there is not much to be gained by plotting this particular segment. In our implementation, we insist that the next line segment be of length between $\epsilon/2$ and ϵ, just to make them all roughly the same size.[1] The table shows the number of steps $N(\epsilon)$ plotted for varying values of ϵ. Looking for the moment only in the first two columns, you can see that $N(\epsilon)$ is increasing much faster than the L/ϵ you might expect. Following the calculation in the note, the Hausdorff dimension should be roughly approximated by calculating $\log N(\epsilon)/\log\frac{1}{\epsilon}$ which we have listed in the last column. You can see that the result does indeed seem to be approaching some value between 1 and 2.

In point of fact, this method is not a good one for actually measuring Hausdorff dimension, because it takes extraordinarily detailed calculations to home in on an accurate value for d. What this count does do, however, is give us a good feel for the complexity of the plot we are trying to make. Of course, the program will run much, much longer with very small ϵ. Since the running time (measured by the number of operations needed) is roughly proportional to the number of nodes that the depth-first search encounters, it also grows something like $N(\epsilon) \sim (1/\epsilon)^d$, and thus is intimately related to the Hausdorff dimension. The more convoluted the curve, the larger d and correspondingly the longer the running time necessary to make a good plot.

If a quasifuchsian group is actually Fuchsian, then the limit set is a circle. You might well ask whether some relatively 'tame' quasicircles (say like the top right frame in Figure 8.2) might still have Hausdorff dimension 1. In fact it is a result of Rufus Bowen[2], that if a quasicircle limit set is not an actual circle (or line) then it has non-integral Hausdorff dimension strictly between 1 and 2. In other words, if a quasicircle is not a circle it is a real fractal!

[1] Whether or not you make restrictions like this is one of the subtleties in the various definitions of fractal dimension. Strictly speaking, to get the Hausdorff dimension, we should allow line segments of *any* length less than ϵ.

[2] The American mathematician Rufus Bowen (1947–1978) was appointed to Berkeley in 1970, but died suddenly and tragically at the age of only 31. Already famous for his definitive work on chaos theory, shortly before his death he was pioneering applying his ideas to Kleinian groups. Had he lived, he would have undoubtedly have made many contributions beyond the one mentioned here.

Box 22: Jørgensen's recipe

(1) Choose complex values for t_a and t_b.

(2) Choose one of the solutions x of the quadratic equation

$$x^2 - t_a t_b\, x + t_a^2 + t_b^2 = 0$$

and set $t_{ab} = x$.

(3) Compute the generator matrices:

$$a = \begin{pmatrix} t_a - \dfrac{t_b}{t_{ab}} & \dfrac{t_a}{t_{ab}^2} \\[2ex] t_a & \dfrac{t_b}{t_{ab}} \end{pmatrix}, \qquad b = \begin{pmatrix} t_b - \dfrac{t_a}{t_{ab}} & -\dfrac{t_b}{t_{ab}^2} \\[2ex] -t_b & \dfrac{t_a}{t_{ab}} \end{pmatrix}.$$

Even though the numerical work exhibits fractal behaviour, Hausdorff dimension is difficult to work with theoretically. Recent advances, made in particular by Peter Jones, Chris Bishop and Curt McMullen, have shown that for groups of Möbius maps various different methods of defining and computing dimension all give the same answer.

Some alternative recipes

It is of course not necessary to confine programs for exploring parabolic commutator groups to Grandma's recipe. Other normalizations also have their advantages. In this section, we describe a few alternatives of this kind.

Jørgensen's recipe. Many of the pioneering explorations of the groups we have been discussing were made by the Danish mathematician Troels Jørgensen.[1] He worked with a different normalization from Grandma's which has several desirable attributes. He begins by conjugating so that the parabolic commutator $abAB$ is the horizontal translation $z \mapsto z+2$. He then produces matrices a and b with prespecified traces t_a and t_b and such that $abAB = \begin{pmatrix} -1 & -2 \\ 0 & -1 \end{pmatrix}$. To do this is a challenging exercise; we simply report the answer in Box 22.[2] To gain confidence in the recipe, all we have to do is compute $\det a = \det b = 1$ and then multiply out to verify we get the correct answer for $abAB$. The interested reader will probably need to sacrifice a good ten sheets of paper to this cause. In the meantime, the rest of us may play!

There is a slight problem. The fixed point of $abAB$ is infinity, always

[1] Jørgensen's work contains many seminal discoveries. He kindly made his article *On pairs of once-punctured tori* available to the authors in the early days of their project. We were impressed that his drafting tools rivalled those of an architect or engineer. Despite its wide influence, this work has never been published.

[2] In writing down Jørgensen's recipe, we have only specified two normalization conditions, not three. The third condition comes from arranging that the bottom righthand entry in the matrix ab is 0.

a difficult value for computation. Difficulty turns to advantage, however, when we realise that since *abAB* is translation $z \mapsto z + 2$, the limit set must be symmetrical under translation by 2. In other words, the limit set repeats every two units horizontally. We only have to compute a section of width 2 and repetitions will give a full picture. The same periodicity can actually be seen hidden in our previous pictures in which the fixed point of *abAB* is always 1. Looking carefully at Figures 8.1 and 8.13, for example, you can see a periodically repeating pattern of spirals which would stretch out into the distance if the point 1 were transported to ∞.

To obtain a basic 'period' of the limit set, start at any limit point, for example the attractive fixed point z_{aBAb} of *aBAb*, corresponding to the infinite word $\overline{aBAb}$. Since *abAB* is translation by 2, applying *abAB* to any limit point z yields the new limit point $z + 2$. Therefore the part of the limit set between $\overline{aBAb}$ and $abAB(\overline{aBAb})$ must be just the kind of section we need.

The picture in Figure 8.19 shows just this strip. We used the same values of the traces as in Figure 8.1. The blue and yellow regions are the 'inside' and 'outside' of the ordinary set. Because the limit set goes through the point ∞, you may prefer to think of them as being 'above' and 'below'. It is rather remarkable that the limit set is symmetric not only under translation by 2, but also by 1. The explanation is connected to Grandma's special symmetry discussed in Note 8.1, and can be explored in Project 8.9.

Figure 8.19. The limit set of the group with $t_a = 1.87 + 0.1i$, $t_b = 1.87 - 0.1i$, the same traces as in Figure 8.1. Try comparing the two pictures to see how to get from one to the other with a Möbius transformation. In this group, *abAB* is translation by 2. The frame of this picture is $-2.1 \leq \text{Re}\, z \leq 2.1$. Only the stretch of limit points running from $\overline{aBAb}$ to $abAB\overline{aBAb}$ was plotted, using the modification of the DFS algorithm described on p. 186.

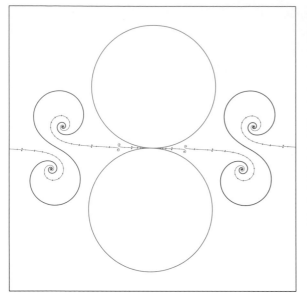 

Riley's recipe. One comparatively simple one-parameter family of groups has received a lot of attention: those generated by two *parabolic* matrices a and b. The free parameter is just the third trace $\text{Tr}\, ab$. You can make simple examples of **Riley groups** by taking two disjoint pairs of tangent Schottky circles.[1] One set of tangent circles is paired by the parabolic a, and the other by the parabolic b. Limit sets of these groups are not connected (because not all of the circles touch), but nor are they are as completely fragmented as Schottky group limit sets (because not all of the circles are disjoint). Groups like this are often associated with Robert Riley, whose systematic explorations led in the late 1970's to some remarkable computer pictures of this family of discrete groups.[2] It is not hard to prove by the methods of Chapter 7 that *any* group both of whose generators are parabolic with different fixed points is conjugate to one whose generators have the simple and special form $a = \begin{pmatrix} 1 & 0 \\ c & 1 \end{pmatrix}$ and $b = \begin{pmatrix} 1 & 2 \\ 0 & 1 \end{pmatrix}$. If $c = 2$ we get the modular group whose limit set, as we have seen, is just the extended real line.

As usual, the real fun begins when you let go of the Schottky circles and let the parameter c become complex. In the left frame of Figure 8.20 we have moved c gingerly away from 2 and plotted the limit set when $c = 1.91 + 0.05i$. Like badly laid linoleum, the limit set has peeled up and you can clearly see tiny spirals forming amidst the fractal mist. This should not come as a complete surprise: spirals appear when traces go complex. Although this picture is not particularly interesting, we can rather clearly see the manifestation of fractal effects. If you look closely you can see

Figure 8.20. Left: Hire a new floorman! The limit set of a Riley group with $c = 1.91 + 0.05i$ is near the triply-punctured sphere group whose limit set is the whole real line. For the Riley group the 'floor' starts peeling in spirals. The flow lines of b, Ba and aB are coloured green, red, and blue, respectively. Right: A Riley limit set with $c = 0.05 + 0.93i$. Notice the fractal structure indicated by repeating spirals at all levels, framing circular shaped holes which are starting to appear.

[1] One way to do this is to set $s = 0$ in the special Schottky family discussed in Project 7.8.

[2] Sadly on the day we write, March 7th 2000, we heard news of Riley's premature death.

small spirals repeated everywhere at ever-decreasing scales. In fact there is one pair of spirals for the fixed points of each conjugate of the element ab. Although the limit set is not a connected curve, neither is it fractal dust. It has a rather delicate structure in which some but not all gaps in the limit set have been sealed.

Varying the parameter c more boldly the pictures get more exotic. In the right frame of Figure 8.20 we took $c = 0.05 + 0.93i$ and got a limit set looking like a series of capital I's with the fanciest serifs ever traced by a medieval scribe. If we take $c = i$, we find $\mathrm{Tr}\, ab = 2 + 2i$. Sound familiar? The three traces are 2, 2, $2 + 2i$. Disguised by this new normalization, we have arrived at the Apollonian gasket, but by a completely different route! Not content with being on the edge of quasifuchsian groups, the gasket must also be on the edge of Riley groups, all the gaps which could cause fragmentation in its limit set having been sealed up, albeit in a completely different way.

In the normalization given, the transformation b is just horizontal translation by 2. Just as with the Jørgensen normalization, we may as well only draw one 'period' of the limit set. The whole picture is obtained by repeatedly translating this section by 2.

Maskit's recipe. Another nice family of groups are the **Maskit groups**, which are always singly- and sometimes doubly-cusped. In these groups, the commutator is parabolic, and so is one of the two generators, which we choose to be b. This reduces the free parameters down to one trace t_a. The formulas simplify if we replace t_a by another variable μ related to t_a via the formula $t_a = -i\mu$.[1] Bernard Maskit used groups like this to implement results in what is called **Teichmüller theory**, which we shall discuss briefly in Chapter 12. We shall be having a lot more to do with Maskit's groups in the next chapter.

You get a nice normalization for these groups by putting the fixed point of b at ∞ to bring out its translation symmetry. Written in terms of the new variable, the generators then look very easy:

$$b(z) = z + 2 \quad \text{and} \quad a(z) = \mu + 1/z.$$

You can check that $ABab$ is parabolic by direct calculation. In fact this particular commutator doesn't depend on μ, and has fixed point at -1.

Because a Maskit group is always a cusp group, there are always lots of circles in the limit set. Figure 8.21 explains why, if $\mathrm{Im}\,\mu > 2$, the group can be made by pairing Schottky circles.

Sometimes we refer to *any* group in which b and the commutator are parabolic as a Maskit group. Any group like this is conjugate to one in the special normalization here.

[1] For readers unacquainted with the Greek alphabet, μ is the Greek letter for 'm' (for Maskit?), pronounced 'mew'.

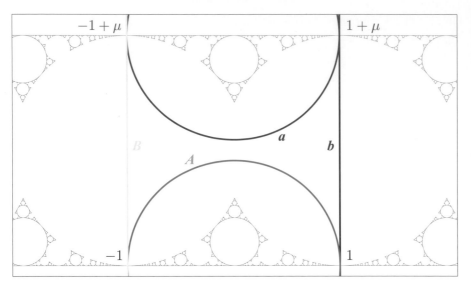

Figure 8.21. The configuration for a Maskit group when μ is on the positive imaginary axis. The four coloured tangent 'circles' are Schottky circles: they don't intersect provided $\operatorname{Im}\mu > 2$. Obviously b pairs the vertical lines through ± 1. Try checking that a pairs the lower and upper circles C_A and C_a. The region inside these circles is a basic tile for the 'inner' half of the ordinary set. When $\operatorname{Im}\mu = 2$ the circles C_A and C_a are tangent, giving us a double-cusp group at $\mu = 2i$. You can check that at this value a is parabolic, in fact
$$a = \begin{pmatrix} 2 & -i \\ -i & 0 \end{pmatrix}.$$

Grandma's four-alarm special. In the recipe used thus far for two-generator groups, we restricted the three natural complex parameters t_a, t_b and t_{ab} to two by insisting that the trace of the commutator be -2. We did this to ensure a reasonable chance that the limit set would be a connected curve. In Box 23 you can see Grandma's prize winning special recipe for arbitrary two-generator groups. In the revised step 2 she let go of the requirement that the three traces satisfy the Markov identity, so the third parameter t_{ab} can also be varied at will.

As you can check, referring back to p. 189 in Chapter 6, the polynomial function of t_a, t_b and t_{ab} in step 2 is just the trace of the commutator $C = abAB$. Once we let go of the condition that $\operatorname{Tr} abAB = -2$, there is no reason to expect that the limit set will be anything more than a swirling cloud of dust. This is the recipe to be used to make a picture of a general Schottky group as in our Road Map on the final page.

The special word algorithm revisited

To conclude this chapter, we want to run through the special word algorithm one more time, adapted not to the gasket as we did in Chapter 7, but to the slightly more complicated situation in which the word aaB is parabolic. If we had plotted Figure 8.16 without proper use of this algorithm, there would have been ugly gaps between the parabolic fixed points just like the ones in Figure 7.14. To close the gaps between the pinch points this time, we need to inform the program that the word aaB is parabolic, and plot not only points with repetends which are cyclic

---

Box 23: **Grandma's special four-alarm two-generator groups**

(1) Choose three complex numbers t_a, t_b and t_{ab}.

(2) Calculate

$$t_C = t_a^2 + t_b^2 + t_{ab}^2 - t_a t_b t_{ab} - 2.$$

(3) Calculate $Q = \sqrt{2 - t_C}$. Choose either square root; it won't terribly matter which.

(4) With $i = \sqrt{-1}$ as usual, determine which sign $\pm$ makes the following a true inequality:

$$|t_C \pm i Q \sqrt{t_C + 2}| \geq 2.$$

If both choices work (in which case $\geq$ will be replaced by $=$), pick one at random. Define $R = \pm\sqrt{t_C + 2}$ with that same sign.

(5) Compute the complex number

$$z_0 = \frac{(t_{ab} - 2)(t_b + R)}{t_b t_{ab} - 2t_a + i Q t_{ab}}.$$

(6) Compute the generator matrices:

$$a = \begin{pmatrix} \dfrac{t_a}{2} & \dfrac{t_a t_{ab} - 2t_b + 2i Q}{(2t_{ab} + 4)z_0} \\[2ex] \dfrac{(t_a t_{ab} - 2t_b - 2i Q) z_0}{2t_{ab} - 4} & \dfrac{t_a}{2} \end{pmatrix},$$

$$b = \begin{pmatrix} \dfrac{t_b - i Q}{2} & \dfrac{t_b t_{ab} - 2t_a - i Q t_{ab}}{(2t_{ab} + 4)z_0} \\[2ex] \dfrac{(t_b t_{ab} - 2t_a + i Q t_{ab}) z_0}{2t_{ab} - 4} & \dfrac{t_b + i Q}{2} \end{pmatrix}.$$

---

permutations of the basic commutator *abAB*, but also those whose repetends are any of the cyclic permutations of *aaB*. This has the effect that we plot the images of all parabolic fixed points of any element in the group.

For the group shown in Figure 8.16, neither of the generators, with traces $t_a = 2 - i$ and $t_b = 3$, are particularly close to parabolic; the element that *is* parabolic is *aaB*. That means that in our list of special words, we should dispense with *a* and *b* but add instead *aaB*. Since this has three letters, we have to permute it cyclically and consider *aBa* and *Baa* as well. (This is because the infinite pattern $\cdots aaB \cdot aaB \cdot aaB \cdots$, at the end of a word can also be grouped as $\cdots a \cdot aBa \cdot aBa \cdot aB \cdots$ or

$\cdots aa \cdot Baa \cdot Baa \cdot B \cdots$.) In Note 8.6 we have organised all the possible repetends in a chart just as we did for the Apollonian gasket on p. 218.

Figure 8.22 shows the plotting algorithm at work making a slow motion pass through part of the limit set of the group in Figure 8.16. We have 'ghosted' the actual limit set light grey underneath. As you can see, we have traced a segment of the limit set in which the words have prefix *aaBa*. The same sort of 'budlike' shape appears throughout. The 'roots' of the buds are points where the limit curve bends on itself though a very sharp (actually 180°) angle meeting another section of limit set bent in the opposite way. These 'roots' are precisely the points where the infinite word ends with the repetend *aaB*, its inverse *bAA*, or one of the cyclic permutations of the same. Notice how the limit curve passes twice through limit points with repetends *aaB*, *aBa*, *Baa* or their inverses.

Let's consider the limit points numbered 3, 4, and 5 in the figure, which belong to the segment of limit set of words with prefix *aaBaab*. Look for the actual section of true limit set in grey underneath. The point 5 is at the end of a black segment which sticks out into a very small grey circle, after which our coarse plot doubles back on itself, so the segment from 4 to 5 is actually the *same* as the segment from 5 to 6.

The prefix *aaBaab* ends in *b*, so referring to our chart, we have to consider, in order, the three points corresponding to this prefix with the

---

Note 8.6: **Repetends for the special word** *aaB*.

Here are the possible repetends for the special word *aaB*. Notice that the two ending tags *a* and *A* have two extra repetends to account for, as opposed to only one for *b* and *B*. Therefore, we need to slightly modify our program to keep track of the number numfp[i] of endings for prefixes ending gens[i]. In this case:

$$\text{numfp}[1] = \text{numfp}[3] = 4$$
$$\text{numfp}[2] = \text{numfp}[4] = 3.$$

gens[1] = a $\Rightarrow$
- repet[1,1] = $bABa$
- repet[1,2] = aBa
- repet[1,3] = Baa
- repet[1,4] = $BAba$

gens[2] = b $\Rightarrow$
- repet[2,1] = $ABab$
- repet[2,2] = AAb
- repet[2,3] = $aBAb$

gens[3] = A $\Rightarrow$
- repet[3,1] = $BabA$
- repet[3,2] = AbA
- repet[3,3] = bAA
- repet[3,4] = $baBA$

gens[4] = B $\Rightarrow$
- repet[4,1] = $abAB$
- repet[4,2] = aaB
- repet[4,3] = $AbaB$

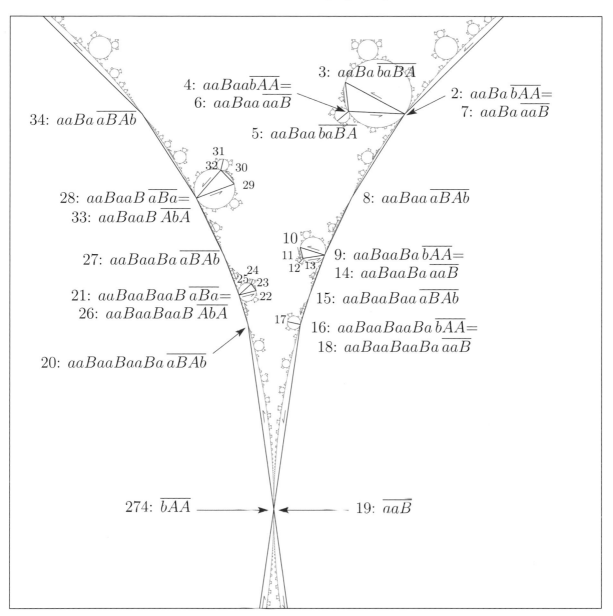

Figure 8.22. Slow motion cusp. Part of Figure 8.16 plotted with a large value of the cut-off level epsilon. The actual limit set is ghosted behind our coarse slow motion plot in grey. The frame shows the segment of limit set corresponding to words with prefix $aaBa$. The initial point 1 and the final point 35 are slightly off screen.

repetends $ABab$, AAb and $aBAb$. This gives us

$$3 : aaBaab\,\overline{ABab} \qquad 4 : aaBaab\,\overline{AAb} \qquad 5 : aaBaab\,\overline{aBAb}.$$

This is not how 3 seems to be labelled in the figure; however remember that since the commutator is parabolic, $ABab$ and its inverse $BAba$ have the same fixed point, so that $aaBaab\,\overline{ABab} = aaBaab\,\overline{BAba} = aaBa\,\overline{baBA}$. This is how the label in the figure actually appears.

The algorithm considers the three points 3, 4 and 5 and decides to terminate because the distances from 3 to 4, and from 4 to 5, are both less than the cut-off value epsilon. In this case the 'outlier' 4 is actually not stretched very far away from 3 and 5; however the limit set has a sharp bend at 4 and if you look closely you can see how 4 is the point at which the segment of limit set from 3 to 5 is pulled out to meet the next segment of limit set which runs from 5 to 8.

Figure 8.23. Raising a crop of spirals. We cultivated all these spirals by arranging that $t_{aaB} = 1.91 + 0.05i$. The input traces are $t_a = 1.965557377 - 0.9953573493i$ (approximately) and $t_b = 3$. Notice the circular structures in the outside part of the ordinary set,

This next section of the limit set corresponds to words with prefix $aaBaaa$. This ends in a, so referring to the chart, there are four words to consider along this segment. Here they are; see how we have had to rearrange the way the words are written by cycling the repetend in order to make them look the same as they appeared in our automatic print-out in Figure 8.22:

$$5 : aaBaaa\,\overline{bABa} = aaBaa\,\overline{baBA} \qquad 6 : aaBaaa\,\overline{aBa} = aaBaa\,\overline{aaB}$$

$$7 : aaBaaa\,\overline{Baa} = aaBa\,\overline{aaB} \qquad 8 : aaBaaa\,\overline{BAba} = aaBaa\,\overline{aBAb}.$$

Now the algorithm checks the three distances from 5 to 6, 6 to 7, 7 to 8 and terminates because they are all sufficiently small. On this section of limit curve, it becomes clearer that plotting the outlier 7 is a good move because it captures the sharp angle in the limit set where 7 gets pulled towards 2 in a previous section of the limit set. As you can see, 2 and 7 both end in the repetend *bAA* which corresponds to our new parabolic special word *aaB*. In a similar way, every stretch of limit set we examine between two consecutive infinite words ending in commutators will correspond to one of the four lists in our chart above.

As we have already seen, the special words algorithm is also useful for plotting limit sets when the special word is not precisely parabolic, but 'nearly so'. For the especially 'spirally' limit set in Figure 8.12 on p. 243, we used traces $1.91 + 0.05i$. Perhaps, if we arrange that $t_{aaB} = 1.91 + 0.05i$, something similar will happen here. We called in a specialist in messy formulas and algebraic calculations, and informed her that we need the value of t_a that produces $t_{aaB} = 1.91 + 0.05i$, given that $t_b = 3$ and $t_{abAB} = -2$. Explaining to her that it's just another equation to solve (technical experts will grouse at everything, so we must be firm), the final answer turned out to be approximately $t_a = 1.965557377 - 0.9953573493i$. With this trace as input, our program produces Figure 8.23. The overall arrangement of the limit set is roughly the same as usual, but the hint of loxodromy causes spirals to bloom. The use of the special words algorithm with *aaB* singled out as nearly parabolic gives a very accurate picture. Unsurprisingly, the spirals bear a close resemblance to their cousins in Figure 8.12.

Projects

8.1: Grandma's recipe and the Apollonian gasket
To get the hang of Grandma's recipe, try using it to recover the formulas

$$a = \begin{pmatrix} 1 & 0 \\ -2i & 1 \end{pmatrix} \quad \text{and} \quad b = \begin{pmatrix} 1-i & 1 \\ 1 & 1+i \end{pmatrix}$$

for the generators of the Apollonian gasket from p. 201. The generators a and b are both parabolic with $\operatorname{Tr} a = \operatorname{Tr} b = 2$, while $\operatorname{Tr} ab = 2 - 2i$, which as you can check is one of the two solutions to the quadratic equation $x^2 - 4x + 8 = 0$. (The other solution is $2 + 2i$, which as you may like to check is the trace of aB.) Start by calculating z_0 (it has a simple value). With a little patience, you should now be able to verify that Grandma has got it right!

8.2: Grandma's recipe and the $\pi/4$ group
Another group we already met which fits Grandma's recipe is the $\pi/4$-Schottky group (see Project 4.2) with generators

$$a = \begin{pmatrix} \sqrt{2} & i \\ -i & \sqrt{2} \end{pmatrix} \quad \text{and} \quad b = \begin{pmatrix} \sqrt{2}-i & \sqrt{2} \\ \sqrt{2} & \sqrt{2}+i \end{pmatrix}$$

belonging to frame (vi) of Figure 6.10. In this case, $t_a = t_b = 2\sqrt{2}$. Try working through the calculations. First check that the equation $x^2 - t_a t_b x + t_a^2 + t_b^2 = 0$ has just one solution giving $t_{ab} = 4$. Then check $z_0 = (1 - \sqrt{2}i)/3$. Now calculate the entries in a and b.

8.3: Schottky circles and real trace

Study the Markov equation when $\text{Tr}\, a = \text{Tr}\, b = t$ for some real number t. Show that the trace of ab is also real exactly when $|t| \geq 2\sqrt{2}$. Use this to explain why the limit sets in the first three frames of Figure 8.2 are circles.

Now fix a real value $s > 2$ for $\text{Tr}\, a$, and let $\text{Tr}\, b = t$ vary. Find a formula for the moment at which the group generated by a and b changes from being Fuchsian to quasifuchsian.

8.4: Traces determine groups up to conjugacy

In order to show that the three traces t_a, t_b and t_{ab} determine the generators a and b up to conjugacy, we use the fact proved in Project 6.1 that a and b share a common fixed point if and only if $t_{abAB} = 2$. Now let a, b and c, d be two pairs of Möbius transformations such that $t_a = t_c$, $t_b = t_d$ and $t_{ab} = t_{cd}$. Prove that, if one of t_{abAB} or t_{cdCD} is not 2, then there is a Möbius transformation e such that $eaE = c$ and $ebE = d$.

8.5: The circular outer boundary

You may notice that the unit circle fits perfectly round the limit set in Figure 8.4. If the trace t_b is real, check that both b and

$$ abAB = \begin{pmatrix} -1 + t_b i & -t_b i \\ t_b i & -1 - t_b i \end{pmatrix} $$

belong to the unit circle group (see p. 88). That means that every transformation in the subgroup generated by b and abA fixes the unit circle, so that its limit set (which is contained in the limit set of the group generated by a and b) will be a sizeable part of the unit circle (all of it, in case $t_b = 2$).

Now give an alternative explanation of this phenomenon using the result of Project 6.6. If you look closely at Figure 8.16, you may spot some 'ghost circles' lurking round the outer boundary of this limit set as well.

8.6: A second cusp group

Go through the process in Note 8.3 again but using the word $aaaB$. You may be able to verify computationally that there is a cusp group for which $\text{Tr}\, aaaB = 2$ at the value $t_a = 1.823932 - 0.595342i$ and $t_b = 3$. The central part of the limit set now gets pinched into four tangent disks.

8.7: Swapping traces conjugates groups

A rather surprising feature of Figure 8.14 is that the two frames actually show conjugate groups! The relationship between the two pictures is that we have interchanged the *traces* of a and b. Obviously, a group generated by two specific matrices S and T is the same as that generated by T and S. In the present case, we have interchanged only the traces, and the reasoning is a bit subtler. The traces of the two generators describe a parabolic commutator group *up to conjugation*. Having specified the traces, Grandma's recipe tells us which specific normalization, that is which conjugate, to pick. Thus if we interchange the traces of the two generators, the two groups we get must be conjugate. More specifically, if the generators of the first group G_1 are a_1 and b_1, and of the second G_2 are a_2 and b_2, then there must be a conjugating map h such that $ha_1h^{-1} = b_2$ and $hb_1h^{-1} = a_2$. Can you see why this tells us that $hG_1h^{-1} = G_2$?

You can think of the righthand picture in Figure 8.14 as the same as the lefthand one turned inside out. Try using what you know about Grandma's normalization to actually calculate the conjugating map h.

8.8: Fixed point of $aBAb$

Grandma organised her recipe so that the fixed point of $abAB$ is at 1 while the fixed point of $aBAb$ is the symmetric point -1. It is interesting to see where the limit points for certain infinite words occur using Jørgensen's recipe. Of course, $\overline{abAB} = \overline{baBA}$ is the point at infinity. Also, for any infinite word w with limit point $\overline{w}$, we have $abAB(\overline{w}) = \overline{w} + 2$. In Jørgensen's recipe, the fixed point of $aBAb$ also doesn't depend on the traces. Where is it?

8.9: The symmetry of order two

Any parabolic commutator group has a special symmetry which comes by interchanging a and A and b with B. It turns out that there is always a Möbius map r such that $r^2 = 1$ (so that $r = r^{-1}$) and such that $rar = A$ and $rbr = B$. Many of Jørgensen's calculations are based on his observation that this symmetry can always be computed from the matrix $r = ab - ba$. This is how it works:

- Compute the determinant of the matrix $ab - ba$ for arbitrary $a = \begin{pmatrix} a_1 & a_2 \\ a_3 & a_4 \end{pmatrix}$ and $b = \begin{pmatrix} b_1 & b_2 \\ b_3 & b_4 \end{pmatrix}$ of determinant 1. Express your answer in terms of t_a, t_b and t_{ab}. You should see something familiar.
- Calculate the trace of $ab - ba$.
- Check that if r is any Möbius transformation with trace 0 then r^2 is the identity transformation I.
- If r is the transformation obtained by 'normalizing' the determinant of $ab - ba$ to be 1, show that $r^2 = 1$ and then check that $rar = A$ and $rbr = B$.
- Grandma's recipe is symmetric with respect to the rotation $r(z) = -z$. What is the analogue of r for Jørgensen's recipe?
- The limit sets for Jørgensen's groups are symmetric under horizontal translation by 2. Actually they are also symmetric under horizontal translation by 1. Explain.

Grandma's recipe is designed so that the symmetry $r = a(ab) - (ab)a$ turns out to be $r/(z) = -z$. Hence, $rar = A$ and $r(ab)r = BA$. That explains why the map r maps all limit points beginning with a to all the limit points beginning with b, A or B, and vice versa.

CHAPTER 9

Accidents will happen

He [Al Gore] was captivated by the metaphorical power of a phenomenon scientists have called the 'edge of chaos'.

John F. Harris, Washington Post

Our progression through the book has been the investigation of more and more remarkable ways in which two Möbius maps a and b can dance together. Figure 9.1 shows another level of complexity, an array of interlocking spirals which literally took our breath away when we first drew it. It results from creating a double cusp group in which the generator b and the word $a^{15}B$ are both parabolic. Surely it cannot be coincidence that there are exactly 16 coloured circles forming a chain across the centre of the picture? Let's pick apart the dynamics of a and b, using the diagrammatic version Figure 9.2 for notation. In particular, let's try to see from the picture why $a^{15}B$ is parabolic and where its fixed point is located.

The action of b is quite easy. It is parabolic with fixed point at the bottom of the picture at $-i$. It pushes points out from its fixed point along clockwise circular trajectories. (You may like to compare with Figure 8.4 on p. 233 to help follow this.) One trajectory lies along the boundary of the outer unit circle framing the picture (note $b(-1) = +1$), and another is the boundary of the white circle tangent to the unit circle at $-i$. The large red disk on the left of the picture (number C_{15} in our schematic version in Figure 9.2) gets moved exactly half way round so it lands on the green disk C_0 on the right. It also pushes the smaller disk C_{16} around to land on the disk C_1.

The action of a is more interesting. It is loxodromic and it pushes the disks in the coloured spiral chain from left to right across the picture, gradually shrinking and spiralling into the sink on the right. This means that in our schematic picture, C_{16} goes to C_{15}, C_{15} goes to C_{14}, and so on, right down to C_1 which goes to C_0.

Now we can trace the effect of $a^{15}B$ on some of the disks. Focussing on

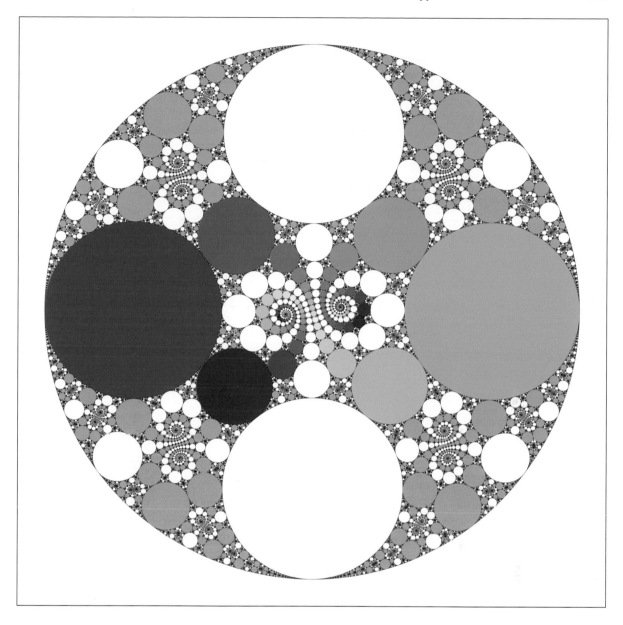

Figure 9.1. A perfect accident! For the group in this picture, both the generator b the word $a^{15}B$ are parabolic, making this a picture of a double cusp. See Figure 9.2 for labels. The parameter values are $t_a \doteq 1.95859 - 0.01128i, t_b = 2$.

C_0, the transformation B carries it to C_{15} and then a^{15} carries it back to C_0. In a similar way, B carries C_1 to C_{16} and then a^{15} carries it back to C_1. This means that $a^{15}B$ maps both the disks C_0 and C_1 to themselves! In consequence, the fixed points of $a^{15}B$ must be on both these circles. Since the two circles are tangent, they have only one point in common. Thus $a^{15}B$ can have only one fixed point, so it must be parabolic. Its fixed point is marked by a red dot in Figure 9.2.

How did we find an accident like this? Surely, you couldn't just cause

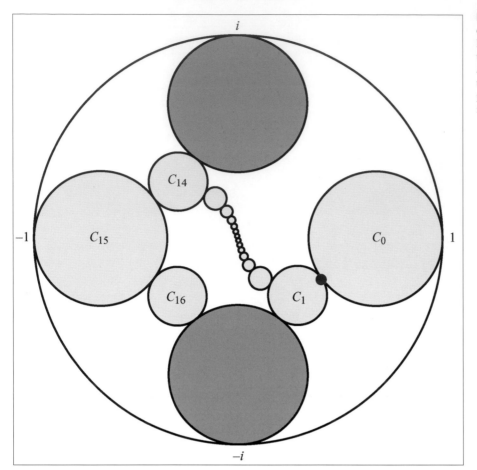

Figure 9.2. Picking out certain of the circles from Figure 9.1 helps us follow the action of the maps *a* and *b* as they push circles along the central chain. The fixed point of the parabolic $a^{15}B$ is marked by a red dot.

such a perfect accident by chance! Well, first of all, we chose the trace of *b* to be 2, because parabolic elements have a tendency to cause circles to emerge in the limit set. Since we have fixed the traces of *b* and the commutator *abAB*, that leaves just one free parameter t_a. We then craftily decided to choose t_a in such a way that the word $a^{15}B$ would also be parabolic. There is a nice way to see how to do this using the trace identities (see Note 9.1) because the trace of $a^{15}B$ is a polynomial in $t = t_a$:

$$t_{a^{15}B} = t^{15} + 2\,it^{14} - 15\,t^{13} - 26\,it^{12} + 90\,t^{11} + 132\,it^{10} - 275\,t^9$$
$$- 330\,it^8 + 450\,t^7 + 420\,it^6 - 378\,t^5 - 252\,it^4 + 140\,t^3$$
$$+ 56\,it^2 - 15\,t - 2\,i.$$

Now all we have to do is solve the equation $t_{a^{15}B} = 2$ to find the correct value for t_a, and BANG! we have our perfect accident.

Actually, like all the best things in life, nothing is ever quite that simple.

An equation of degree 15 has 15 complex number solutions, which we should really multiply by 2 since we don't know whether to go for $t_{a^{15}B} = 2$ or $t_{a^{15}B} = -2$.[1] The trouble with most solutions to these equations is that you will get either a non-discrete or non-free group, and either of those situations will cause problems for our limit set algorithm. For just the right solution $t_a = 1.958591030 - 0.011278560i$, however, we discover the group responsible for the limit set in Figure 9.1.

French-kissing Schottky curves

Figure 9.3 is another picture of an accident. This time, the parabolic words are b and a^3Ba^2B. Why choose such a funny word? To explain, we have again picked out a rather special chain of numbered circles, which we call the **core chain**, whose progress we are going to trace under a and b, shown in a stripped down version in Figure 9.4. Note in Figure 9.3 that the core chain separates the parts of the limit set given by words beginning in a, b, A and B.

All these circles are wonderfully arranged so that the generators a and b move them in a tight pattern. The map a moves each circle back 2 relative to our numbering, so C_5 goes to C_3, C_3 goes to C_1 and so on. Since b is parabolic with fixed point $-i$, we can see from Figure 9.4 that the map B moves the circles forward by 5, so for example C_0 goes to C_5 and C_1 goes to C_6. (Fortunately we shall never need to answer the question of what happens when one of these moves takes us off the end.) Following

---

Note 9.1: Solving $t_{a^{15}B} = \pm 2$

According to our work on p. 189 in Chapter 6, the trace of any word can be expressed as a polynomial in t_a, t_b and t_{aB}. In our case, this means that $t_{a^{15}B}$ should be expressible as a polynomial in the one free parameter t_a.

To work this out explicitly, first we use the Markov identity to find t_{aB} in term of t_a. Since $t_b = 2$ we know

$$t_a^2 + 4 + t_{aB}^2 = 2t_at_{aB}.$$

Solving this as a quadratic for t_{aB} gives $t_{aB} = t_a \pm 2i$, and with the conventions of Grandma's normalization, the correct root to choose is $t_{aB} = t_a + 2i$.

Now we repeatedly apply the grandfather identity

from p. 192:

$$t_B = 2$$
$$t_{aB} = t_a + 2i$$
$$t_{a^2B} = t_at_{aB} - t_B = t_a^2 + 2it_a - 2$$
$$t_{a^3B} = t_at_{a^2B} - t_{aB} = t_a^3 + 2it_a^2 - 3t_a - 2i$$
$$t_{a^4B} = t_at_{a^3B} - t_{a^2B} = t_a^4 + 2it_a^3 - 4t_a^2 - 4it_a + 2$$

and so on. Notice how this calculation is recursive: at each new step, to find the value of t_{a^kB}, we will need to know the value of $t_{a^{k-1}B}$ which we found at the stage before. In this way, $t_{a^{15}B}$ can be expressed as a polynomial (of degree 15!) in t_a.

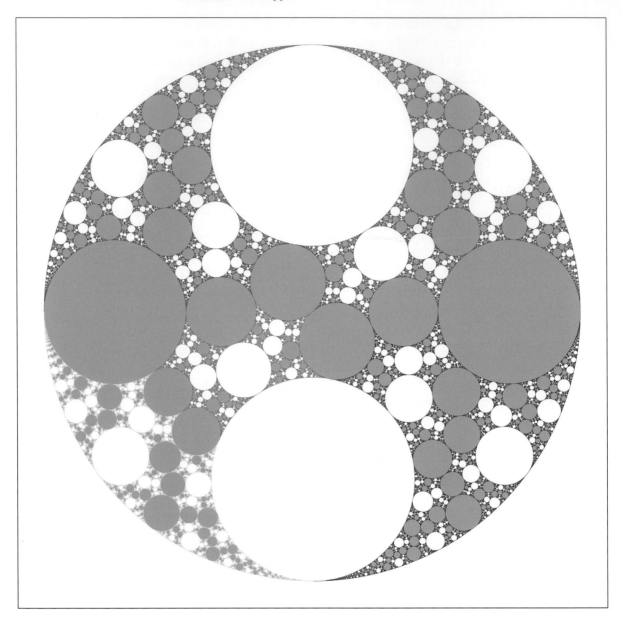

this action on circles, we are taken along the route

$$C_0 \xrightarrow{B} C_5 \xrightarrow{a} C_3 \xrightarrow{a} C_1 \xrightarrow{B} C_6 \xrightarrow{a} C_4 \xrightarrow{a} C_2 \xrightarrow{a} C_0$$

which means (remember that you have to read transformations backwards) that the circle C_0 is fixed by *aaaBaaB*. The rules force us to follow the same path starting from C_1:

$$C_1 \xrightarrow{B} C_6 \xrightarrow{a} C_4 \xrightarrow{a} C_2 \xrightarrow{B} C_7 \xrightarrow{a} C_5 \xrightarrow{a} C_3 \xrightarrow{a} C_1,$$

implying that *aaaBaaB* also fixes the circle C_1. Just as before, we argue

Figure 9.3. In this accident, the transformation a^3Ba^2B is parabolic. The value of t_a for which this happens is $1.64213876 - 0.76658841i$. The transformation b is also parabolic, so $t_b = 2$. We have coloured the limit points according to the first letter of their infinite word: red for a, blue for b, green for A and yellow for B.

Figure 9.4. A less cluttered view of the core chain of circles in the limit set for the a^3Ba^2B cusp group. There are also two grey flowlines showing the action of the transformation a.

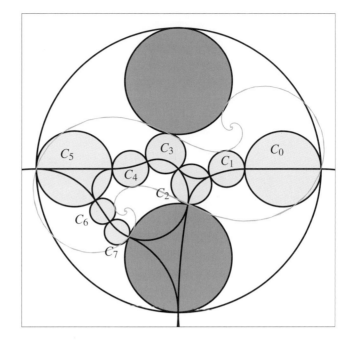

that *aaaBaaB* can have only one fixed point, the point of tangency between C_0 and C_1. In other words, the dynamics of the circle chain alone shows that *aaaBaaB* is parabolic as we claimed.

Notice that the word *aaaBaaB* contains 2 B's and 5 a's. Because of the pattern, we shall call this group the **2/5 cusp**. In fact you can work out this sequence of moves, and hence the word *aaaBaaB*, just starting from the two numbers 2 and 5. Perhaps we would get other interesting patterns and words if we replaced 2 and 5 by some other pair of integers p and q?

There is another remarkable arrangement to be seen in this picture. In Figure 9.5 we have superimposed four slightly wavy 'circles' on the limit set, their insides marked by the generators a, b, A and B. (The region marked a contains the point at infinity, which explains why it appears to have infinite extent. You should think of the roughly horizontal red line in the figure as enclosing the 'blob' above it, in just the same way as we have become used to thinking of the real axis as a circle bounding the upper half plane.) These four magic blobs are so organised that the generator a maps the 'outside' of the blob marked A to the 'inside' of the blob a, and similarly b maps the outside of blob B to the inside of blob b. This is analogous to the way the four Schottky disks were moved in Chapter 4, what we called the 'Schottky dynamics'. The Schottky curves which bound these regions are something special: unlike Schottky circles which can do no more than kiss, these curves not only kiss, they are 'smushed'

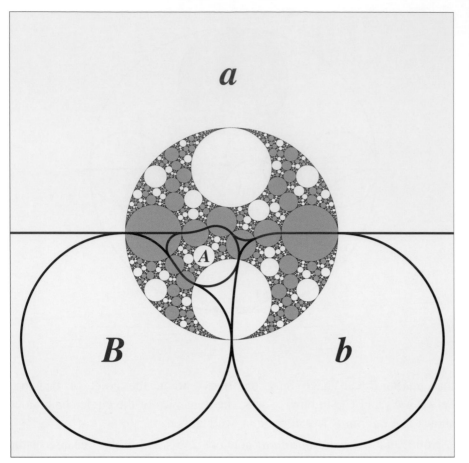

Figure 9.5. Schottky curves. Another view of the 2/5 cusp. The smushed loops are Schottky curves paired by the generators *a* and *b*. The yellow region is the remnant of the outside of the quasicircle for the quasifuchsian group and the pink is the remnant of the inside. Thus *a*, *b*, *A* and *B* map yellow circles to yellow circles and pink to pink.

against each other.[1] The blobs really look almost circular, getting smushed only where they touch.

The three 'triangular' shaped regions between the blobs are remnants of the original Schottky tile. (Actually there is a fourth triangular region on the lower outside region.) Topologically, this picture works just like the arrangement of kissing circles for the gasket in the right frame of Figure 7.4.

As soon as we started drawing pictures of limit sets with spirals like Figure 8.5, it became clear that it would surely be impossible to find four correctly paired tangent Schottky circles defining a basic tile. Here, then, is the solution! Once we let go of the requirement that we are looking for actual circles to pair, and allow more elaborate shapes instead, all is well. The core chain of numbered circles gave us the clue for finding these 'Schottky blobs'. As you can see most clearly in Figure 9.4, within each disk, the Schottky curve is just a circular arc joining the two tangency points where the disk meets its two neighbours in the chain. These arcs

[1] French kissing perhaps?

join up smoothly because we arranged them at right angles to the boundary of the disks which contain them, making a snaky curve wending its way through the core chain of disks.

It is interesting to look at the tangency points between neighbouring circles in the core chain. They are all fixed points of parabolic elements in the group. Notice that they are exactly the tangency points on the boundary between the different coloured regions in Figure 9.3, in which the 'anatomy' of the limit set is shown using the same colouring as in the figure on p. 233. In both pictures, the red points are those which can be represented by an infinite word beginning with *a*. But you will see a dramatic difference between the two pictures if you look at the contact points between the different coloured regions. In Figure 8.4, the limit set is a loop and the red part is a segment which meets the rest of the limit set only at its two endpoints, the parabolic fixed points $\overline{abAB} = \overline{baBA}$ (on the right side) and $\overline{aBAb} = \overline{BabA}$ (on the left). For the 2/5 cusp, there are 5 additional points of contact with outer parts of the limit set, at the other tangency points of the core circle chain. Written as infinite words, here are the contact points from left to right:

$$\overline{aBAba}, \quad \overline{aBaaBaaa}, \quad \overline{aBaaaBaa}, \quad \overline{aaBaaBaa},$$

$$\overline{aaBaaaBa}, \quad \overline{aaaBaaBa}, \quad \overline{abABa}.$$

Aha! Since we are at the 2/5 cusp, to get a good plot we should organise that the word *aaaBaaB* is special. We have just listed precisely the repetends which would have to be considered in the special word algorithm following prefixes ending in *a*. This is the reason why the special words algorithm (see p. 260) was formulated: to be sure we get all these points of contact between different parts of the limit set when we make our plot.

Accident-prone words

Locating cusp groups is very important in our probing, because they hover on the boundary between being quasifuchsian and non-discrete. We cross over into non-discrete groups at the exact moment when a word becomes accidentally parabolic and we have reached the cusp. Double rather than single cusp groups seem an especially good target to search for, because having two extra parabolics gives two equations between our trace parameters t_a and t_b. This means that, as in the examples we have just given, there is a good chance of solving to locate a small number of parameter values to try. Besides, as we have just seen, the limit sets of double cusp groups look as if they hold some fascinating patterns in store.

So what words in the generators might be suitable targets for creating accidents by being parabolic? Let's play about a bit and experiment with the pattern that gave us the word *aaaBaaB*, which as we saw was closely related to the numbers 2 and 5. We already hinted you might try the same thing for any pair of numbers p and q. Schematically it would go like this. Starting from 1 (short for the circle C_1), at every stage we are going to either add q or subtract p. Keep adding unless adding would get to a number larger than $p + q$. Then change to subtracting p, and continue unless you get to a number less than 1, in which case revert to adding q. Keep going and see how long it takes to get back to the start. For example, if $p = 3$ and $q = 5$ we would get

$$1 \xrightarrow{+5} 6 \xrightarrow{-3} 3 \xrightarrow{+5} 8 \xrightarrow{-3} 5 \xrightarrow{-3} 2 \xrightarrow{+5} 7 \xrightarrow{-3} 4 \xrightarrow{-3} 1.$$

Now for every +5 write a 'B', and for every -3 write an 'a'. Then the word we get, *read from right to left*, is *aaBaaBaB*. The reason that we read off the word backwards is because this is what we did when we were thinking about how generators acted on the circle C_1. If you prefer to read words from left to right, you can achieve this by changing the rule to subtracting q and adding p. Then we would get

$$1 \xrightarrow{+3} 4 \xrightarrow{+3} 7 \xrightarrow{-5} 2 \xrightarrow{+3} 5 \xrightarrow{+3} 8 \xrightarrow{-5} 3 \xrightarrow{+3} 6 \xrightarrow{-5} 1.$$

Check you will read off the same word, this time from left to right, by putting 'B' for -5 and 'a' for +3.

We can do exactly the same thing for any values of p and q. We call the word we get the **p/q word** and write it $w_{p/q}$.[1] You may like to amuse yourself reading some off – here are some of the smaller ones displayed in a way which may remind you of the 'Farey addition' of fractions we met in Chapter 7.

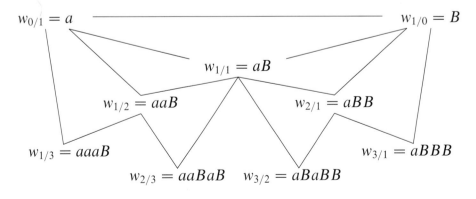

If you arrange your words on a chart like this, you will soon discover there is an easier way to get them than following round our cycle of adding p's

[1] Notice we are writing the numbers p and q as a fraction p/q. This is because there is not much to be gained by considering the case in which p and q have a common factor d, because the pattern you get will be exactly the same as if you cancelled and worked out the pattern for p/d and q/d instead.

and subtracting q's. For example:

$$w_{3/5} = aaBaaBaB = (aaB)(aaBaB) = w_{1/2}\, w_{2/3}$$

exactly mirroring the Farey addition formula

$$\frac{1}{2} \oplus \frac{2}{3} = \frac{1+2}{2+3} = \frac{3}{5}.$$

Let's play a bit more with these word patterns before we get to explaining why they are exactly the right kind of words to use to make cusps. Believe it or not, these peculiar sequences of a's and B's are familiar to every reader of this book! The fact is that these patterns appear in any piece of written text printed by a computer, which, nowadays, means almost anything in print at all. To a computer, a page to be printed is just a vast array of tiny squares each of which may be coloured entirely black or left entirely white. (For colour printing one needs several different arrays all of the same kind.) The tiny squares are known as 'pixels', short for 'picture elements'. Just as atoms are the fundamental building blocks of all matter, all pictures are made up of a collection of pixels. The process of realizing geometric objects as a set of darkened pixels is known as **scan conversion**. Although to Euclid a straight line was an ideal with no width at all, such ideals would be rather hard to see on a printed page. In the real world, if a computer is told to draw a sloping straight line, it must decide what combination of filled-in pixels will be the best fit.

So what has this got to do with p/q words? The left frame of Figure 9.6 is a picture of a line of slope 3/5 overlaid on a square grid. If the squares were in fact the minuscule pixels of a printed page, then a good procedure for the computer would be to colour black the pixels which the line meets and leave the rest white. If you write down a whenever the line crosses a vertical line, and B whenever it crosses a horizontal one, then, reading from left to right you get exactly the 3/5 pattern $aaBaaBaB$, where we stop because after this point the pattern repeats. If we assign a dynamical meaning to the symbols a and B, so that a means move one pixel to the right, while B is the instruction to move one pixel upward, then the word $aaBaaBaB$ may be interpreted as a program to draw the straight line sloping upward 3 units for each 5 units horizontally right. The bottom left square of the picture shows copies of all the different segments of the line as they fall into different yellow squares, transported back under various combinations of the moves A (that is, a^{-1}) and b (that is, B^{-1}).

This pixel version of straight lines explains the pattern

$$w_{3/5} = w_{1/2}\, w_{2/3}.$$

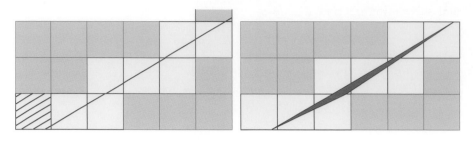

Figure 9.6. Left: The straight line of slope 3/5 passing through a square grid. The line passes through eight yellow squares: these would be the pixels that a computer printer would darken to depict this particular line. Each of the stripes appearing in the lower left square has a copy in exactly one of the eight yellow squares which the line now meets. The line begins in the lower left yellow square and extends far enough to return to its 'beginning' in the light green square.
Right: This shows lines of slopes 3/5, 1/2 and 2/3 forming a very thin triangle shaded in light blue. The long straight side and the other 'broken' side of the triangle pass through the same sequence of vertical and horizontal sides, thereby demonstrating the concatenation law in this example.

The right frame of Figure 9.6 is exactly the same as the left one, except that we have superimposed the lines $p/q = 1/2$ and $p/q = 2/3$. The space between these three line segments is a thin blue triangle. The $1/2$ line repeats after travelling one horizontal square upward, while the $2/3$ line repeats after two. Moving upwards, first along the $1/2$ line and then the $1/3$ one, picks out *exactly the same sequence* of squares or pixels as the yellow ones which highlighted the $3/5$ path. This explains why $w_{3/5}$ is exactly the word you get by 'chaining together' or **concatenating** the $1/2$ and $2/3$ words $w_{1/2}$ and $w_{2/3}$.

Exactly the same concatenation procedure works whenever we have neighbouring words. In other words, if p/q and r/s are neighbours, and if $p/q < r/s$, then

$$\frac{p}{q} \oplus \frac{r}{s} = \frac{p+r}{q+s}$$

and

$$w_{\frac{p+r}{q+s}} = w_{\frac{p}{q}}\, w_{\frac{r}{s}}.$$

The full pattern of words is shown in Figure 9.7, where rational numbers are joined whenever they are Farey neighbours. This is just the modular group tessellation which we saw in Figure 7.11, repositioned using the Cayley transform. The vertex p/q has also been labelled by the word $w_{p/q}$, so the diagram on p. 276 is a small fragment of this picture. It is easy to read off the concatenation formula: for example $2/5$ is between $1/3$ and $1/2$ and

$$(aaaB) \cdot (aaB) = (aaaBaaB).$$

(Remember that rational numbers are Farey neighbours whenever they are the endpoints of a common edge.)

Notice that on the diagram we have extended the definition of p/q words to negative fractions. We did this by decreeing that exactly the same concatenation rule should hold for *any* two positive or negative fractions $p/q < r/s$ which are Farey neighbours. To get started, we only needed to make sure the rule works for all the negative integers. The correct formula for $-p/1$ (where $p > 0$) is $w_{-p/1} = ab^p$. (This is because we have to think

Figure 9.7. Patterns in p/q words. This is the modular group tessellation from Chapter 7, repositioned inside the unit disk. The vertices of each triangle in the figure are a triple of neighbours $\frac{p}{q}, \frac{r}{s}, \frac{p+r}{q+s}$. Every rational number is assigned to a vertex, and every pair of neighbouring fractions have an arc stretching between them. Each vertex is also labelled with the corresponding p/q word in accordance with the concatenation formula explained in the text.

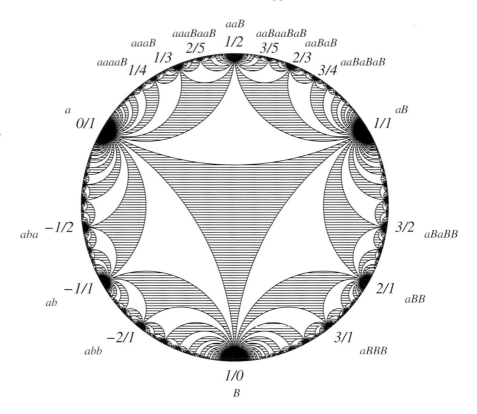

of $0/1$ as *between* $-1/1$ on its left and $1/0$ on its right, as in the formula $0/1 = (-1 + 1)/(1 + 0)$. If $w_{-1/1} = x$, then our concatenation rule will work provided $w_{0/1} = w_{-1/1}w_{1/0}$, in other words if $xB = a$ so that $x = ab$.)

An important property of accident-prone words is that they come in families: if a Möbius transformation c in a group is parabolic, so are all its conjugates, so are all its powers and so are the conjugates of its powers! So the specific p/q word $w_{p/q}$ is a somewhat arbitrary choice. For instance, instead of $w_{2/3} = aaBaB$, we could have used any of its cyclic permutations $aBaBa$, $BaBaa$, $aBaaB$ or $BaaBa$ or any of their inverses $bAbAA$, $AbAbA$, $AAbAb$, $bAAbA$ or $AbAAb$. When one suffers an accident, so do they all! And if the order of the summands in $w_{p/q} w_{r/s}$ is changed, then $w_{3/5}$ would be replaced by one of its cyclic permutations. We made the particular choice we did to ensure that the concatenation formula works nicely. In Project 9.4, we describe another elegant property of our special choice: whenever $p/q < r/s$ are Farey neighbours, the commutator of $w_{p/q}$ and $w_{r/s}^{-1}$ exactly equals $abAB$.

Notice that the p/q word (for positive p/q) contains p letter B's and q letter a's, distributed just about as 'evenly' as possible through the pattern.

You might well ask why we are confining our attention to the special patterns. Why not, for example, study *aaBBB*? The answer to this lies in some topology, which we discuss in the next section.

Gift-wrapping a torus

Now we have explored the p/q word patterns, let's try to see what they have to do with pinching curves and finding cusps. We have already hinted that there is a close connection between words in a group of symmetries and loops on the glued up surface associated to a basic tile. (We saw this phenomenon in the Schottky group in Figure 6.15 with the commutator *abAB*.) The connection works like this. Look at an example of a quasi-fuchsian group from Chapter 6, for example the nice square Fuchsian one in Figure 6.5. Each red Dr. Stickler is labelled by exactly one reduced word. Imagine him standing on the central tile, connecting himself by a path to the image of himself on another tile labelled by his favourite word. As we have seen, the effect of the generating transformations is to glue up the opposite sides of the central tile to make a torus. When you do this, the path connecting Dr. Stickler to his image across the tiles winds around and around the glued up surface, finally meeting itself at the initial point from which it began. Equally, any path on the torus can be 'unwound' to produce a word.

Obviously there are many different paths connecting two different points, but since we shall be tightening everything up anyway, we may as well only look at the 'shortest' or 'straightest' ones. Actually we can say more. We are aiming to find out which paths it makes sense to shrink to zero length. If we choose a word at random, there is no reason why the corresponding curve on the torus shouldn't intersect itself several times; usually, it does. If you cut the curve at each of its intersection points, you will get a number of separate loops. One would expect that shortening the whole curve should automatically lead to shortening of these individual parts. So it should be enough to think only about what happens when you shrink the **simple** or non-self-intersecting loops. From the mathematical theory[1] this turns out to indeed be the case. So now the question: 'Why are the p/q words the ones which it is sensible to try to make parabolic?' becomes the question 'Why are the p/q words the ones which correspond to simple loops on the glued-up once-punctured torus?' The clue is that the scan conversion instructions in the last section do indeed seem to have something to do with a two-generator group which makes a torus. In fact the two moves *a* and *B* described in the last

[1] Due to Bers and his former student Maskit, this uses so-called Teichmüller theory which will be touched on in Chapter 12.

section exactly generate the symmetries of our tiling of the Iowa plain by a grid of identical square fields in Chapter 1. The basic square which is transported around by these symmetries glues up into a torus (as in the picture on p. 24). Figure 9.8 illustrates the same gluing process, and we have drawn the same set of line segments that we had in the bottom left square of Figure 9.6. As you see, the segments glue together into a spiralling barber pole loop round the torus. Notice that the loop on the torus never crosses itself, because the line segments across the square in the left frame don't cross. The glued-up curve on the torus is a simple loop!

Figure 9.8. Gluing the B edge to the b edge in the marked square on the left gives the cylinder in the middle of this three-part picture. Then gluing the a edge to the A edge forms the torus on the right. The chocolate striping on the square becomes a 'barber pole' pattern on the cylinder, gluing up to a delicate spiral of icing on the doughnut.

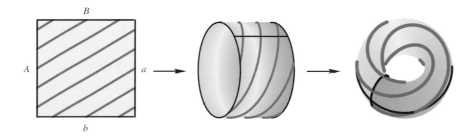

So far we are talking about an ordinary (unpunctured) torus, which goes with the Iowa plane and the commutative group of translations in which $ab = ba$. But suppose we think of the square as a schematic representation of the basic half-tile for the kissing Schottky group in Figure 6.5. We must imagine that the puncture is at the corners: the point where the four corners of the square meet on the glued-up torus on the right. The vertical sides are marked with the transformations a and A, and the horizontal sides marked by b and B. The labels are just the same as if the sides of the tile were parts of the four Schottky circles C_a, C_A, C_b and C_b. Thus the generator b carries the side labelled B to the side labelled b, and so on, and the glued-up torus-with-a-puncture is made by gluing in the same pattern. We could draw the same striped pattern of arcs[1] on the Schottky tile, and they would glue up into a spiralling non-self-intersecting loop on the torus-with-a-puncture in exactly the same way.

[1] It makes no difference to what we say if these arcs aren't straight lines, as long as they join edges of the tile in the same pattern and never cross each other.

The two pictures in Figure 9.9 illustrate the reason behind our rules for the p/q words. All you need to do is to read off, in order, the labels of the sequence of sides which the joined-up curve cuts. We explain the details of how to do this in Note 9.2, and you can find a detailed explanation of why our recipe for reading off labels in Figure 9.9 gives the correct word in Note 9.3.

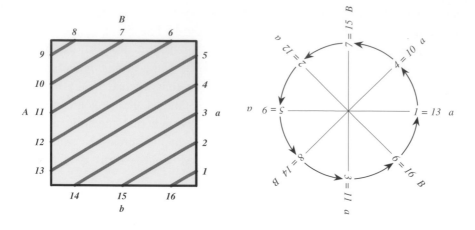

Figure 9.9. Left: The two possible arc patterns of a simple closed curve on a torus. In both frames, the number of endpoints on the top and bottom is $p = 3$, while the number on the left and right sides is $q = 5$. The endpoints are numbered from 1 to $2p + 2q = 16$, starting at the bottom righthand corner.
Right: How the word in a and B is read off the pattern of arcs on the left. Following the arcs upward is equivalent to tracing this circle in anticlockwise fashion; see Note 9.2 for details.

So here finally is the punchline: the words which correspond to the simple loops on the once-punctured torus are exactly the p/q words! It turns out that *every* simple loop on the torus comes about in exactly this way: there is exactly one such word for each value of p and q. Actually we can also take negative values for p. This gives another curve in which the straight line segments across the square slope in the opposite direction with negative slope $-p/q$. All you need to do is to specify the numbers p and q of endpoints which lie on the vertical and horizontal sides,

---

Note 9.2: **Reading off the 3/5 word**

Here's how to use Figure 9.9 to reconstruct the 3/5 word, represented by the arc pattern in the left frame, in which the arcs slope upwards with slope 3/5. By following the arcs along the loop, paying close attention to the way in which they leave and re-enter the square, we should be able to read off the word which belongs to this curve. As you can see, we have numbered the endpoints from 1 to 16 in anticlockwise order starting from the lower right corner.

Start at point on the segment between numbers 1 and 16, and then follow the arcs upward. We first encounter the point numbered 1, and since it is on the side marked with a, we record the generator a. Next, we have to use the inverse transformation A to teleport ourselves to the matching point number 13

on the left edge. From there we may continue our upward journey along the next arc to point number 4. Again we record a, and then teleport to point 10 on the left edge by A. The next arc carries us to point 7 on the top edge; here we record the generator B and the transformation b carries us to point 15 on the bottom edge. Eventually we will return to our starting position. Our entire journey is summarized by the wheel of numbers in the right frame. The pattern in the wheel (limiting our attention to the numbers from 1 to 8) brings us right back to our original recipe:

$$1 \xrightarrow{+3} 4 \xrightarrow{+3} 7 \xrightarrow{-5} 2 \xrightarrow{+3} 5 \xrightarrow{+3} 8 \xrightarrow{-5} 3 \xrightarrow{+3} 6 \xrightarrow{-5} 1$$

giving the 3/5 word *aaBaaBaB*.

respectively. Given these numbers, there are only two possibilities. It is an interesting puzzle to work out why.

There are wonderful connections between the p/q words and the Markov conjecture (see p. 191), described by the middle author in *The Geometry of Markoff Numbers*, Mathematical Intelligencer, 7, 1985.

Calculating traces

Now that we have found out which curves or words are good targets for creating accidents, we need an efficient method for working out their traces. According to what we said about the trace identities on p. 189, the trace of the p/q word $w_{p/q}$, which we shall write for short as $T_{p/q}$, should be a polynomial in the basic three traces t_a, t_b and t_{aB}. Once we know the **trace polynomial**, we shall be set to try solving equations like $T_{p/q} = 2$ to find cusps.

There is a nice way to find $T_{p/q}$, which rests on the concatenation

---

Note 9.3: **Labelling paths by words**

Here is the detailed explanation of our recipe for reading off labels in Figure 9.9 to reconstruct a path on a surface. We have to see why reading off the labels as we describe gives exactly the word W which connects a point P on a tile F to its image point $W(P)$ on tile $W(F)$.

Imagine F is a four-sided tile for a punctured torus, bounded by four tangent Schottky circles C_a, C_b, C_A and C_B. Follow the path in F until it meets one of the four bounding circles, say for example the point X on the circle C_a. At this point the path crosses over into the next tile, which, because of the way we have set up the labelling, is the tile $a(F)$. The map A has the effect of hauling back everything that happens in tile $a(F)$ and letting us view it instead in tile F. In particular, since X is also a point on a side of $a(F)$, it follows that $A(X)$ must be a point on the edge C_A of F. Thus the pulled-back path re-enters F from its new

entry point $A(X)$, creating a new segment running across F.

Now follow this new stretch of path, hauling it back to F when it crosses one of the four boundary circles by applying the appropriate one of the symmetries A, B, a or b. If the path leaves across the side C_B, then it would be hauled back by the transformation b to re-enter across C_b, and so on. Continuing in this way, the whole path can be viewed entirely on F. It has become a collection of separate segments, each of which is a copy of a part of the original continuous path, transported back to F by a symmetry in the group. Everything works out exactly so that when the circles are glued up in pairs, the segments glue to make an unbroken loop on the punctured torus.

The word W can now be read off just by looking at the path segments in F. All you have to do is write down, in order from left to right, the labels of the circles across which the path exits F.

formula

$$w_{\frac{p+r}{q+s}} = w_{\frac{p}{q}} w_{\frac{r}{s}}.$$

To use this, we are going to build up the traces *recursively*, following the exact same right–left road map as we would to form the fraction $\frac{p+r}{q+s}$ starting from the initial fractions $\frac{0}{1}$ and $\frac{1}{1}$. At each stage along our path, we calculate the trace of the word belonging to the current fraction, and use it for our calculations in the next few steps.

The calculations are based on the grandfather trace identity from p. 192 in Chapter 6 which, slightly turned around, gives the formula

$$\operatorname{Tr} w_{\frac{p+r}{q+s}} = \operatorname{Tr} w_{\frac{r}{s}} \operatorname{Tr} w_{\frac{p}{q}} - \operatorname{Tr} \left(w_{\frac{p}{q}}^{-1} w_{\frac{r}{s}} \right).$$

(Assume here $p/q < r/s$.) So far so good, but what about that awkward term $\operatorname{Tr} \left(w_{\frac{p}{q}}^{-1} w_{\frac{r}{s}} \right)$?

Because of the way our special words build up, the neighbouring words $w_{r/s}$ and $w_{p/q}$ will begin with a longish common string. So the expression $w_{p/q}^{-1} w_{r/s}$ involves some cancellation. If $s > q$ it turns out to equal $w_{(r-p)/(s-q)}$, while if $s < q$, then instead $w_{p/q} w_{r/s}^{-1}$ equals $w_{(p-r)/(q-s)}$. No matter: $w_{p/q} w_{r/s}^{-1}$ has the same trace as $w_{p/q}^{-1} w_{r/s}$. If you have calculated previous words along the right–left path already, the trace of this word will be already on your list.

For example, suppose we want to find the trace for

$$w_{\frac{8}{21}} = aaaBaaaBaaBaaaBaaaBaaBaaaBaaB.$$

Here is the list of words we shall have passed on our left–right journey. See how the fractions alternate in size and how every successive pair are neighbours:

p/q:	$w_{p/q}$
1/1:	aB
0/1:	a
1/2:	aaB
1/3:	$aaaB$
2/5:	$aaaBaaB$
3/8:	$aaaBaaaBaaB$
5/13:	$aaaBaaaBaaBaaaBaaB$
8/21:	$aaaBaaaBaaBaaaBaaaBaaBaaaBaaB$

If we want to take $(p + q)/(r + s) = 8/21$, then $p/q = 3/8$ and $r/s = 5/13$. So the Grandfather identity tells us

$$\operatorname{Tr} w_{\frac{8}{21}} = \operatorname{Tr} w_{\frac{5}{13}} \operatorname{Tr} w_{\frac{3}{8}} - \operatorname{Tr} w_{\frac{3}{8}}^{-1} w_{\frac{5}{13}}.$$

Now the nice thing about this is that

$$w_{\frac{3}{8}}^{-1} w_{\frac{5}{13}} = (aaaBaaaBaaB)^{-1}(aaaBaaaBaaBaaaBaaB)$$

$$= aaaBaaB = w_{\frac{2}{5}}.$$

In other words,

$$\text{Tr}\, w_{\frac{8}{21}} = \text{Tr}\, w_{\frac{5}{13}}\, \text{Tr}\, w_{\frac{3}{8}} - \text{Tr}\, w_{\frac{2}{5}}$$

and not only $w_{5/13}$ and $w_{3/8}$ *but also* $w_{2/5}$ are already on our list. Notice that $2/5 = (5-3)/(13-8) = (r-p)/(s-q)$.

In this way we can recursively calculate the traces of all the words on our list. At each stage there is a nice easy formula telling us how to get from one level to the next. There is just one slight twist: the pattern is slightly different depending on whether $q > s$ or $s > q$. Never mind, this shouldn't be too much for a good computer. The choice to make just depends on which way you turned in the left–right path at each stage. The particular pattern of words we wrote down is unusually regular, because it simply alternates left–right–left–right–$\cdots$. The general scenario is explained in Figure 9.10.

Figure 9.10. The left–right turning pattern. This schematic picture shows a bit of the Farey tessellation. The 3 vertices of each triangle give a triple of neighbouring words. Turning left or right corresponds to replacing the triple (u, v, uv) by (u, uv, uuv) or (uv, v, uvv), respectively (writing the outer pair of words first).

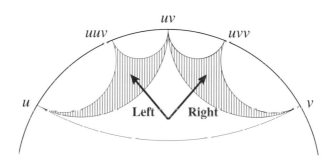

Let's assume that at the stage we have reached we have already worked out the three traces t_u, t_v and t_{uv} corresponding to a triple of neighbouring fractions $p/q, r/s$ and $(p+r)/(q+s)$. Left and right turns will mean replacing those three values t_u, t_v and t_{uv} by two possible choices: (t_u, t_{uv}, t_{uuv}) if turning left and (t_{uv}, t_v, t_{uvv}) if turning right. Using the Grandfather trace identity gives the formulas

$$L : (t_u, t_v, t_{uv}) \longrightarrow (t_u, t_{uv},\ t_u t_{uv} - t_v) \qquad \text{A left turn,}$$

$$R : (t_u, t_v, t_{uv}) \longrightarrow (t_{uv}, t_v,\ t_{uv} t_v - t_u) \qquad \text{A right turn.}$$

You can see some code which implements this in Box 24.

Where are all the accidents?

Now we know which words it makes sense to try to make parabolic, and we know how to find their traces, let's try locating all the cusps.

To make sure our equations only have a limited number of solutions, we shall confine our search to double cusp groups. In fact we are going to insist that the 1/0-word *b* be parabolic and look for double cusp accidents in which another *p/q* word is parabolic at the same time. (We shall

---

Box 24: Trace recursion

This routine has as its input the fraction `p/q` and the three traces `tr_a`, `tr_B` and `tr_aB`. We were a little creative with the pseudocode with such testing expressions as `while p3/q3 != p/q`, which means do the commands inside the while loop while the fraction `p3/q3` is not equal to `p/q`. Some programming languages do have a rational number data type, making this a little more convenient. In general though, one has to replace this expression with something that tests the equality of `p3` with `p` and `q3` with `q`.

```
trace_poly( p/q , tr_a, tr_B, tr_aB) {
   if p/q = 0/1 then RETURN(tr_a)
   else if p/q = 1/0 then RETURN(tr_B)          Deal with easy cases.


   p1/q1 := 0/1; p2/q2 := 1/0; p3/q3 := 1/1;    The fractions p1/q1 and p2/q2 will be Farey
                                                neighbours zooming in on p/q, and p3/q3
                                                will be the next step.
   tr_u := tr_a; tr_v := tr_B; tr_uv := tr_aB;
   while p3/q3 != p/q do {                       The loop terminates when p/q = p3/q3.
      if p/q < p3/q3 then {
         p2/q2 := p3/q3; p3/q3 := (p1+p3)/(q1+q3);
                                                 During the execution of the loop, tr_u,
         temp := tr_uv;                          tr_v, tr_uv will be the trace of the words
                                                 for p1/q1,p2/q2,p3/q3.

         tr_uv := tr_u * tr_uv - tr_v;
         tr_v := temp}
      else {
         p1/q1 := p3/q3; p3/q3 := (p3+p2)/(q3+q2);
         temp := tr_uv;
         tr_uv := tr_v * tr_uv - tr_u;
         tr_u := temp}
      }
   RETURN(tr_uv)
}
```

explain later in the chapter why we wouldn't get anything really different by choosing a word other than 1/0.) As we have seen, the restriction that $t_b = 2$ cuts down our parameters to just one free variable t_a. The formulas look a bit neater if we replace t_a with the Maskit variable $\mu = t_a/i$ as on p. 259. If we arrange for the fixed point of b to be at ∞, then the groups we are studying are exactly the ones we met in Maskit's recipe, with generators

$$a = \begin{pmatrix} -i\mu & -i \\ -i & 0 \end{pmatrix} \quad \text{and} \quad b = \begin{pmatrix} 1 & 2 \\ 0 & 1 \end{pmatrix}.$$

Being single cusp groups, one half of the ordinary set still glues up into a once-punctured torus and the other half breaks up into disks, each of which glues up into a triply-punctured sphere (midway between the two once-punctured tori on p. 190 and the two triply-punctured spheres on p. 205). The set of these groups is known as **Maskit's slice**.[1]

Figure 9.11 is a plot in the μ plane of *all* the solutions to $T_{p/q}(\mu) = \pm 2$, for all possible values of p/q between -1.25 and 1.25 with q less than 57. The solutions are coloured according to the size of p/q. What an amazing picture! The pattern repeats with a horizontal symmetry of translation. All the solutions appear to be confined to a horizontal strip, above and below which is a white solution-free zone. The upper boundary between the two regions has been traced in black.

This picture took a LOT of computing! We had to run a routine called Newton's algorithm for finding roots to polynomial equations with sufficient accuracy to find all q solutions of each equation $\text{Tr}\, w_{p/q} = \pm 2$ for each p/q in the required range. The roots of a polynomial can be widely

Note 9.4: **Finding cusps and the fundamental theorem of algebra**

Way back in Chapter 2 we mentioned the *Fundamental Theorem of algebra* which says that every polynomial equation has a complex number solution. In fact, a polynomial of degree n (the degree means the highest power of the unknown you are seeking, in our case μ) has exactly n solutions. (Some of these may be 'multiple solutions', so for example quadratic equations usually have exactly two solu-

tions, but if $b^2 - 4ac = 0$ then they have only one.) That means that we should expect the equation $\text{Tr}\, w_{p/q} = 2$ to have q solutions. There are another q solutions to the equation $\text{Tr}\, w_{p/q} = -2$. We are telling you that there is exactly one which is actually on the black boundary in Figure 9.11. How are we ever going to tell which one?

scattered around the plane. Never mind; it's only computer time. We'll move on to Newton's method later. The animated version of this book will do the picture greater justice, for as the fraction p/q moves continuously, wave after wave of ghostlike cuspy curves come crashing through the swirling dust.

Actually this picture was not the first thing we plotted. We were looking for the double-cusp groups which hover on the boundary between discrete and non-discrete. Theory confirms that among all possible solutions to the

Figure 9.11. Parabolic dust. This cloud of dust shows all the points in the μ-plane at which the p/q word is parabolic, with $-1.25 \le p/q \le 1.25, q \le 57$.

[1] In the picture, you actually see each double-cusp twice, see Note 9.5. This result was proved by Keen, Maskit and Series, J. Reine u. Angew. Mat., 1993.

trace equation (for a fixed p/q), *exactly one* will be discrete.[1] These were the transition points we sought. Theory also told us that no value of μ with imaginary part between -1 and $+1$ could give a discrete group, while values with imaginary part more than $+2$ or less than -2 would always be discrete. Moreover the Maskit slice had certain symmetries, see Note 9.5 and Project 9.2. But what did it look like?

Our original plots (after a huge amount of stumbling around in the dark) looked more like Figure 9.12, which shows the stretch of the boundary corresponding to cusps from $0/1$ to $1/1$. To get oriented, we have marked the first three cusps which you can easily calculate by hand. A few more cusps are marked on the slightly closer up picture in the left frame of Figure 9.12. See how the fractions increase in their proper order from left to right along the curve. The cusp for the p/q word is marked $\mu(p/q)$. You should imagine this boundary extending out to infinity in both directions, repeating with a horizontal period of 2. You can also see an extra symmetry: reflection in the vertical line through the mid-point of the picture corresponding to the $1/2$ cusp at the value $\mu(1/2) = 1 + \sqrt{3}i$.

[2] Some isolated μ-values in the grey region give discrete groups, but such groups are never free (see p. 98). We go into this further in Chapter 11.

The points in the white upper region all represent discrete groups, while 'most' points in the shaded region are not discrete.[2] The black cuspy curve is called the **Maskit boundary**; μ-values on the boundary give groups right on the transition between order and chaos.

'Cuspy' now takes on two meanings. We got the Maskit boundary by plotting what we called 'cusp groups' in which a certain word became

---

Note 9.5: **Symmetries of the Maskit slice**

A nice feature of Maskit's slice is its translational symmetry. This means that if the group at a value μ is in the slice, then so are the groups at $\mu+2$, $\mu+4$, $\mu+6$ and also $\mu-2$, $\mu-4$ and so on. In fact, these values all really give the same group, because the transformation 'ba' in the μ group is identical with the transformation 'a' in the group for $\mu+2$. The slice also has a reflectional symmetry in the imaginary axis Re $\mu = 0$. This can be expressed by saying that if μ is in the Maskit slice, then so is $-\overline{\mu}$.

Actually, for each group at a value μ in the upper half plane, there is a 'complex conjugate group' at the complex conjugate value $\overline{\mu}$. The group at $\overline{\mu}$ is the 'mirror image' of the group at μ. If one of these two groups is discrete then the other is; the only difference between their two limit sets is that one is a reflection of the other so that their spirals curl in opposite directions. This is the reason why we concentrate our attention entirely on μ-values in the upper half plane. Whatever map we find of the Maskit slice above the real axis, the same picture will be exactly mirrored by reflection into the lower half plane below.

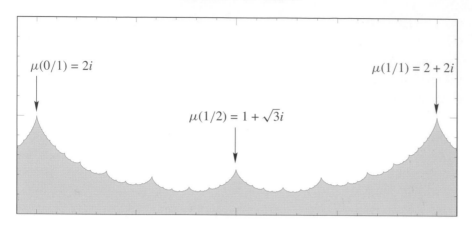

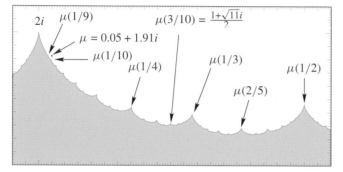

Figure 9.12. Maskit's slice. This is a map of μ values showing the complicated cusped boundary between discreteness and chaos. Each complex number μ determines a two-generator group with traces $t_a = i\mu$, $t_b = 2$ and $t_{abAB} = -2$. The white regions are single-cusp groups while the shaded area is the zone of non-discreteness. The word corresponding to the fraction p/q becomes accidentally parabolic at the value $\mu(p/q)$. On the bottom left, more cusps are labelled and the bottom right is a closeup of the principal cusp at $\mu(0/1) = 2i$. The red dot at $\mu = \pm 0.05 + 1.91i$ corresponds to trace $1.91 - 0.05i$, used in some of our plots in Chapter 8.

accidentally parabolic, and, miraculously, the boundary near a value corresponding to a cusp group actually looks like a cusp! This coincidence of terminology was quite unexpected, but the technical term 'cusp group' seems to have been a fortunate choice indeed. In fact it is possible to work out the behaviour of the boundary near the cusps in great detail, so that for example the very regular looking stretch of boundary in the right frame of Figure 9.12, near the principal cusp at $\mu(0/1) = 2i$, has exactly the same shape as the cusp in a classical Greek curve called a *cardioid* (for 'heart-shaped'). You can see a complete cardioid in the Mandelbrot set, shown for comparison in Figure 9.13.

If you were wondering about the introduction of fractions and Farey addition, this picture of the Maskit boundary may begin to convince you of their relevance. The peaks convey a strong visual image of Farey addition; the most prominent peak between two Farey neighbours p/q and r/s corresponds to the Farey sum $(p+r)/(q+s)$. Notice the direct 'line of sight' between neighbouring cusps $\mu(p/q)$ and $\mu(r/s)$, meaning that the line segment between these two cusps seems to lie entirely in the white region, although this remains unproved.

The cusp groups seem to be scattered thickly (densely, in mathemati-

Figure 9.13. The World's Most Famous Fractal. There had to be at least a cameo appearance of the *Mandelbrot set*, a picture of the complex numbers c for which the images of the point c under iteration of the map $z \mapsto z^2 + c$ never head off to ∞. The main blob is a cardioid, in which the trademark cusp can be seen pushing in from the right. In fact, our cuspy Maskit boundary itself may look familiar to devotees of complex iteration. See the similarity to the Julia set of the quadratic map $z \mapsto z^2 + \frac{1}{4}$, shown here on the right. In the geography of the Mandelbrot set, $c = 1/4$ is the exactly the parameter value at the cardioid cusp! Pictures courtesy of C. McMullen.

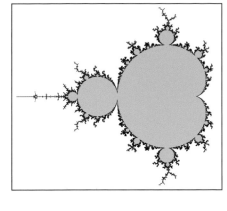

[1] Ann. Math., 1991

[2] *The classification of punctured-torus groups*, Ann. Math., 1999. Among other things, Yair's results show that the Maskit boundary is 'homeomorphic' to the real line.

cal language) along the boundary, in just the same way that the rational numbers (alias repeating decimals) are scattered thickly along the real line. Bers had conjectured in the 1960's that this would be true, and although our pictures bore it out, Bers' conjecture remained unproved until a paper by Curt McMullen called *Cusps are dense.*[1] It was not until 20 years after our experimental discoveries that Yair Minsky finally succeeded in proving that the boundary is actually a genuine boundary with no holes or cracks.[2] We shall return to this in the next chapter.

The cuspy Maskit boundary doesn't look as wild as some of our spirally limit sets, but it *is* decidedly fractal. Pointy cusps seem to appear at all scales, and as you can see from Figure 9.13, it is tantalizingly similar to the **Julia set** obtained by iterating the quadratic map $z \mapsto z^2 + \frac{1}{4}$. Based on Minsky's work, Hideki Miyachi has recently proved that the cardioid shape really does appear at every single cusp. In short, the Maskit boundary is very intricate, although this only really becomes apparent with a program capable of making deep zooms.

Heading for the boundary

[3] We liked the sound of that.

In Chapter 8, we played (rather irresponsibly) with traces around 2. One especially spiralliferous[3] spot was near trace $1.91 + 0.05i$, exactly the complex conjugate of the trace corresponding to the μ-value $0.05 + 1.91i$, the red dot in Figure 9.12. (Complex conjugate traces, remember, give mirror image groups.) Looking back at Figure 8.17, perhaps you can see how we might have predicted that the μ-value for this group should be near the 1/9 and 1/10 cusps. The spiral head is largest around words with prefix ab^{10} and ab^9, and in fact the most extreme point seems to be about ab^9ab^{10}. The roles of a and b are reversed because of the mirror symmetry, which more or less explains why in our present set-up it is a good idea to declare $a^{10}B$ a special word.

Figure 9.12 is like a road map, delineating the boundary between order and chaos. How about using it to drive right up to one of these two nearby cusps? With any luck, it should exhibit both the beautiful spirals of Figure 8.17 *and* the delicate lacework of the Apollonian gasket. Imagine how those two effects will be combined. The results are – wait one moment, let's not get ahead of ourselves. Perhaps we should look very closely at the boundary near $\mu = 0.05 + 1.91i$. Zoom in to the red dot in Figure 9.12 to get the very small (0.03×0.03) frame in Figure 9.14.[1] If we were to head straight towards the boundary from the red dot, we would reach it very close to the 2/19 cusp. On the other hand, the 1/9 cusp is perhaps a tad closer as the crow flies, and the 1/10 cusp is not that far off either. All three cusps may exert an influence over the groups in this neck of the woods.

[1] As usual, the units are set in terms of the parameter, in this case μ.

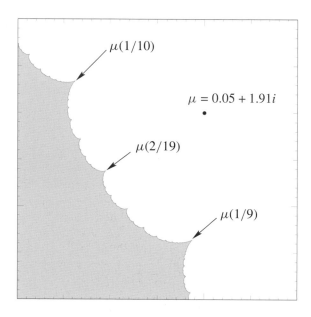

Figure 9.14. An extremely close-up view of the region round the red dot ($\mu = 0.05 + 1.91i$) in Figure 9.12.

You can see exactly what does happen when we plot the limit sets in Figure 9.15. We have used Grandma's recipe rather than Maskit's, so that the limit set is nicely enclosed by the unit circle. The pink region inside the limit set for the group for $\mu = 0.05+1.91i$ is very spirally. As usual in a Maskit group, the outer yellow disk gets carried into each of the infinitely many other yellow disks in the picture by the different elements in the group. All the disks have all been drawn precisely, because we told our program to use the correct special words: a^9B for the 1/9 cusp, $a^{10}B$ for

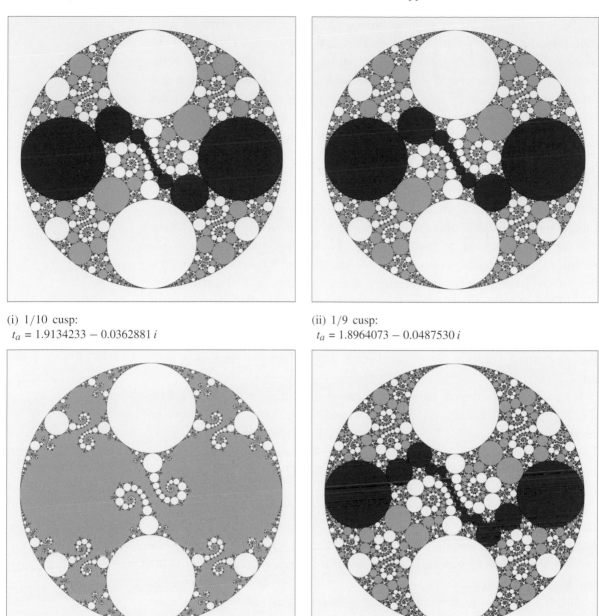

(i) 1/10 cusp:
$t_a = 1.9134233 - 0.0362881\,i$

(ii) 1/9 cusp:
$t_a = 1.8964073 - 0.0487530\,i$

(iii) Nearby group:
$t_a = 1.91 - 0.05i$

(iv) 2/19 cusp:
$t_a = 1.90378 - 0.03958\,i$

Figure 9.15. Four nearby limit sets. See text for discussion.

the 1/10 cusp and $a^{10}Ba^9B$ for the 2/19 one. For $\mu = 0.05 + 1.91i$, any of these choices would work.

Watching the limit set evolve as we follow μ-values out to the various cusps, the pink region gets severed into infinitely many disks as well. Most of the new disks in the three cusp groups are shown in pink, but we have highlighted a special chain in red. It stretches from a disk tangent to the outer yellow disk at -1 (the limit point $\overline{aBAb}$), to one tangent to the outer circle at 1 (the limit point $\overline{abAB}$). The pink and red disks are all transported one to another by elements in the group as well. The pattern of the 'core' red chain depends on the cusp. Just as in our examples at the beginning of the chapter, you can read the fraction directly from the chain by counting. Try it!

Truthfully, when we were first examining these limit sets, we delighted in bludgeoning ourselves with pictures like this for all the p/q cusps we computed before acknowledging such a simple pattern. As we go to more complicated fractions the patterns become extraordinary, and here we can only mention a few highlights. Notice that the 2/19 limit set seems to be a splitting or melding of the 1/9 or 1/10 ones in which the circles have been approximately subdivided into pairs of smaller circles. These circle twins are very noticeable because they are oriented somewhat at right angles to the overall path of the chain. The pattern isn't really related to the numerator of 2/19. It has more to do with the **continued fraction** $2/19 = 1/(2 + 1/9)$ which we met in Chapter 7.

Is this just coincidence, or are we seeing a significant effect? Let's take the plunge and look at a more complicated cusp 7/43. The boundary program finds that $\mu(7/43)$ is approximately $0.136998688 + 1.80785524i$. The associated parabolic word is $a^7B(a^6B)^6$; written out in full this would have 50 letters! After only a mild outflow of steam from our computer, the limit set is shown in Figure 9.16.

The complexity of the fraction 7/43 is reflected in the increased wildness of the limit set, a little reminiscent of the probe we carried out in Figure 8.15, slicing through a strip of non-discrete groups. If we use our boundary program to view a close-up chart of Maskit's slice in this vicinity, we may be able to find some dominant nearby cusp. The plot is in Figure 9.17, and sure enough the line $\mu = 0.05 + xi$ (where x is a variable real number) passes right by $\mu(3/26)$.[1] The point marked by the green crosshairs corresponds to the frame in Figure 8.15 where the limit set program went haywire. The vertical green line was the path of our probe. We should show the 3/26 cusp, but this time let's plot it using Maskit's recipe. Thus b is translation by 2 in the horizontal direction, and the fixed point of b has been moved to infinity. The limit set is a periodic frieze of circles shown in Figure 9.18.

The inlet in the boundary at the 3/26 cusp is so slender that the reader may wonder how the authors were so lucky as to strike it. Well, hindsight docs have its virtues. Initially, though, we really did experiment with the

[1] In Chapter 8, we actually probed the line $t_a = x + 0.05i$, which would correspond to $\mu = -0.05 + xi$. This is a mirror image of the line discussed here.

Figure 9.16. The $7/43$
cusp group. The red chain
contains some very sharp
turns, between which are 6
gentle spirals of 7 disks.
Surely it cannot be
coincidence that

$$\frac{7}{43} = \frac{1}{6 + \dfrac{1}{7}}?$$

limit set program, producing plenty of smoking computer screens as the
entwining tentacles only whetted our appetites for ever more tangled limit
sets teetering on the edge of non-discreteness. Eventually, we developed
a rather delicate algorithm for tracing the boundary to make pictures like
Figure 9.12, which we shall be explaining on p. 302 ff. Before we come to
that, though, let us pause briefly to contemplate some yet more elaborate
pictures of double cusps.

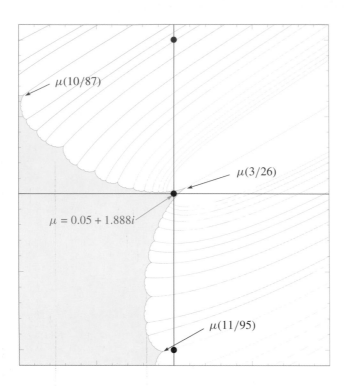

$\mu(10/87)$

$\mu(3/26)$

$\mu = 0.05 + 1.888i$

$\mu(11/95)$

Figure 9.17. Slicing a cusp. The points marked by blue dots correspond to the probe carried out in Figure 8.15. The line we travelled sliced right through a cusp of nondiscreteness. We have added some curves which indicate where p/q words have real trace; these we call the **real trace rays**. For example, the 3/26 word has trace 2 at the point marked $\mu(3/26)$, and along the ray emanating from that point the trace of the 3/26 word remains real and increases from 2 upwards. This is a useful charting technique which will be expanded on in the final chapter.

Changing generators and Grandma's party trick

Thus far, we have confined our search to double cusp groups in which the 1/0 word b is parabolic. What would happen if we decided instead to look for double cusp groups in which a pair of words $w_{p'/q'}$ and $w_{p/q}$ were parabolic, where there was no particular relation between the two fractions p'/q' and p/q? It turns out that the groups you will get this way will always be conjugate to a Maskit group in which $w_{1/0}$ and $w_{k/l}$ are parabolic for some other fraction k/l (depending of course on p'/q' and p/q).

Figure 9.19 shows a picture of this. In this example, the two words $w_{3/5}$ and $w_{-5/3}$ are parabolic. As we shall shortly explain, this group is conjugate to the Maskit cusp group in which the words $B = w_{1/0}$ and $w_{21/34}$ are parabolic. Notice the two green and pink 'core chains' corresponding to the two cusps which separate the four parts of the limit set corresponding to limit points whose infinite words begin with a, A, b and B. In place of the 1/0 chain of length $0 + 1 + 1 = 2$, we now have a pink $-5/3$ chain of length $5 + 3 + 1 = 9$.

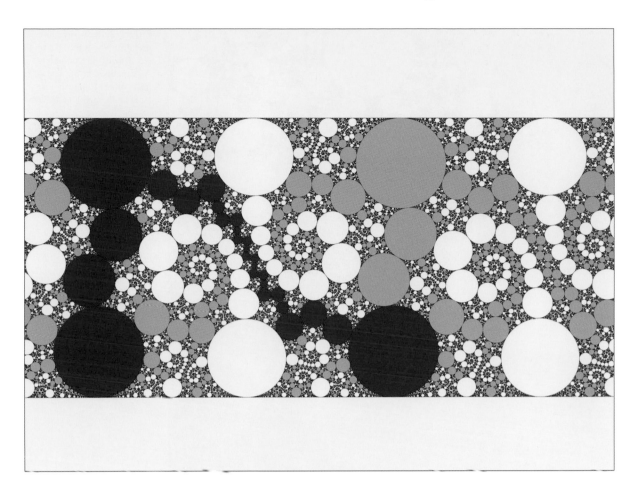

Figure 9.18. The 3/26 cusp. This is the 3/26 cusp in Maskit's slice with generators $a(z) = \mu + \frac{1}{z}$ and $b(z) = z + 2$. Can you read the continued fraction from the core red circle chain?

To understand what is going on here, we first need to look at another nice property of the p/q words. The fact is that if p/q and r/s are neighbours, that is, the fractions at the two ends of some edge of Figure 9.7, then the words $w_{p/q}$ and $w_{r/s}$ generate the same group as a and b. This means that not only are $w_{p/q}$ and $w_{r/s}$ expressible as words in a and b (and their inverses A and B), but also that a and b can be expressed as words in the new generators $w_{p/q}$ and $w_{r/s}$. Thus, by substitution, words in a and b and words in $w_{p/q}$ and $w_{r/s}$ amount to the same thing.

For example, the pair (a, aB), corresponding to the neighbouring fractions $(0/1, 1/1)$ are generators, because any word you can write in terms of (a, B) you can also rewrite in terms of the pair (a, aB). A more complicated example is the pair of words $u = w_{1/2} = aaB$ and $v = w_{2/3} = aaBaB$. In fact, using our usual upper case convention (that $U = u^{-1}$ and $V = v^{-1}$),

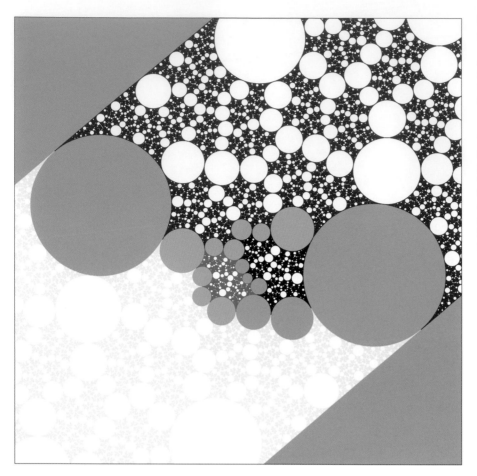

Figure 9.19. A double-cusp group in which the 3/5 and −5/3 words are parabolic. As explained in the text, the trace values to put in Grandma's recipe to make this picture are $t_a \doteq 1.5306639 - 0.8501047\,i$, $t_b \doteq 1.5306639 + 0.8501047\,i$. This group is conjugate to the Maskit double-cusp group in which $w_{1/0}$ and $w_{21/34}$ are parabolic. Notice that the traces of a and b are complex conjugates! We shall come back to this interesting fact in the next chapter.

you find that:

$$uVu = aaB(bAbAA)aaB = aaBbAbAAaaB = a$$
$$\text{and } UvUUv = (bAA)(aaBaB)(bAA)(bAA)(aaBaB) = B.$$

Thus any word you can write in terms of a and b can also be expressed using these two formulas in terms of the new generators u and v. It is a beautiful fact that *every possible* pair of generators (u, v) is a pair of words coming from a pair of neighbours in just this way.

Getting from one pair of generators to another is not as hard as it might seem. Figure 9.7 actually encodes the recipe. Suppose you are on an edge labelled by a pair (u, v) which you already know are generators. Face towards the boundary as in Figure 9.10 and you see in front of you two more pairs of generators, the pair (u, uv) (corresponding to a Left turn) and (uv, v) (corresponding to a Right turn). It is not hard to see that if u and v generate the group, then so do both the pairs (u, uv) and (uv, v). We can write the rules in symbols:

$$L : (u, v) \longrightarrow (u, uv)$$
$$R : (u, v) \longrightarrow (uv, v).$$

Now here is a neat party trick that Grandma plays with her recipe. It can be used to prove our claim that *any* double-cusp group can be directly created from some $1/0$, p/q Maskit cusp group which we know how to find. (A word of warning: what happens from here on really is a bit tricky and we shall really only want to make use of it some way into the next chapter.) Suppose you have a particular group generated by two matrices a and b. Compute the traces of the words $u = w_{p/q}$ and $V = w_{r/s}$ in this exact group. (We used V not v because in our pattern of words, $B = w_{1/0}$.) Then plug those traces into Grandma's old recipe and get new matrices which we'll call $\hat{u}$ and $\hat{v}$. The way Grandma's recipe works, we know that $\mathrm{Tr}\,\hat{u} = \mathrm{Tr}\,u$, $\mathrm{Tr}\,\hat{v} = \mathrm{Tr}\,v$ and $\mathrm{Tr}\,\hat{u}\hat{v}\hat{U}\hat{V} = -2$. This is like sleight of hand: out of the a, b-group of matrices, we have created the $\hat{u}, \hat{v}$-group by playing games. The trick is that it's not a truly new group, in fact, as explained in Note 9.6, the new group is actually a conjugate of the old. (Although the traces of the new generators $\hat{u}$ and $\hat{v}$ have no resemblance to those of a and b.)

Grandma's trick is very useful for constructing new cusp groups: if some third word $w_{m/n}(a, B)$ in the original group was parabolic and causing accidents, then we can rewrite $w_{m/n}$ as a word in u and V and we find that the word written in terms of $\hat{u}, \hat{V}$ is also parabolic. The general rule for rewriting special words in new generators is explained in Project 9.4.

Let's work through our $3/5$, $-5/3$ example. Our aim is to produce a group in which the words $w_{3/5}$ and $w_{-5/3}$ are parabolic and which is conjugate to the Maskit double-cusp group in which b and $w_{21/34}$ are parabolic. For the Maskit group, the trace value is $t_a(21/34) = 1.61799 - 1.29170\,i$. If we apply the generator change $RLRLRLRL$ to the generator pair (a, B), we are making four left turns and four right turns in the Farey diagram, starting from the fraction $(0/1, 1/0)$. It's fun to check that

$$RLRLRLRL(a, B) = (w_{21/34}, w_{13/21}).$$

Now let's focus on the generator pair right in the middle, that is $RLRL(a, B) = (w_{3/5}, w_{2/3})$ and let's call $u = w_{3/5}(a, B)$, $V = w_{2/3}(a, B)$. With the value $t_a = 1.61799 - 1.29170\,i$, $t_b = 2$, these have traces

$$\mathrm{Tr}\,w_{3/5} \doteq 1.53066 - 0.85010\,i, \quad \mathrm{Tr}\,w_{2/3} \doteq 1.53066 + 0.85010\,i.$$

Now in Grandma's trick, she wants to make $(w_{3/5}, w_{2/3})$ her new generators, so she is going to use her recipe to make matrices $\hat{u}$ and $\hat{v}$ with $\text{Tr}\,\hat{u} = \text{Tr}\,u, \text{Tr}\,\hat{v} = \text{Tr}\,v$. In this group, she says, the two words $w_{3/5}(\hat{u}, \hat{V})$ and $w_{-5/3}(\hat{u}, \hat{V})$ are parabolic. Why does it work?

The trick is that $w_{21/34}(a, B)$ is equal to the '3/5 word' in the new generators u and V while the beginning word B is equal to the '$-5/3$ word' in u and V. The first is not so hard to show:

$$(w_{21/34}(a, B), w_{13/21}(a, B)) = (RL)^4(a, B) =$$
$$RLRL(u, V) = (w_{3/5}(u, V), w_{2/3}(u, V)).$$

Now because $\hat{u}$ and $\hat{V}$ have the same traces as u and V, every word which we can make from $\hat{u}$ and $\hat{V}$ has the same trace as the same word made from u and V. (Actually we also need to know that $\text{Tr}\,uv = \text{Tr}\,\hat{u}\hat{v}$. The reason that this is true is explained in Note 9.6.) In particular:

$$\text{Tr}\,w_{3/5}(\hat{u}, \hat{V}) = \text{Tr}\,w_{3/5}(u, V) = \text{Tr}\,w_{21/34}(a, B) = \pm 2$$

so $w_{3/5}(u, V)$ is parabolic, in other words, it has a 3/5 cusp!

Note 9.6: **Changing generators leads to conjugate groups**

Let's call $G(t_a, t_b)$ the group you get by feeding in t_a and t_b to Grandma's recipe. Take this group $G(t_a, t_b)$, and choose a different pair of generators say u and v. Each of these are words in a and b, so in the specific group $G(t_a, t_b)$ they are just actual matrices with specific traces, say the complex numbers t_u and t_v.

Make use of Grandma a second time and construct the group $G(t_u, t_v)$. It will be generated by two specific matrices $\hat{u}$ and $\hat{v}$ with traces t_u and t_v. What is the relationship between $G(t_a, t_b)$ and $G(t_u, t_v)$? We claim they are just conjugate groups! We saw a special case of this in Project 8.7 in which we found two conjugate groups just by swapping in Grandma's recipe the complex number which was the trace of a with that which was the trace of b.

For short, let's write G for $G(t_a, t_b)$ and $\hat{G}$ for $G(t_u, t_v)$. Also a and b will mean the specific pair of generator matrices churned out by Grandma's recipe putting in parameters t_a and t_b, while $\hat{u}$ and $\hat{v}$ will mean the specific generators you get putting in parameters t_u and t_v.

The reason the two groups are conjugate is that $\text{Tr}(u \text{ in } G) = \text{Tr}(\hat{u} \text{ in } \hat{G})$ and $\text{Tr}(v \text{ in } G) = \text{Tr}(\hat{v} \text{ in } \hat{G})$. For conjugacy we also have to check that $\text{Tr}(uv \text{ in } G) = \text{Tr}(\hat{u}\hat{v} \text{ in } \hat{G})$. Now, as you can see in Project 9.4, $aBAb = uVUv$ and so $\text{Tr}\,uVUv = \text{Tr}\,aBAb = -2$. Thus not only $\text{Tr}\,a$ and $\text{Tr}\,b$, but also $\text{Tr}\,u$ and $\text{Tr}\,v$ satisfy the Markov identity. Comparing the Markov identity for $\text{Tr}\,\hat{u}$ and $\text{Tr}\,\hat{v}$ with the same formula for $\text{Tr}\,u$ and $\text{Tr}\,v$, we conclude that either $\text{Tr}\,\hat{u}\hat{v} = \text{Tr}\,uv$ or $\text{Tr}\,uV$. Project 8.4 now tells us that we can conjugate the pair $\hat{u}, \hat{v}$ to either u, v or u, V. However we have to preserve what is called the **orientation** of the two generators, which controls the order (anticlockwise or clockwise) in which the fixed points of the four cyclic permutations of the commutator appear round the boundary of the inside part of the ordinary set. This leads us to conclude that the right choice is that $\hat{u}, \hat{v}$ must map to u, v.

How about that negative trace value? To find this we have to use the rules for negative words. All you have to do is convince yourself that $w_{1/0}(a, B) = w_{-5/3}(u, V)$. Well – er – oops – this isn't quite exactly true. As you can check out by solving explicitly, we get $B = VUVVUVUV$. This isn't the word $w_{-5/3}(u, v) = uvvuvvuv$, but it is part of the same family, being the inverse of some cyclic permutation. In other words, it is parabolic at the same time as $w_{-5/3}(u, v)$.

Tracing the boundary

We promised up above to explain our algorithm for tracing the cuspy Maskit boundary in Figure 9.12. The plan goes like this. We start at the 0/1 cusp, which as we already know is at $\mu(0/1) = 2i$. We then aim to move along a steadily increasing sequence of fractions using the previous solution as the starting guess for Newton's method (to be explained shortly) to find the solution to the next trace equation.

Of course, glibly mentioning a 'sequence of fractions' overlooks one glaring problem: what sequence? What's the next fraction after 0/1? We must choose a reasonable spacing of fractions that will work. We tried several possibilities before latching onto one natural candidate which works very well. We simply choose an upper bound denom for the denominator and write down in order all the fractions with denominator *at most* denom. Note that the sequence we get is not simply

$$\frac{1}{\text{denom}}, \quad \frac{2}{\text{denom}}, \quad \frac{3}{\text{denom}}, \quad \frac{4}{\text{denom}}, \quad \frac{5}{\text{denom}}, \quad \frac{6}{\text{denom}}, \quad \dots$$

which is certainly a possible choice. Our choice includes more than these fractions. For instance, the sequence of all fractions of denominator at most 6 from 0/1 to 1/1 is

$$\frac{0}{1}, \frac{1}{6}, \frac{1}{5}, \frac{1}{4}, \frac{1}{3}, \frac{2}{5}, \frac{1}{2}, \frac{3}{5}, \frac{2}{3}, \frac{3}{4}, \frac{4}{5}, \frac{5}{6}, \frac{1}{1}.$$

This is known as the **Farey sequence** of order 6. Farey sequences have the interesting feature that every pair of consecutive fractions $\frac{r}{s} < \frac{p}{q}$ are 'Farey neighbours'. In other words, you can check that for each adjacent pair in the sequence, $ps - qr = 1$[1]. The fractions in the Farey sequence are not at all uniformly spread out over the interval, but it will emerge that they *are* well spread out with respect to the boundary we are trying to compute.

The one trouble with the Farey series is that it is not quite straightforward to find the next term in the sequence. We shall shortly explain how Euclid's algorithm can be adapted to do just what we want, enabling us

[1] This was the sequence originally considered by Farey (see p. 211), who noticed that successive terms were always neighbours. It was the mathematician Cauchy who first seems to have written down a proof of this interesting fact.

to create a subroutine whose input is the maximum denominator `denom` and the current fraction p/q, and whose output is the next fraction in the sequence, `nextpq( p/q , denom)`.

The other ingredient we need is the `newton solver`. Let's imagine that we have somehow already found the solution $\mu(r/s)$ to the cusp equation $T_{r/s}(\mu) = 2$ for some fraction r/s that is 'near' to p/q. In other words, suppose we know a value $\mu(r/s)$ which we think is on the boundary for which $T_{r/s}(\mu(r/s)) = 2$. In a kind universe, $T_{p/q}(\mu(r/s))$ (that is, the value of $\mathrm{Tr}\, w_{p/q}$ at the point $\mu(r/s)$) will be 'close' to 2. With a bit of luck, there will be a solution to the p/q equation which is 'close' to $\mu(r/s)$. Sir Isaac Newton gave the quintessential method for finding solutions of equations given a sufficiently good initial guess.

Usually Newton's method is applied with real variables, however the same idea works if the variable is *complex*, so we shall write our variable as z. (In our application, the variable will be the Maskit parameter which we have symbolized by μ.) It starts with an arbitrary function $f(z)$ and a guess z_0 for a solution to the equation $f(z) = 0$. For example, $f(z)$ might be the trace polynomial

$$T_{3/8}(z) = z^8 - 6z^7 + 20z^6 - 44z^5 + 68z^4 - 76z^3 + 60z^2 - 32z + 10$$

in which case we would be looking for the solutions to the equation

$$z^8 - 6z^7 + 20z^6 - 44z^5 + 68z^4 - 76z^3 + 60z^2 - 32z + 10 - 2 = 0.$$

In general, unless we make a fabulously good guess, $f(z_0)$ will not be 0. However we may be able to work backwards to a better guess for the

---

Note 9.7: **How does Newton's method work?**

Usually, Newton's method is explained in terms of calculus. However, we used a variant sometimes called the **secant method** in which (rather than taking derivatives) we estimate the rate of change of our function f near the point of interest z_0. Take a very small increment, which we call `smidge`, and then define the `rate` of f to be the quantity

$$\mathrm{rate}(z_0) = \frac{f(z_0 + \mathrm{smidge}) - f(z_0)}{\mathrm{smidge}}.$$

As you can see, this measures the average rate of increase of f over the very small interval between z_0 and $z_0 + \mathrm{smidge}$.

Now suppose z_1 is another nearby point where we are interested in the value of f. Supposing that the rate of increase of f is roughly constant over the interval between z_0 and z_1 (this is where we are using linear interpolation) then, replacing `smidge` by the other small quantity $z_1 - z_0$, we get:

$$\mathrm{rate}(z_0) \sim \frac{f(z_1) - f(z_0)}{z_1 - z_0}.$$

If we now set $f(z_1) = 0$ (to get a better approximation to the equation we are trying to solve), after a little mental gymnastics we find the formula

$$z_1 = z_0 - \frac{f(z_0)}{\mathrm{rate}(z_0)}$$

for our improved guess.

root, which we call z_1. This is done by linear interpolation, as explained in Note 9.7. Newton's method is meant to be iterative. This means we now have to repeat the process over and over, giving a list of new values $z_2, z_3 \ldots$. Hopefully, this sequence of improved guesses will home in on the solution we desire. In the next section we explain in more detail how to build our newton solver which you can think of as a black box `newton(mu_0,trace_eqn(p,q))` which takes in a fraction p/q and a starting guess μ_0, and outputs a solution to the equation $T_{p/q}(\mu) = 2$ which (all being well) is very close to μ_0.

Putting all our algorithms together, and supposing that `denom` has been set, the program for drawing part of the boundary of our slice is shown in Box 25.

Building the newton solver. As it stands, Newton's method will probably take infinitely many steps to arrive at the exact solution. Worse, it may take infinitely many steps and never even come close. Both of these eventualities tend to aggravate users waiting for answers. Since we can expect a computer to complete only a finite (preferably smallish) number of operations, we have to come up with a termination criterion. The method produces a stream z_0, z_1, z_2, z_3, $\ldots$ of allegedly improved guesses. We preselect a small threshold ϵ, for example $\epsilon = 10^{-10}$ for ten decimal places

---

Note 9.8: **How can Newton's method fail?**

Here are several ways in which Newton's method can go wrong:

Convergence to the Wrong Root The iteration converges to a solution but it is not the one we wanted.

Near a Double Root This is a rather nasty case where the rate is almost 0, which happens when two solutions have nearly (or even exactly) merged into one. If this happens, the formula for z_{n+1} involves $f(z_n)/(\text{tiny rate})$, so it's huge, or even infinite. This leads to a bad case of 'Divergence'.

Divergence The iterates z_n may grow larger and larger in absolute value without bound; we say they 'diverge to infinity'.

Oscillation The iterates do not diverge, but neither do they approach any limit. Instead, they 'oscillate'. They may stay in a finite region, or even worse their absolute values can erratically (and possibly only after a very long period of almost convergent behaviour) get extremely big. The full variety of possibilities is almost too large to catalogue.

Knowing that we are applying Newton's method to polynomials restricts some of the possible behaviours. For example, if $f(z)$ is a polynomial of degree m, and if z_n is extremely large, then z_{n+1} will be approximately $\frac{m-1}{m} z_n$, which is slightly smaller. Thus in our case it is impossible for divergence to occur.

of accuracy. Discounting the unlikely event that one of our guesses will actually be a solution $f(z_n) = 0$, we decide to declare z_n close enough to a solution if $|f(z_n)| \le \epsilon$.

Of course life is never quite so simple, and in reality there are various ways in which Newton's procedure can go wrong. The most obvious form of failure is that the iteration may converge, but not to the solution we want. We have listed some problems you might run into in Box 9.8. To get a program which really works, you will have to make provision for them all.[1]

To see how badly convergence to roots of a polynomial can fail in Newton's algorithm you have to go no further than the equation $z^3 - 1 = 0$. A beautiful fractal picture of which points converge to which of the three roots is to be seen in the book *The Beauty of Fractals*, H.O. Peitgen and P.H. Richter, Springer, 1988 (and also on the web). In fact, for some cubics, there is non-zero probability that from a randomly chosen starting point, oscillation will occur.

A very important feature of the routine is that the polynomials we shall

[1] There are many possible sources of information about programming and Newton's method; one is *Numerical Recipes in C* already referred to on p. 46.

Box 25: Boundary tracing: part I

In this algorithm, Newton's method is applied consecutively to all fractions p/q in a Farey sequence, seeking the solution μ to the equation $T_{p/q}(\mu) - 2 = 0$, taking as the starting value $\mu(r/s)$, the position we found for the previous cusp. The procedure `newton` is passed the function `trace_eqn` which should return the value

```
trace_poly( p , q , -I*mu, 2, -I*mu + 2*I ) - 2,
```

of the polynomial $T_{p/q} - 2$ at the input value μ. Several possible errors may occur during the running of Newton's algorithm, and some decisions should be made about how to handle these. How you do this depends on your programming environment, so we omit this.

```
boundary(denom) {
    p/q := 0/1                              Start at known cusp 0/1
    mu := 2*I                               and use known value of mu here.
    plot(mu)
    while p/q < 1 do {                      Loop through Farey sequence,
        oldmu := mu                         save previous mu,
        p/q := nextpq( p/q, denom )         find next cusp,
        mu := newton( oldmu, trace_eqn(p,q) )    use oldmu to start newton.
        plot( mu ) }
}
```

be working with may have HUGE degree. (We have taken q up to well over 100,000 in some of our plots.) So it would rapidly become impossible to store information about all this vast array of coefficients as we move along. Instead, our routine calls not for the polynomial itself but only its value, and the value of its `rate`, at various points. Because our trace recursion works with any values plugged in for the traces, the values of a trace at a point z may be found directly from our recursive formula without having to know the trace polynomial itself at all.

This idea is easy to handle in the program because in many programming languages, there is a way to pass a function or procedure as well as a variable to a subroutine. If there isn't, then one has to 'hardwire' the function into the Newton's method subroutine. Thus, rather than being a record of all the coefficients in the polynomial f, `funct` will be a subroutine which calculates the value $f(z) = $ `funct(prm)` given the parameter value $z = $ `prm` by the above method. The argument `rate` will be another similar subroutine. Our newton solver will take the function `funct` and the `rate` as arguments, as well as the initial guess `prm0` for the solution to `funct(z)=0`.

Our routine for finding the rate function is in Box 27. There is a trick involved which will be appreciated by the experts: we use a fancy improved estimate coming from the fact we are dealing with *complex differentiable* functions to give an improved value, see Project 9.5. The process of numerically calculating a rate like this can be quite tricky and there are other approaches.[1]

The 'bad' phenomena listed in Note 9.8 are detected in the program by setting limits. Here are some possible choices:

maxitr This is be the maximum number of iterations allowed before declaring that oscillation has occurred. In practice, we commonly set this at 100.

[1] One reason for choosing the secant method was that since we have not stored the coefficients of the trace polynomial, the usual calculus formula for reading off the derivative from the coefficients will not work. In the interest of speed, we opted to use numerical approximation. This is notoriously dangerous, particularly, when the function in question tends to have strong oscillations. Still, it seemed to be the most practical method.

Box 26: Boundary tracing II: the rate formula

```
rate(prm, funct) {
   smidge := 0.1 * prmeps
   hrate:= (funct(prm+smidge)-funct(prm-smidge))      Find rate in real direction.
           /(2*smidge)
   vrate:= (funct(prm+smidge*I)-funct(prm-smidge*I)) Rate in imaginary direction, I denotes i.
           /(2*I*smidge)
   RETURN((hrate+vrate)/2)                             Average the two estimates.
   }
```

valeps This is how small we shall require $|f(z_n)|$ to be in order to declare the process convergent with answer z_n. We keep `valeps` rather small, say about 10^{-13}.

prmeps While our primary goal is that $f(z)$ be small, we also want to be sure that the parameter z is calculated accurately. Therefore, it is usually advisable to require that the difference from z_{n-1} to z_n also be small, so that it does indeed appear that the method is converging. Thus, to declare victory, we also require that $|z_n - z_{n-1}|$ be less than `prmeps`. Since we are asking for visual accuracy, we usually set `prmeps` to be about 10^{-4}, depending on the frame of reference.

Having got all this set up, it is easy to write our newton solver routine in Box 27.

Finding the next term in a Farey sequence. The input to the routine for calculating successive terms in the Farey sequence will be the maximum denominator `denom` and the current fraction p/q. The output will be the next fraction in the sequence.

So how *are* we going to find the next fraction? Suppose we wish to calculate the fraction p/q which comes after the fraction r/s in the Farey series of denominator d. Any neighbour a/b of r/s on its right gives a solution $x = a$, $y = b$ of the equation $xs - yr = 1$. Because successive terms in the Farey sequence are neighbours, it follows that the next fraction p/q should be the solution $x = p$, $y = q$ to the equation $xs - yr = 1$ with the largest denominator q not exceeding d.

This is where Euclid's marvellous algorithm comes to the rescue.[1] It exactly finds the 'simplest' or 'smallest' solution $x = a, y = b$ to our

[1] See Project 7.5. Detailed accounts of Euclid's algorithm and related matters like Farey sequences and continued fractions can be found in G.H. Hardy and E.M. Wright, *An Introduction to the Theory of Numbers*, Oxford University Press, 1980. (First published 1938).

Box 27: Boundary tracing III: the newton solver

```
newton(prm0,funct,rate) {
    for i from 1 to maxitr do {
        rate := rate(prm0)                    Find rate at prm0.
        prm:= prm0 - funct(prm0) / rate       Update guess for root.
        if |funct(prm0)| <= valeps and        Test new guess.
            |prm-prm0| <= prmeps then RETURN(prm)
        else prm0 := prm                      Try again with new prm0.
        }
    Print "Newton's method failed."           If no root found after maxitr tries, abandon
    }                                          all hope.
```

equation $xs - yr = 1$. Having done this, we just need to adjust our solution a/b to get as near as we can to the previous fraction r/s. We do this by Farey addition: if $x = a, y = b$ is a solution, then the fractions $(a + kr)/(b + ks)$ are also solutions $x = a + kr, y = b + ks$ for any positive integer k. The solution we desire is the largest positive integer k such that $b + ks \leq d$.

Box 28 shows a routine which does all this work. It may be easier to follow if you realise that it is actually the same routine that we used to find trace polynomials, but with several steps combined into one. In the trace routine in Box 24, we ran one pass of our algorithm for each new subdivision or turn in the process of homing in on our goal p/q. Now, because when we do each pass we only want to Farey add fractions rather than invoke our complicated trace formula recursion, it is much faster to combine the whole block of left steps into one single addition, then the whole block of right steps, and so on.

Projects

9.1: Extra circles

The $3/10$ cusp on the Maskit slice has some very unusual properties. First, even though the trace equation has degree 10, the actual value of $\mu(3/10)$ is the very

---

Box 28: Boundary tracing IV: the next Farey fraction

The input to this routine is the maximum denominator `denom` and the current fraction p/q. The output `nextpq(p/q ,denom)` is the next fraction in the sequence.

```
nextpq( p/q , denom) {
   p1/q1 := 0/1; p2/q2 := 1/0;
   r/s := p/q; sign:=-1;
   while s !=0 do {
      a := floor( r/s )
      r/s := s/(r-a*s)
      temp:=p2/q2
      p2/q2 := (a*p2 + p1)/(a*q2 + q1)
      p1/q1 := temp
      sign := -sign }
   k := floor( (denom -sign*q1)/denom )
   RETURN( (k*p + sign*p1)/(k*q + sign*q1) ) }
```

`floor` is largest integer less than r/s.
We are finding the continued fraction of p/q.

$p1/q1, p2/q2$ are Farey neighbours zeroing in on p/q.

simple number

$$\mu(3/10) = \frac{1 + \sqrt{11}\,i}{2} \sim 0.5 + 1.658312\,i.$$

Moreover the continued fraction is very symmetric:

$$\frac{3}{10} = \cfrac{1}{3 + \cfrac{1}{3}}.$$

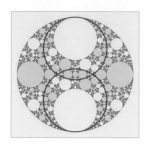

The surprise is in the plot of the limit set. There are some *extra* circles, two of which are marked in red. Why do these circles occur for this value? Do extra circles appear for any other fraction p/q?

9.2: Lower limit for Maskit's boundary

We said that groups with $\mathrm{Im}\,\mu < 1$ are definitely outside the Maskit slice. This is because for any group in the slice, the image of the lower half plane under any element of the group should be strictly contained in the upper half plane. Using the formulas for the image of a circle or a line under a Möbius map, check that:

• The image of the lower half plane under $a(z) = \mu + 1/z$ is the half plane $\mathrm{Im}\,z > \mathrm{Im}\,\mu$.

• The image of the lower half plane under $A(z) = 1/(z - \mu)$ is the disk centre $i/(2\,\mathrm{Im}\,\mu)$ and radius $1/(2\,\mathrm{Im}\,\mu)$.

The condition that the image of the lower half plane under a^2 be disjoint from the lower half plane is the same as requiring that these two disks should be disjoint. Boil this all down to an inequality which shows that $\mathrm{Im}\,\mu \geq 1$.

9.3: Leading terms

There is an interesting pattern in the top two terms of the trace polynomial $T_{p/q}(\mu)$. Can you work out what it is?

9.4: Identities for the p/q words

There are many beautiful relationships among the infinite family of special words $w_{p/q}$. Here are two of them:

• If $p/q < r/s$ and $qr - ps = 1$, then $w_{p/q} w_{r/s}^{-1} w_{p/q}^{-1} w_{r/s} = abAB$.

• If, in addition, $p/q \geq 0$, then when we substitute $w_{p/q}$ for a and $w_{r/s}$ for B, the special word $w_{k/l}$ becomes another special word, namely $w_{(rk+pl)/(sk+ql)}$.

The second looks mysterious but it's not so complicated if you just count the number of a's and b's in the different words. Check that they agree on the two sides.

A good way to make friends with identities like this is to take a special case and see them work. Take, for instance, $p/q = 3/5$, $r/s = 2/3$, $k/l = 1/3$. Once you begin to believe them, the tried and true method for verifying them in *all* cases is to use induction. In our case, this means induction on the R and L moves! In other words, build up the more complicated cases of the identities from the simpler ones and show, just by looking at the effect of these two moves, that if the simple cases hold, then so must the more complicated.

9.5: Calculating derivatives

Readers who know some complex analysis may like to try to work out the meaning of the expression we have used for calculating the `rate` of the function f used in the newton solver. In mathematical notation the expression in Box 26 is

$$\frac{f(z+h)-f(z-h)}{4h} + \frac{f(z+ih)-f(z-ih)}{4ih}.$$

Unlike more obvious approximations you might make, this is accurate up to terms of order h^4.

9.6: Hausdorff dimension

The limit set for the p/q cusp on the boundary of Maskit's slice seems to be increasingly complicated for increasingly complicated fractions. It would be interesting to compute the Hausdorff dimension of cusp groups for various values, revealing the dependence on the fraction p/q. This is a real research project!

CHAPTER 10

Between the cracks

'I thought of a labyrinth of labyrinths, of one sinuous spreading labyrinth that would encompass the past and the future and in some way involve the stars.'

The Garden of Forking Paths, Jorge Luis Borges[1]

In the last chapter, we investigated a special collection of groups we called 'accidents'. These were the beautiful double cusp groups in which two symmetries are forced to be parabolic, corresponding to two different rational numbers of our choice. These groups lay right on the borderline between the relatively well-behaved quasifuchsian regime, and the total disorder of non-discreteness. As we are about to see, however, this is not the full tale. There are other yet stranger groups hovering on this same boundary between order and chaos.

Two millennia before the hesitant introduction of imaginary numbers, came the ancient discovery of another class of numbers whose name has acquired over time an even more negative [*sic*] connotation: *irrational numbers*, numbers, that is, which cannot be expressed as a ratio p/q. Legend has it that Pythagoras, founder of a religious cult circa 500 BC, discovered that $\sqrt{2}$ is irrational, and that the revelation was so unnerving the fact was kept secret within the brotherhood.[2] The irrationals lurk in the cracks between the rationals, a kind of invisible glue without which the line would fall apart.

We drew the Maskit boundary in the last chapter by locating the double-cusp groups corresponding to the words $w_{1/0}$ and $w_{p/q}$. The boundary looked like a continuous cuspy curve. In between the rational cusp points, must lurk some similar invisible glue.

What do these 'irrational' boundary groups look like? We can recognise a group in the quasifuchsian regime because its limit set is a quasicircle which divides the Riemann sphere into two pieces, its inside and its outside. As we move through our parameter space towards a 'rational'

[2] The sources for this apocryphal story are Iamblichus and Proclus – neither a mathematician – writing nearly a millennium later. No one however disputes that the Pythagoreans knew the connection between harmonious chords and small integer ratios. Plato has the Pythagorean Theodorus (465–398 BC) discussing irrational square roots, and Aristotle (384–322 BC) took the non-commensurability between the diagonal and sides of a square as his prime example of the insights mathematics has to offer.

---

[1] Extract from LABYRINTHS by Jorge Luis Borges. Copyright © New Directions Publishing Corporations, 1962, 1964, used by permissions of The Wylie Agency (UK) Limited.

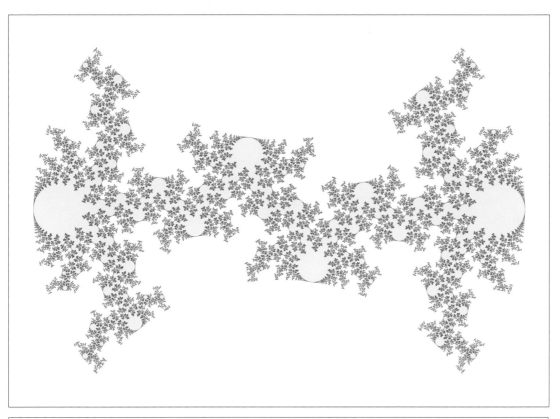

Figure 10.1. Degeneracy.
See text.

boundary point or cusp group, various points on the limit set come together. At the moment we reach the boundary, one or possibly both sides of the ordinary set fragment into a collection of tangent circles. This is what produced the beautiful spiralling chains of circles we met in the last chapter.

However, if we aim instead towards an irrational boundary point, something quite new and startling happens. As we approach the boundary, *one or both of the pieces of the ordinary set shrivels up into a wildly convoluted affair with smaller and smaller area*, until at the point at which we actually reach the boundary, on that side of the limit set, *nothing remains of the ordinary set at all*. Groups whose ordinary sets shrivel away like this are called **degenerate**. The group is called **singly-degenerate** if the quasicircle has swallowed one half of the ordinary set and **doubly-** or **totally degenerate** if it has swallowed both.

An example of a singly degenerate group is shown in Figure 10.1. The group in the top frame corresponds to trace values t_a = 1.785 − 0.73852i, t_b = 3 and is just quasifuchsian. The inside of its extremely convoluted limit set is yellow. Underneath is a singly degenerate group smack on the borderline of non-discreteness at values t_a = 1.7846648 − 0.73852i, t_b = 3. Notice the wild zig zagging of the limit set. If the resolution were finer, you would see that the bottom limit set also has microscopic zigs and zags and that, despite appearances, it actually has area 0. The white disks 'inside' the limit set in the lower picture are due to incomplete searching of the tree of words. Notice the outlines of circular structures still present on the outer part of the ordinary set. They occur because t_b = 3 is real.

The existence of groups like this was discovered by Lipman Bers.[1] His argument ran something like this. Each cusp group is defined by a number of polynomial equations. There are only a countable number of equations and therefore only a countable number of possible cusps. But a continuous boundary has to contain more points than this – it needs some irrational 'glue'. (This is closely connected to Cantor's diagonal argument explained in Note 5.4.) Therefore, there must be groups on the boundary which are *not* double cusps. He proved that for any such group, one side of the ordinary set would completely disappear, so he named these groups degenerate. Although Bers proved that every group on the boundary which didn't contain an accidental parabolic would be degenerate, he didn't actually find matrices which would generate even one single example! This was done by Troels Jørgensen, who first worked out how to locate examples like the ones shown here.

You might have been wondering how Hausdorff's ideas about dimension fit into all this. It turns out that your typical limit set on the

[1] *On boundaries of Teichmüller spaces and on Kleinian groups*, Ann. Math. 91, 1970. Teichmüller theory will be discussed briefly in Chapter 12.

Lipman Bers, 1914–1993

Lipman Bers, known to his students and friends affectionately as Lipa, was born in Riga, Latvia. Already having lived in Petrograd (St Petersburg) and Berlin, he studied briefly in Zurich. On his return to Riga, in addition to his mathematical studies, he began a lifetime of social and political activism. Fleeing to Prague, he continued his studies under Loewner where he received his doctorat in 1938. In great danger from the Nazis, Bers fled once again with his wife Mary, this time to Paris and then, finally obtaining an American visa, to New York. Through all of this turmoil Bers continued his work on mathematics developing ideas which were later used in his work on Teichmüller theory and Kleinian groups. In 1942 he obtained a position at Brown University where he was soon joined by Loewner. His work there on fluid flows contributed to the war effort but also laid the foundations for techniques central to his later work. From 1951 Bers moved to New York, as professor first in the Courant Institute and later at Columbia. Until long after his retirement he inspired and sustained a large mathematical circle in and around New York.

Throughout his life, Bers strove courageously to live up to his ideals as a social and humanitarian activist. He made great contributions to the American Mathematical Society, broadening its horizons to include social and political concerns. He was outspoken in defence of human rights, yet ever courteous, admired and loved by all who came into contact with him. He had in all 48 students, of whom, remarkably and probably uniquely, exactly one third were women. He had legendary talents as a teacher and his love of mathematics inspired a tradition which continues strongly to this day.

Bers felt the irony of devoting much of his life to elaborating the work of such a virulent Nazi as Teichmüller, but quoted Plutarch 'It does not of necessity follow that, if the work delights you with its grace, the one who wrought it is worthy of your esteem.'

boundary of quasifuchsian space is pretty craggy, mathematically speaking. Christopher Bishop and Peter Jones proved that the limit sets of singly degenerate groups like this have the pathological properties of having zero area while their Hausdorff or fractal dimension is 2![1]

[1] *Hausdorff dimension and Kleinian groups*, Acta Math., 179, 1997.

The groups in Figure 10.1 form part of the $t_b = 3$ slice of quasifuchsian space. We drew a picture of this slice using the same method as in the last chapter, just putting $t_b = 3$ in place of $t_b = 2$. The picture obtained from our newton solver is shown in Figure 10.2. The picture has

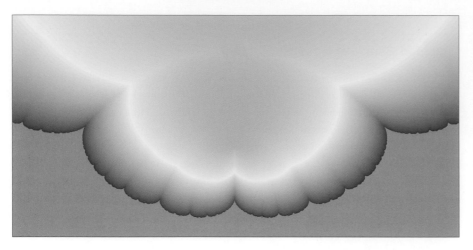

Figure 10.2. The trace 3 slice of quasifuchsian space. This is the slice of groups for which $t_b = 3$ and the commutator has trace -2. The coloured region indicates μ-values for which $t_a = -i\mu$ gives a quasifuchsian group. The boundary bends up slightly, but the cusped effect remains the same as in Maskit's slice in which $t_b = 2$. The recipe which produces the colouring is explained in Note 12.3.

been coloured by a method discovered by the middle author and Linda Keen relating to three-dimensional hyperbolic geometry, and closely connected to the real trace rays shown in Figure 9.17. The idea is explained at the end of Chapter 12.

The most prominent cusps in the picture correspond, from left to right, to the fractions $-1/1$, $0/1$ and $1/1$. The degenerate group in Figure 10.1 corresponds to the irrational number $(\sqrt{3}-5)/2$ and is located near one of the lowest points on the boundary in Figure 10.2. In the next section, we shall explain how to find the analogous group in the slice $t_b = 2$.

Jørgensen's singly-degenerate group

The first person to explicitly locate a singly-degenerate group was Troels Jørgensen who, as we have already mentioned, discovered many definitive facts about quasifuchsian punctured torus groups. His group lies in the Maskit slice where $t_b = 2$, so while one side of the quasicircle collapses, the other side will be a union of disks. To reach Troels' group, we shall try to get down to the boundary of the Maskit slice avoiding all the 'rational' points which belong to the double-cusp groups we studied in the last chapter. On the boundary, the rational points show up as the inward-pointing cusps, just like the ones in Figure 10.2. To avoid them, we are going to aim for the *lowest* point on the boundary, on the soft, round underbelly where no peaky rational cusp group can possibly lurk. The three frames in Figure 10.3 show a series of zooms heading for what appears to be the very bottom of the boundary. In each frame we shrink to a much finer scale, the last frame being a mere 0.00007 times the width of the first.

Figure 10.3.
Self-similarity in the
Maskit boundary. These
three frames show
progressively deeper zooms
into the part of the Maskit
boundary which appears to
be the lowest point on the
curve. The top frame shows
the part of the boundary
between $\mu(0/1) = 2i$ and
$\mu(1/2) = 1 + \sqrt{3}i$. The
minimum appears to be
between the 3/8 cusp and
the 2/5 cusp. The second
frame shows only the part of
the boundary between
$\mu(3/8)$ and $\mu(5/13)$, which
is only 0.03774 wide. (The
scale, as usual, being set by
the parameter μ.) The
bottom frame is the part of
the boundary from
$\mu(144/377)$ to $\mu(233/610)$,
which measures only
0.000071 across. The
colouring, still chosen by the
Keen–Series method, is
suggestive of descent, but is
not intended to be consistent
from one frame to the next.

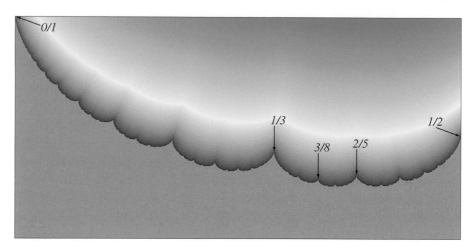

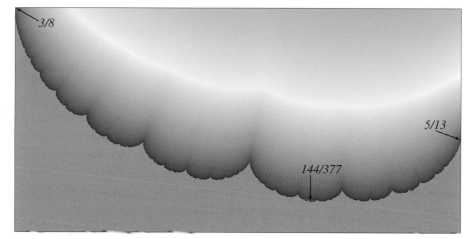

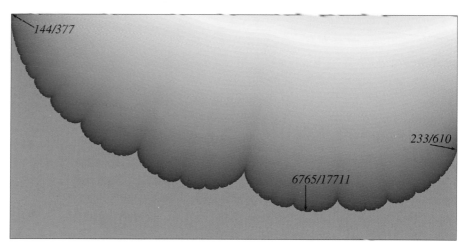

As we head deeper into the boundary, numerical calculations give a suggestive pattern. At each stage, we zoom in to a pair of peaked cusp points which look as if they surround the lowest point in sight. The first frame shows all the cusps between $0/1$ and $1/2$. Then, since $1/3$ is a lower peak than $1/2$, we zoom into the right 'half' from $1/3$ to $1/2$, then $2/5$ looks still lower so we rescale to $1/3$ to $2/5$, then go to the right half of the next interval from $3/8$ to $2/5$, and so on. The 'halving' process is carried out by Farey addition, so the next 'midpoint' would be $(3 + 2)/(8 + 5)$. Notice the pattern in which the moves go: Left Right Left Right Left Right $\cdots$. We have shown only selected stages in the pictures. The second frame shows all cusps between $3/8$ and $5/13$, and the third all those between $144/377$ and $233/610$.

The fractions coming here up are closely connected to another ancient pattern: the **Fibonacci numbers**. The Fibonacci numbers are defined by a recursive process. Starting with $F_0 = 0$ and $F_1 = 1$, we proceed to

$$F_2 = 1 + 0 = 1, \ F_3 = 1 + 1 = 2, \ F_4 = 2 + 1 = 3, \ F_5 = 3 + 2 = 5$$

and so on. At each stage, the next number is the sum of the previous two. The first sixteen Fibonacci numbers are:

$$0, \ 1, \ 1, \ 2, \ 3, \ 5, \ 8, \ 13, \ 21, \ 34, \ 55, \ 89, \ 144, \ 233, \ 377, \ 610, \ 978, \ 1597.$$

Taking the ratio of successive Fibonacci numbers,

$$1/1, \ 2/1, \ 3/2, \ 5/3, \ 8/5, \ 13/8, \ \ldots, \ F_{n+1}/F_n, \ \ldots$$

and so on brings you ever nearer the famous **golden ratio** or **golden mean**:

$$\frac{F_{n+1}}{F_n} \longrightarrow \frac{1 + \sqrt{5}}{2} \doteq 1.618033989.$$

In terms of continued fractions:

$$\frac{3}{2} = 1 + \cfrac{1}{1 + \cfrac{1}{1}}, \qquad \frac{5}{3} = 1 + \cfrac{1}{1 + \cfrac{1}{1 + \cfrac{1}{1}}}, \qquad \frac{8}{5} = 1 + \cfrac{1}{1 + \cfrac{1}{1 + \cfrac{1}{1 + \cfrac{1}{1}}}}$$

and so on. The way to continue the pattern is obvious! As explained in Project 10.1, the golden mean is exactly the *infinite* continued fraction

$$\frac{1 + \sqrt{5}}{2} = 1 + \cfrac{1}{1 + \cfrac{1}{1 + \cfrac{1}{1 + \cfrac{1}{1 + \cdots}}}}.$$

Now the fractions belonging to cusps which bound our zooms come in the pattern:

$$0/1, \ 1/2, \ 1/3, \ 2/5, \ 3/8, \ 5/13, \ \ldots, \ F_{n-2}/F_n, \ \ldots$$

whose limiting value can be found as:

$$\frac{F_n}{F_{n+2}} = \frac{F_n}{F_{n+1}} \frac{F_{n+1}}{F_{n+2}} \longrightarrow \left(\frac{2}{1+\sqrt{5}}\right)^2 = \frac{3-\sqrt{5}}{2}.$$

We are led to conjecture that the minimum point on the Maskit boundary occurs at the μ-value which corresponds to the irrational number $\frac{3-\sqrt{5}}{2}$! Because the Maskit boundary is symmetrical under translation by 2 and reflection in the imaginary axis, we could just as well have done the same thing at the symmetrical point corresponding to the number $-(\frac{3-\sqrt{5}}{2})+2 = \frac{\sqrt{5}+1}{2}$. That is, experimentally, *Maskit's boundary reaches a minimum at the μ-value corresponding to the golden mean*. Pythagoras would have been proud of us!

When we first made these calculations, Troels eagerly anticipated the actual value of μ(golden mean), especially its imaginary part. The answer (drumroll, please!) was

$$\mu\left(\frac{1+\sqrt{5}}{2}\right) \doteq 3.2943265032 + 1.6168866453\,i$$

correct to 10 places. At least, let's say our Newton's algorithm stabilized at this value. The imaginary part is just a shade below the golden mean itself. No explanation for this strange coincidence has yet been offered.

The reader can hardly have failed to spot another phenomenon which the three frames in Figure 10.3 were designed to suggest. There is hardly any difference between them! Indeed they are scaled so that if the third frame were overlaid on the second, the two peaked boundaries would match almost exactly. This phenomenon was discovered by Curt McMullen. It is called **asymptotic self-similarity**. Let's phrase it in terms of scaling. Suppose we just consider the part of Maskit's boundary between cusps belonging to two successive Fibonacci fractions F_{2n}/F_{2n+2} and F_{2n+1}/F_{2n+3}. Call that part $M(n)$, just to give it a name. Thus $M(0)$ is the part from the $0/1$ cusp to the $1/2$ cusp, $M(1)$ is the part from $1/3$ to $2/5$, and so on. The second frame in our picture, from $3/8$ to $5/13$ is $M(3)$, and the bottom frame from $144/377$ to $233/610$ is $M(6)$.

The ratio of the size of $M(n)$ to that of $M(n+1)$ (the next piece down) is approximately (close your eyes and guess first):

$$\frac{5+\sqrt{21}}{2} \doteq 4.791287848.$$

For those of you who guessed the golden mean and are now scratching your heads in bewilderment, this extraordinary rabbit was pulled out of a hat by Curt at the International Congress of Mathematicians, Kyoto, 1990, see Note 10.1 and Project 10.2.

So what does the limit set at Jørgensen's point look like? We just found a numerical value of μ to use in our limit set program. The traces of the words a, b and $abAB$ are $-i\mu$, 2 and -2, respectively. Let's enter them in. Rather than using the golden mean itself, we'll choose the group on the same horizontal level corresponding to the value $\frac{3-\sqrt{5}}{2}$ for which

$$\mu \doteq 0.7056734968 + 1.6168866453\,i.$$

The result is Figure 10.4.

As usual, the figure has been plotted as a closed curve with one side coloured white and the other black. The white disks of the ordinary set all represent the same triply-punctured sphere. What was the other half of the ordinary set has completely disappeared! The jagged yellow lightning streaking across the picture is also part of the limit set; it is the remnants of the special core chains of circles which we met in Chapter 9. It separates the part of the limit set given by words beginning with a from the parts beginning with A, b and B: we will discuss this decomposition further later in the chapter.

Both pictures are actually very crude representations of the limit set. One problem is that it looks as if there are substantial black regions, notably at the two extreme ends of the yellow lightning, but these only appear because of the imperfect convergence of our algorithm and the difficulty of plotting a limit curve of zero thickness! In a true picture,

---

Note 10.1: **Checking the self-similarity factor**

Let's check Curt's prediction that the self-similarity factor for our zooms into Jørgensen's group is

$$\frac{5+\sqrt{21}}{2} \doteq 4.791287848$$

against some good old-fashioned numerical calculation. The ratio of the size of successive zooms can be computed as the ratio of the horizontal distances between the appropriate Fibonacci cusps. Our newton solver furnishes a table:

p/q:	Re(μ)	Im(μ)
55/144:	0.7054471339427	1.6170362741583
89/233:	0.7057877733739	1.6169347201325
144/377:	0.7056262552364	1.6169178809595
233/610:	0.7056973462763	1.6168966786925

This gives the ratio of $M(5)$ to $M(6)$ as

$$\frac{0.7057877733739 - 0.7054471339427}{0.7056973462763 - 0.7056262552364} \doteq 4.7916$$

which is pretty darn close!

Figure 10.4. Troels' Point.
Both these pictures show the limit set of Jørgensen's group. The 'true' picture consists of a dense set of white disks – the ordinary set – plus a tree-like black web of zero area but Hausdorff dimension 2 – the limit set. Drawing it is problematic and the two figures here differ only by allowing the white disks to include one more pixel on their boundary than the black ones. The jagged yellow curve is what remains of the 'circle chain' in the limit set; it separates the *a*-part of the limit set from the *b*, *A*- and *B*- parts. The μ-value used to make this picture was $0.70567 + 1.61688\,i$.

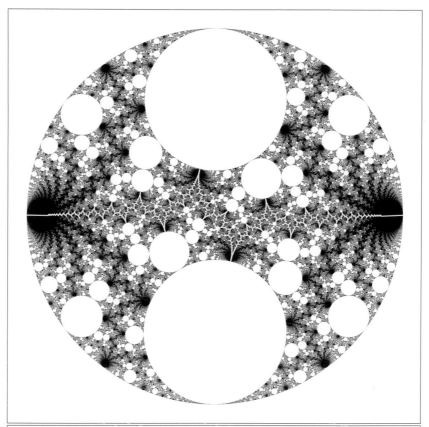

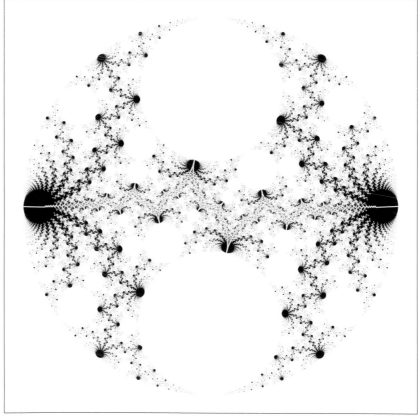

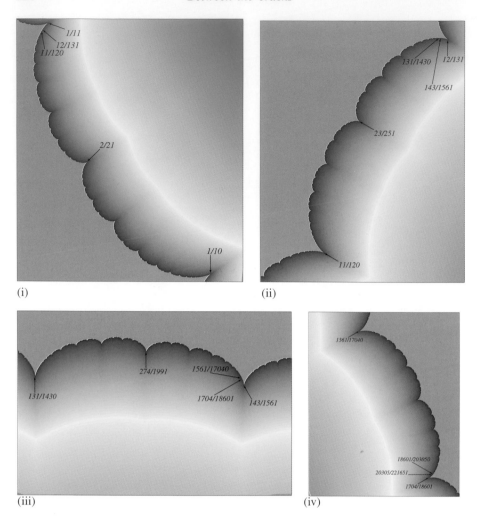

Figure 10.5. Overhang in the Maskit boundary. Four successive zooms into a small part of the boundary very near the principal cusp at 0/1. Each frame zooms in to the arrows marked in the previous one. Already at $\mu(1/11)$ in frame (i), we can see a slight overhang in the boundary. In frame (ii) the boundary is at a very steep angle and in frame (iii) we are completely upside-down so that the region of discrete Maskit punctured torus groups is below us. Frame (iv) has spiralled almost back to the horizontal. Continuing this zooming, we realize that the boundary spirals forever. To give an idea of the success of the Newton's method approach to tracing the boundary, the width of frame (iv) is 0.0000000023; the width of one period of Maskit's boundary is 2.

the black region would completely disappear except for a very intricate and twisted web, which is the limit set. Tendrils of white circles reach out and touch other parts of the limit set in a completely interlocking way.

A spirally degenerate group

The large grey pointed peak just at the left side of the top frame of Figure 10.3 is the cusp corresponding to 0/1. High on its slopes are the smaller cusps $\mu(1/11)$ and $\mu(1/10)$. Let's put on our crampons and climb up there, which is another way of saying we shall adjust the parameters of our boundary-drawing program to plot just that stretch (or slightly more). The result is the first frame of Figure 10.5.

We shall need technical climbing gear if we wish to scale right up to the rocky grey underside of the 1/11 cusp. The result is shown in the second frame, where we have blown up the nearby stretch from 11/120 to 12/131. Along this stretch, the grey rocks above us are nearly pointing upside down! We are also shrinking ourselves dramatically as we climb into these overhangs. We shall need Batman's suction-cups to climb even further, aiming for the very roof underneath the 12/131 cusp as we do in the third frame. The grey stalagmites in the first frame have now become stalactites, or vice versa, since we always have trouble remembering which is which.[1] Let's carry on and shrink ourselves even smaller. By the time we reach the fourth frame, we have travelled in a very tight spiral so that the small piece marked by arrows towards the bottom right is practically level ground.

[1] Actually, we have researched the matter, and we have them correctly distinguished. (We think.)

What is the pattern we used to make these zooms? In the language of making left and right turns in the Farey road map used in the discussion on p. 283, each stage in our approach to Jørgensen's point in the last section required precisely one left turn and one right turn (or *LR* for short). The bottom of the boundary was at the end of a forced march of 'Left! Right! Left! Right! Left! Right! ...' forever chanted by a sadistic drill sergeant.

Our new approach has been conducted by taking ten 'left turns' followed by one 'right turn'. More precisely, starting at the pair of fractions 0/1, 1/0, we took ten left turns by means of Farey addition to arrive at 0/1, 1/10. Then a right turn leads us to the pair 1/11, 1/10. Applying the same steps again leads to the prominent cusps in the second frame corresponding to 12/131, 11/120. Carrying out ten lefts and one right again leads to the third frame, and one more time still into the fourth.

The number corresponding to our new pattern is the infinite continued fraction

$$\cfrac{1}{10 + \cfrac{1}{1 + \cfrac{1}{10 + \cfrac{1}{1 + \cdots}}}}$$

which by a little trickery (see Project 10.1) works out as equal to

$$\frac{\sqrt{35} - 5}{10} \doteq 0.0916079783.$$

The limit of this process should also be an irrational point on Maskit's boundary – another degenerate group. McMullen's asymptotic self-similarity theorem works for this boundary point too. In this case the scaling factor is a complex number, *which also accounts for the spiralling!*

This is just like our discussion about the effects of multiplication by a complex number way back in Chapter 2. McMullen's theory calculates the factor by which μ is multiplied or **renormalized** as you move from one frame to the next. In this case, an approximation to this scaling factor is $86.214 + 164.198\,i$.

Our series of pictures makes it possible to test this value. The renormalization represented by $L^{10}R$ should carry the piece marked out by arrows in the second frame to that in third. This leads us to the calculation (according to our newton solver)

$$\frac{\mu(12/131) - \mu(11/120)}{\mu(143/1561) - \mu(131/1430)} \doteq 86.158 + 164.247i,$$

again accurate to three decimal places.

Let's draw the limit set for the degenerate group at the end of all this spiralling. Remarkably, the spiralling in our Maskit road map is mirrored by spiralling in the limit set. The sequence of zooms already suggests to some accuracy the value of μ that determines this group; by the time we reach the last frame of Figure 10.5 that value is known to within 10^{-8}. With double precision arithmetic, the newton solver can extract even finer information; the value we use to make Figure 10.6 is $\mu = 0.0273817919 + 1.9264340533\,i$. Again the black part is degenerating to nothing, although the convergence is so slow there is still a substantial black part visible. This is due to the human imperfection of our drawing algorithm, and does not reflect the state of grace this degenerate group has achieved. Indra has flung his thunderbolt, and the lightning flashes back and forth across the white $1/10$ circle chains which can be seen reproducing themselves throughout the picture at ever diminishing scales.

How it all collapsed

If we start with a quasifuchsian group and then gradually change trace parameters approaching a degenerate group, the regular set gets crushed to nothing and the points in the limit set come together in a wildly convoluted pattern. The easiest way to understand this collapse is not to approach Jørgensen's group from the interior of quasifuchsian space as we have done up to now, but through double-cusp groups given by the fractions F_n/F_{n+2}, like the $55/144$ group shown in Figure 10.7. These groups are very close to Jørgensen's group, and we know quite a lot about them.

The story is a continuation of the one we started in Chapter 9, especially on p. 274, where we discussed the anatomy of the $2/5$ double cusp.

Figure 10.6. Spirally degenerate. This limit set corresponds to the group on Maskit's boundary at the end of the spiralling pattern of zooms in Figure 10.5.

Figure 9.5 is a simpler version of Figure 10.7. Both figures show the division of the limit set into its a-, b-, A- and B-parts, the parts defined by infinite words beginning in a, b, A and B. These four regions are separated by a network of crimson and bright blue circle chains, analogous to the chains we drew in Chapter 9. The 'a-part' and 'A-part' of the limit set (red and green) are surrounded by $q + 1 = 145$ of the crimson disks and one of the bright blue ones, while the b- and B-parts (dark blue and yellow) are surrounded by $p + 1 = 56$ of the crimson ones and both of the bright blue ones. The crimson disks form part of the remnants of the inner part of the ordinary set and you can get from one to another by applying a suitable transformation in the group. The bright blue disks belong to the second class of disks – the remnants of the 'outside'.

Analogous to the decomposition in Figure 9.5, we can use these two chains to define four Schottky regions surrounding the a, b, A and B parts of the limit set in such a way that the transformation a carries

Figure 10.7.
Approximating Jørgensen's group. The parts of the limit set corresponding to infinite words beginning with *a*, *b*, *A* and *B* are coloured just as in Figure 9.5. The chains of circles which separate these four parts are crimson and bright blue.

the interior of the *A*-region to the exterior of *a*-region and so on, which we called Schottky dynamics in Chapter 4. As before, the boundaries of each region are made up of orthogonal circular arcs through the tangency points which snake their way through the chains. Notice that the circles in the chain can be divided into two types: those which contact only two of the four parts of the limit set, and those which contact three different parts. There are precisely two crimson and two bright blue circles of this latter type, representing the junctions of three of the four parts of the limit set. In these, the Schottky regions omit ideal triangles just as in Figure 9.5. The Schottky regions are now very seriously smushed together indeed.

As we move through cusp groups for fractions p/q approximating $\frac{3-\sqrt{5}}{2}$ (the location giving Jørgensen's point on the Maskit slice), the disks in the crimson chain increase in number and shrink steadily in size, until finally we have the continuous jagged lightning curve seen in Figure 10.4.[1]

[1] To prove that the violet circles really do shrink to zero radius posed a severe theoretical challenge. The result is a consequence of recent work by Yair Minsky and Curt McMullen.

It is an interesting puzzle to work out just which infinite words represent the same limit point in a degenerate group. In a Schottky group, each limit point was represented by just one infinite word. When we brought the Schottky circles together in Chapter 6, the fixed points of the commutator came together so that the two words $\overline{abAB}$ and $\overline{baBA}$ represented the same point. When we made the gasket and then other cusp groups, each extra accidental parabolic created more pairs of infinite words representing the same limit point. In our 55/144 cusp, for example, *the points where two infinite words define the same limit point are exactly the tangency points between the circles in the crimson and blue chains.* These points are the fixed points of b, $abAB$, the word $w_{55/144}$, and all their cyclic permutations. For instance, a tangency point where the red region meets the green one is defined by an infinite word ending in a and another infinite word beginning in A.

Our method of approximating an irrational degenerate group enables us to determine which pairs of words define the same limit points in the degenerate case. Sticking to the case of Jørgensen's group, the list on p. 284 shows the words which become parabolic in the approximations by double-cusp groups belonging to successive Fibonacci fractions. If you look back, you will see several patterns visible in this table. As we should expect, the next word in the sequence is a concatenation of the previous two words (in alternating orders, because you always have to put the word belonging to the smaller of the two fractions first). The final three words listed are:

$$w_{3/8} = aaaBaaaBaaB, \quad w_{5/13} = aaaBaaaBaaBaaaBaaB,$$

$$w_{8/21} = aaaBaaaBaaBaaaBaaaBaaBaaaBaaB.$$

The most important observation for our purposes is that these words appear to be 'stabilizing': after a certain distance down the list, the beginning part of the word does not change. That means there is a well-defined infinite word which is the 'limit' of these p/q words. The first 21 terms of it are the last word of our table. We'll call this infinite word $w_{\frac{3-\sqrt{5}}{2}}$, or w_r for short.

Now if one of these words, say $w = w_{8/21}$, is parabolic, then the infinite word consisting of infinite periodic repetitions $\overline{w}$ collapses to the same limit point as the infinite word corresponding to its inverse, that is $\overline{W}$, where

$$W = bAAbAAAbAAbAAAbAAAbAAbAAAbAAA.$$

We should like to say that in the limit, the infinite word w_r collapses to its 'inverse' word – the only snag is, what *is* the inverse of w_r? To find the inverse of a finite word you have to start at the right and read backwards, so how are we going to do this for an infinite word?

To resolve this problem, let's look at another way of getting w_r. In Chapter 9, we explained how the p/q words can be read off from a straight line of slope p/q cutting across a square grid in the plane. Why not let the slope be an irrational r? Figure 10.8 shows a line of slope $\frac{3-\sqrt{5}}{2}$ cutting through a grid in which the vertical and horizontal lines go through the integer points of the plane. As usual, we can read off w_r from the pattern of cuts by interpreting a cut across a vertical line as an 'a' and a horizontal cut as a 'B'.

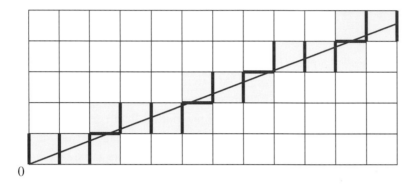

0

Figure 10.8. Line of irrational slope. Each red or vertical cut corresponds to an a and each blue or horizontal cut corresponds to a B. The cuts shown here give *aaaBaaaBaaBaa* at the beginning of the infinite word, corresponding to slope $(3 - \sqrt{5})/2 \doteq 0.381966$. Sliding the initial point up along the initial red vertical segment leads to an uncountable number of different infinite words all representing lines of the same slope.

For a line of rational slope p/q, the pattern of cuts repeats as soon as the line passes through the integer grid point (q, p), giving an infinite repetition of the p/q word. The clue is, that if you read backwards from the same starting point, the word you get will consist of infinite repetitions of the *inverse* word. (This is because we are crossing the grid lines in the opposite direction, so we should read A for a vertical cut and b for a horizontal one.) Since the p/q word is parabolic, these two infinite periodic words represent the *same* limit point. More generally, you don't have to start at a grid point. Take any point on the y-axis and lay out a line of slope p/q and you will get a repeating word, just a cyclic permutation of the first one. The infinite repetitions of each of these forwards and backwards cyclic permutations give a list of the pairs of infinite words which represent the same limit point on the limit set for the group $\mu(p/q)$. (Of course there are other pairs whose limit points coincide as well – the result of prepending the same finite non-cancelling string to the beginning of each of the two words in any of these special pairs.)

If the slope is irrational, on the other hand, the line goes through at most one grid point and we get an infinite sequence which never repeats. If you also read off the backwards word for the same line from the same starting point you will get another infinite non-periodic word. For the golden mean, for instance, the line which starts at the origin gives the two words

$$w_r = aaaBaaaBaaBaaaBaaaBaaBaaaBaaB \cdots,$$
$$W_r = bAAbAAAbAAbAAAbAAAbAAbAAAbAAb \cdots.$$

Thus we have found two infinite words which represent the same limit point on the limit set for the group $\mu(r)$!

Once again, if you slide the initial point up along the initial red vertical segment you will get a different infinite word. However, while in the rational case this only gave a finite number of different words, now we get an *uncountable* number of different infinite words all representing lines of the same slope r. With any slope r and starting at any point on the y-axis, you will get a pair of infinite words which represent the same limit point on the limit set for the group $\mu(r)$!

This describes the pattern of collapsing in the degenerate groups at all but the points on the orbit of $\overline{abAB}$, the flat points on the limit set. Here there is a more complicated pattern (which, incidentally, is quite visible if you are able to run a program plotting the limit set in real time so that you can watch when it returns to plot the same point twice). The key is that in an approximating double-cusp group, for any fraction p/q, both $w_{p/q}$ and $\overline{abAB}$ always belong to the same circle (the one called C_0 in Chapter 9) in the core chain. When the diameter of this circle shrinks to zero, these limit points come together so that the limit point corresponding to the infinite word w_r must also equal the commutator fixed point $\overline{abAB}$.

That gives at least *three* distinct infinite words ($\overline{abAB}$, $\overline{baBA}$ and w_r) corresponding to the *same* limit point in this degenerate group. In fact, the commutator must now fix the infinite word w_r, since the point represented by w_r is actually *equal* to the commutator fixed point 1. This means we may precede w_r by any power of the commutator, as in

$$(abAB)^n \, aaaBaaaBaaBaaaBaaaBaaBaaaBaaB \cdots.$$

All these infinitely many distinct infinite words correspond to the same limit point of the group!

As we explain in Project 10.4, deriving the word associated to an irrational number in this way leads to an algorithm for plotting the 'lightning' in Figures 10.4 and 10.6.

Good old-fashioned 'natural' sphere-filling curves

We next want to look at doubly-degenerate groups. For these groups, both sides of the ordinary set are supposed to shrivel away, so the limit set must fill up the whole Riemann sphere. Wait a minute! The limit set is plotted as a single closed curve by a pen which moves continuously and never leaves the paper. However crinkled up it gets, how can a *curve* possibly go through every single point on the sphere? We had better reflect on this paradox first.

Through the exploration of the world we live in, we have come to some kind of understanding of the *dimension* of space and the objects encountered in it. A one-dimensional thing has only one direction to move in, a two-dimensional thing has two 'independent' directions, and so on. It sounds simple, but as mathematicians began to think about these questions more deeply, they realised they needed to ask more and more subtle questions about what really distinguishes a one-dimensional thing from a two-dimensional thing. We have already seen how Hausdorff's ideas on dimension led to the modern idea of a fractal, but another quite different approach started by Cantor also led to some astonishing revelations.

We have already mentioned Cantor's investigations into countability. His 'diagonal' proof that the real numbers are 'uncountable' (see p. 129) showed there are different sizes of infinity, of all things. At the same time, he sought a proof that the points of the plane formed a 'larger' set than the points of a line, which would make sense since a two-dimensional thing is clearly[1] 'larger' than a one-dimensional thing. The startling realization Cantor slowly came to was that, in the sense of correspondence that he had established as a means of comparing two infinite sets, the line and the plane were precisely the same size!

Then what really is the difference between one and two-dimensional objects? One problem with the correspondence that Cantor found is that it is not 'continuous', in the sense that nearby points on the line may correspond to widely separated points in the plane, and vice versa.

[1] We use the term *clearly* in the strict mathematical sense that we don't know how to prove it so don't want students to bother us with questions.

Note 10.2: **Cantor's correspondence**

The essence of Cantor's proof that the line and the plane contain the same number of points is quite simple. Start with a point on the plane, thought of as a pair of infinite decimals measuring its x, y coordinates. For simplicity, let's assume the two numbers are between 0 and 1. Now 'weave' the two numbers in the pair to form a single number by alternating their decimal digits:

$$\left.\begin{array}{l} 0.\,a_1\,a_2\,a_3\,\ldots \\ 0.\,b_1\,b_2\,b_3\,\ldots \end{array}\right\} \Rightarrow 0.\,a_1\,b_1\,a_2\,b_2\,a_3\,b_3\,\ldots$$

Going backwards, we can 'unweave' a single decimal to give a pair of numbers, so there is an exact one-to-one correspondence between points on the line and points on the plane.

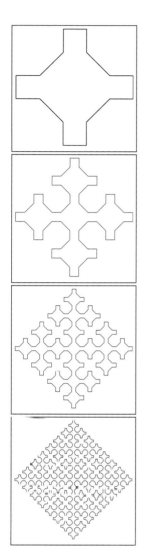

[1] See *The Algorithmic Beauty of Plants* by P. Prusinkiewicz and A. Lindenmayer, Springer, 1990.

It was believed that it would surely be impossible to find a *continuous* way to draw a curve which filled up the whole plane. That belief also was shaken by the discovery by the Italian mathematician Giuseppe Peano (1858–1932) of so called **area-filling** or **Peano curves**, curves so convoluted that they actually fill up the whole plane. (Unlike Cantor's maps, in a Peano curve several different points on the curve can (in fact must) go through the same point on the plane.)

In the margin you can see a variation of Peano's constructions (adapted from Mandelbrot's *Fractal Geometry of Nature*). If the pattern is repeatedly infinitely it produces a continuous mapping of a line segment onto a square. The plots were made with the aid of an *L*-system (short for **Lindenmayer system**) designed by Adrian Marano.[1]

The curves illustrated in the margin are rather like the singly-degenerate limit sets. Each curve is a closed loop and as you progress through the sequence, the region inside the loop fills up with finer and finer squiggles. The region outside the diamond, on the other hand, remains empty. In the limit, the region inside the loop disappears but the region outside remains. With a little thought, one can devise a way to modify this construction to produce a truly sphere-filling curve. The first step is to transfer each loop from the plane to the sphere via stereographic projection. Imagine each curve fitting on the sphere rather like a loosely knitted sock. Because the curves were enclosed in a square on the plane, the sock only covers the southern hemisphere. Now pull suitable strands in each successive loop up towards the North Pole, so that the modified 'socks' evenly cover more and more of the sphere. If you do this with enough care, you can arrange that in the limit you get a continuous map of the line segment whose image fills up the whole sphere (including the North Pole).

Peano created his area-filling curves for no other purpose than to show it could be done. In the 19th century, many intuitive ideas about space and time were cast aside by such ingenious but artificial constructions. The great Karl Weierstrass (1815–1897), himself one of the foremost producers of exotic counterexamples, is said to have referred to them as 'monstrosities'. We had to wait until the last quarter of the 20th century for Mandelbrot to point out that these fractal 'monstrosities' in fact model the normal behaviour of nature – just look at the clouds and the trees!

Mathematicians may disagree about many things, but they are all strongly of the opinion that some of the objects they discover are also 'natural'. The beautiful and intricate limit sets we have been plotting in the last few chapters are a good example. The crinkled and repetitive fractal shape of Peano's curve is not unlike the limit sets in Figures 10.4

and 10.6, but of course our pictures were remnants of Maskit groups so they contain lots of open white holes which the limit set never penetrates. Perhaps, inspired by the sock analogy, we can find a limit set which actually fills the plane.

Here's a way in which this can be done. We are going to approach a degenerate group, but at the same time we are going to *couple* the inside and the outside of the limit set, so that they are forced to behave in a similar way. We saw an example of the kind of coupling we have in mind in the sequence of kissing Schottky groups on p. 231 with $t_a = t_b$ real and varying between $2\sqrt{2}$ and 2. As these groups approach the gasket group, both the inner and outer parts of the ordinary set develop horns. These groups belong to a bigger class in which $t_a = \overline{t_b}, t_{ab} = \overline{t_{aB}}$. As described in Project 10.5, one can write down a pair of matrices a and b with traces satisfying these relations in such a way as to create a group in which the map $z \mapsto i/\bar{z}$ interchanges the inside and the outside parts of the ordinary set. An example is shown in Figure 10.9. On the left, the quasifuchsian limit set is shown as usual in the plane, but the symmetry between the inside and outside is more easily apparent when on the right it is lifted to the sphere. See how the limit set is systematically unfolding its wings as it seeks to fill every nook and cranny – very much the same process as that adopted by the 'man-made' Peano curves.

Figure 10.9.
Approximating a 'natural' sphere-filling curve. On the left, the limit set for $t_a = 1.55 + 0.866i$, $t_b = 1.55 - 0.866i$ in the plane as usual. On the right is the same limit set on the Riemann sphere. Peering through to the far side, see how symmetrical the limit set is around the North and South Poles.

Jørgensen's doubly-degenerate group

[1] *Compact 3-manifolds of constant negative curvature fibering over the circle*, Ann. Math., 1977.

The example in Figure 10.9 is an approximation to a true sphere-filling limit set discovered by Jørgensen.[1] Jørgensen actually discovered a whole category of groups of Möbius maps whose limit sets were later described by Cannon and Thurston as 'natural sphere-filling curves', as if they were suitable for sale in the organic produce section of the grocery. One of the remarkable feature of these groups is that they are still *discrete* in the sense we discussed on p. 239 ff., even though *their limit sets fill up the whole sphere*. When there is no regular set, it is not at all obvious how to tell when a group is discrete, although in fact it is a theorem that any limit of discrete groups is itself discrete. In the case of Jørgensen's groups there are deeper reasons connected with three-dimensional hyperbolic geometry, to be touched on in the final chapter.

One way to understand Jørgensen's doubly-degenerate group is to see it as a limit of double cusp groups, using a form of the sock trick explained in the last section. The idea is that we shall start with the sequence of double cusp groups with cusps $1/0$, P/Q which we used to converge to Jørgensen's singly-degenerate group, and use Grandma's party trick explained on p. 299 in the last chapter to conjugate these groups into groups where the cusps are of the form p/q, $-q/p$ (where q is quite a bit smaller than Q). In this new group, it turns out, the new generators u and v will have traces satisfying the relation $t_u = \overline{t_v}$. Thus the new group we get will satisfy the symmetry property that the inside of the limit set behaves in exactly the same way as the outside. As the unconjugated groups approach a singly-degenerate group, so the conjugated groups will approach a doubly-degenerate group. The inside and the outside of the new limit sets will act alike, one shrinking if the other does.

It turns out that this plan works beautifully if we take P/Q to be the word we get from an even number $2n$ of passes of the Right–Left change of generators. We shall choose the new generators u and v to be the ones we get after exactly n passes of the algorithm; that is, the 'middle' generator pair. In this new group, the words we get from n forward and n backwards RL passes are cusps, which turn out to be always of the form $w_{p/q}$ and $w_{-q/p}$. Moreover it also turns out that $t_u = \overline{t_v}$. In the example in Chapter 9, P/Q is $21/34$, coming from 4 passes of the RL algorithm. We found that the double cusp group with parabolic words $w_{3/5}$ and $w_{-5/3}$ (coming from two forward and two backwards passes of the RL algorithm) was a conjugate of the group with parabolic words $w_{21/34}$ and B. We already saw in this double cusp group that $t_u = \overline{t_v}$.

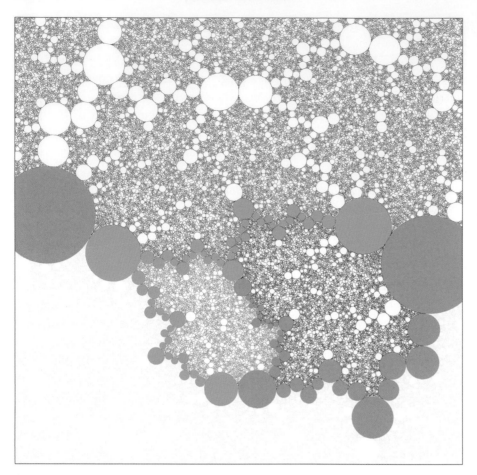

Figure 10.10. A big double cusp. In this group the $\frac{21}{34}$ and $\frac{-34}{21}$ words are cusps. The corresponding circle chains are light blue (for 21/34) and lavender (for −34/21). The a, b, A and B parts of the limit set are, in order, red, blue, green, and yellow. See how these chains of circles delineate the four parts of the limit set. In terms of the μ-parameter, this group is within a few thousandths of the doubly degenerate group in Figure 10.11.

Here is a much bigger example: start with a big double cusp group, say the one with cusps for $B = w_{1/0}$ and $w_{987/1597}$ (coming from eight forward passes of the *RL* algorithm), on the boundary of Maskit's slice. Our newton solver gives the value

$$t_a(987/1597) = 1.616888739343 - i\,1.294321525602.$$

The generator pair right in the middle is

$$(w_{21/34}, w_{13/21}).$$

In our double cusp group, the traces of these middle generators are

$$\operatorname{Tr} w_{21/34} = 1.501347474086 - 0.865385203320\,i,$$
$$\operatorname{Tr} w_{13/21} = 1.501347474169 + 0.865385203218\,i.$$

Grandma's procedure is to plug these trace values into her recipe, resulting in two new generators u and v. In the new group, the (a, B) word

$w_{987/1597}$ equals the '21/34 word' in the new generators u and v, while the word B is the '$-34/21$ word'. This means we have created a group in which the 21/34 and $-34/21$ words are cusps. Its portrait is displayed in Figure 10.10. Notice that the traces of the two generators are conjugate, so (in a suitable normalization) the inside and the outside of the ordinary set are 'the same'.

To find the traces which made our first singly-degenerate group, all we could do was run through ever better approximations to the trace values as we sunk through cusp groups lower and lower towards the bottom of the Maskit slice. A wonderful discovery of Jørgensen was that, just as to find double cusp groups we had to solve a set of equations which said that certain words were parabolic, so there is a quite simple set of equations which can be solved to find the *exact* values of the traces which give his doubly degenerate group. To understand where his equations come from, let us proceed, as the mathematician Dennis Sullivan likes to say, by 'pure thought'.

When we took the sequence of Maskit double cusp groups, with parabolic words $1/0$, p_n/q_n with p_n/q_n tending to the value $r = (3-\sqrt{5})/2$, then the trace values converged to the value for the Jørgensen single cusp group $t_a = i\mu \left(\frac{1+\sqrt{5}}{2} \right) \doteq -1.6168866453 + 3.2943265032\,i$, $t_B = 2$. Now, we are studying a sequence of double cusp groups with cusps belonging to the fractions p_n/q_n and $-q_n/p_n$. If we are lucky, this sequence of groups should converge to a group in which *both* sides of the ordinary set degenerate – corresponding to the irrational numbers r and $-1/r$. If this is true, what can we say about the traces? Well, as we go through this limiting process, the chain of generator moves which gets us from the pair involving $-q_n/p_n$ to the pair involving p_n/q_n gets longer and longer. The idea is that, when we are working with an extremely long sequence, it should matter less and less exactly which are the central pair, so the traces of the generator pairs roughly in the middle of the sequence should all be roughly the same. McMullen has demonstrated that such heuristic reasoning can actually be justified, although it requires a good deal of three-dimensional hyperbolic geometry and topology (as touched on in our final chapter) to do it.[1] We have tested the idea experimentally in Note 10.3.

Anyway, the reasoning continues that if a group were to be defined by a two-sided infinite sequence of generators, then you certainly could no longer tell what would be the 'central pair'. This suggests that in this limit group, the traces for the generator pair (a, B) should actually be *the same* as the traces of the generators next along in the list, namely the one with

[1] See McMullen's book *Renormalization and 3-manifolds which Fiber over the Circle*, Ann. Math. Studies 142, Princeton University Press, 1996.

generators (aaB, aB). How can we find such a pair a and B? In terms of the basic traces (t_a, t_B, t_{aB}), we have $t_{aaB} = t_a t_{aB} - t_B$. Thus these equations turn into two polynomial equations:

$$t_a t_{aB} - t_B = t_a \quad \text{and} \quad t_B = t_{aB}.$$

This plus the Markov equation $t_a^2 + t_B^2 + t_{aB}^2 = t_a t_B t_{aB}$ gives three polynomial equations for the three unknowns (t_a, t_B, t_{aB}), from which it only requires a certain amount of scratch paper (plus maybe an assist from Dr. Stickler) to get the answer:

$$t_a = \frac{3 - \sqrt{3}i}{2} \doteq 1.5 - 0.8660254038i,$$

$$t_b = \frac{3 + \sqrt{3}i}{2} \doteq 1.5 + 0.8660254038i.$$

Note how close the traces of the central generators are to these values. These are the traces that Troels told us years ago should make a group which is doubly-degenerate!

It was with some trepidation that we carefully transferred these traces to our limit set program; after all, we were about to witness the first all-natural area-filling curve. All remnants of the ordinary set should have finally disappeared! With our basic limit set algorithm, we saw the program twist and turn its way through something like the small snapshot in Figure 10.11. The curve fills the whole sphere, and so it leaves the viewing window in this figure to go off to infinity before it returns. This curve was perfectly capable of filling our hard drive and every stitch of memory in the computer if we pushed it with our algorithm to even barely adequate accuracy. That we kept the accuracy fairly low explains why we can see

---

Note 10.3: **Convergence to the double cusp**

The idea that traces for 'big' double cusp groups should be almost fixed or invariant under the *RL* map $a \mapsto aaB$ and $B \mapsto aB$ is borne out by examples. Here is a list of some of the 'central' traces for the $46368/75025$ cusp, corresponding to twelve passes of the *RL* process instead of eight. It is a double cusp group in which the $144/233$ and $-233/144$ words are parabolic. As you can see from our list, the traces do indeed seem to be almost invariant under the *RL* mapping. See how close they are to the values

$t_a = 1.5 - 0.8660254038i, \quad t_b = 1.5 + 0.8660254038i$ of the degenerate group!

$$\text{Tr } a = 1.50005872063 - 0.86599761708\, i$$
$$\text{Tr } B = 1.50005872063 + 0.86599761708\, i$$
$$\text{Tr } aaB = 1.50022957551 - 0.86573837802\, i$$
$$\text{Tr } aB = 1.50006401907 + 0.86617349520\, i$$
$$\text{Tr } aaBaaBaB = 1.50108897866 - 0.86461772278\, i$$
$$\text{Tr } aaBaB = 1.50026132274 + 0.86679372131\, i.$$

its craggy pattern at all in our picture; if drawn perfectly, we would see a solid black square. Although this may remind our reader of our reckless ventures into the domain of non-discrete groups in Chapter 8, there is a very big difference. Even though we have achieved total degeneracy with this group, it was not through careless abandon, but by deliberate fore-thought and calculation. So there!

Figure 10.11. Utterly degenerate. The whole surface has degenerated to nothing and the limit set is a sphere-filling curve. (Which means, er, that an accurate portrayal would be a solid black square, which is slightly easier to draw than the horribly inaccurate picture we provide.)

Of course, we can carry out a similar process for any cusp on the boundary of Maskit's slice. Consider the cusp associated to a large fraction p/q and the sequence of generator pairs leading to the p/q word. Choosing traces of the generator pair in the 'middle' of this sequence gives a double cusp very close to a doubly-degenerate group. Figure 10.12 shows just such a double cusp, which was obtained by using the transformation RL^{10} (that is, R followed by ten L's) in place of RL in our above calculations. To its right is the best approximation we can coax out of our program for the nearby degenerate group which is the fixed point of RL^{10}. (Of course, the ideal plot is a black square; the figure just suggests that there is some rather fine structuring in this particular shade of black.)

 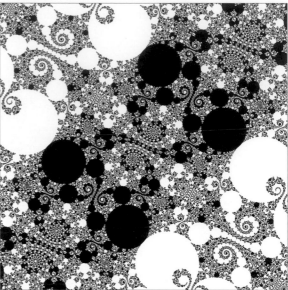

Realpolitik: a fractal partition of the sphere

What can one say about a sphere filling curve? There seems to be no picture to draw except for a black plane. Nonetheless, the points have nationalities: there are the *a*lphans, the *b*etans, the *A*gonists and the *B*lasphemes; or let's just say those with infinite words beginning with *a*, *b*, *A* and *B*. As we have seen, there are some with multiple affiliations; such limit points sit on the precarious fractal edge between the nations. What we want to do in this section is to review some additional features of this convoluted division of the world.[1]

The circles at the 'junctions' of three parts of the limit set in Figure 10.10 are still noticeably large, despite our assurances to the reader that as we approach the degenerate group they will indeed shrink to zero diameter. However, the convergence is almost unimaginably slow. Nonetheless, if we imagine the sky blue and pink circle chains shrinking to zero width, we arrive at the end of our story about Schottky curves. While we started with pristine circles defining a classical Schottky group, at these degenerate points, the Schottky curves have been deformed to fractal jigsaw shapes which fit together to comprise the entire Riemann

Figure 10.12. Spirally degenerate. The right frame shows as good an approximation as we can draw to the sphere-filling limit set of the group fixed by the transformation RL^{10}. The left frame shows a double cusp approximating this degenerate group, with the different parts of the limit set coloured as before.

[1] This intricate partition of the sphere is the subject of an interesting article by Alperin, Dicks and Porti, *The boundary of the Gieseking tree in hyperbolic three-space*, Topology and its Applications, 1999.

sphere. For example, the yellow lightning streak in Figure 10.4 marks the boundary between the *a*- and the other three nations.

Figure 10.13 shows this basic division of the world with some extra information. Let's think of the infinite word representing a limit point as a record of its genealogy. The first letter tells the nationality of the citizen limit point itself, the second the nationality of his father, the third of his paternal grandfather, and so on. It turns out that there has been enormous freedom of movement of the citizen limit points from one nation to another, which has meant that many citizens have ancestors coming from other nations. In fact, there are only four snobs[1] who can claim that all their ancestors come from their beloved home country. (Fortunately, there is only one snob in each country.) The colouring in our figure is based on the grandfathers of our limit points (meaning, the third letter in their infinite words). Those with grandfather *a* are coloured red, those with grandfather *b* are blue, and so on in our usual colouring scheme. Amazingly, this gives a finer partition of the sphere into $4 \times 3 \times 3 = 36$ regions, with no neighbouring regions sharing the same colour. We should advise the reader that the *a*- and *A*-nations have very poor relations with each other, and there is no way that an *a*-citizen would have a father living in the *A*-nation, and vice versa. The same goes for the *b*- and *B*-nations.

We now have the ultimate example of Schottky dynamics. The inhabitants of the *a* part can expand themselves by applying the transformation *A*. This has the effect of knocking the first letter off their infinite words, in other words, it takes them back to their father's homeland which will begin with *a, b* or *B* but not *A*. Miraculously (considering the jaggedy nature of the borders) the nation swells up and *exactly* swallows up the *b* and *B* nations as well. In the language of Schottky dynamics, *a* takes the interior of the *a*lphan country to the exterior of the *A*gonist country. Similarly, *b* takes the interior of the *b*etan country to the exterior of the *B*lasphemes. (If you live on the frontier you have a choice of what to do, because you have affiliations with several states at once. Luckily, whichever choice you make, the map will keep you on the frontier.)

In Chapter 4, the four paired Schottky circles omitted the swiss cheese which was a basic tile for a surface of genus 2. In Chapter 6, when we brought the circles together in a chain, this basic tile split into two halves, an inside and an outside. In Chapter 7, the Schottky circles had yet more points of tangency and the basic tile was shattered into yet more parts. In Chapter 9, by choosing more complicated parabolics, we managed to make Schottky regions which smushed right up against each other, leaving just the barest remnants of empty space in between.

---

[1] You know who they are.

Now, at the natural conclusion of this process, we have fractally smushed the Schottky regions up against each other so tightly that there is nothing left outside any of them at all.

Another way of describing the situation uses the map T which takes a limit point defined by an infinite word w to the new limit point defined by knocking off the first letter of w. This map T is called a **Markov map** and it gives another way of thinking about the infinite word which labels a limit point. Suppose we start at some point on the sphere, and we want to

Figure 10.13. Partition of the world. The limit set shown in Figure 10.11 with limit points coloured red, blue, green and yellow according to the *third* generator in their infinite word.

Figure 10.14. Names of the parts. Only the curves that separate the coloured parts in Figure 10.13 are drawn (in black and orange) here. The black curves are all images of one another under the group, and similarly so are the orange ones. Each part is labelled by the three tags that begin all limit points in that part. Some very small parts are not labelled.

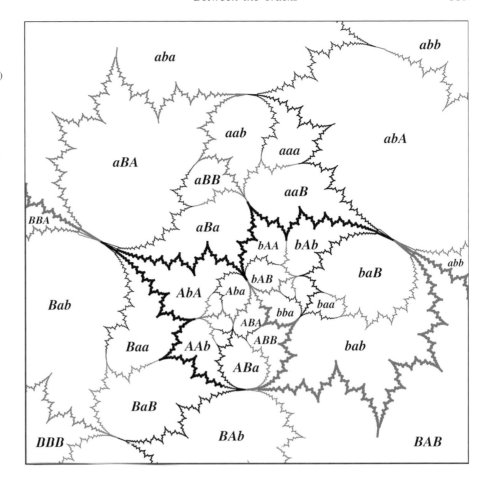

find its label as an infinite word. As long as we know which of the four nations it belongs to, all we have to do is apply T. The nation we find ourself in will be our father's, giving us the second letter of our infinite address. Repeating the process, you can find your address up to any level you please. Plotting the orbit of a point under the Markov map is very closely related to the IFS system for drawing the limit set explained on p. 152.

Another remarkable fact about our doubly-degenerate group is that its limit set has an extra symmetry. This comes about because the trace equations $t_{aaB} = t_a$ and $t_{aB} = t_B$ imply that the RL move which changes the generator pair (a, B) to the new pair (aaB, aB) must actually be implemented by a Möbius map which conjugates the group to itself. As explained in Project 10.6, this symmetry turns out to be a parabolic map c which commutes with the commutator $abAB$ and which has the same fixed point but a different translation direction. This symmetry is reflected in the patchwork of nations in Figure 10.14. The fixed point

of *abAB* is at the juncture of the large *abA* and *baB* regions on the right
of the picture. The commutator pushes these regions round one onto the
other, but the *c* symmetry is seen in the repetition of structure in the thin
pointy parts of the 'maple leaves' as they work their way ever further into
the narrow zero-angled crack between *aBA* and *Bab*.

Monsters in disguise

Early in this chapter, we sailed down into the deeper southern reaches
of the Maskit boundary by carefully avoiding the jutting rocky promontories where the cusp groups sit. To reach Jørgensen's degenerate group
at the furthermost point, we had to head through the slice holding a tight
almost due south course without very much room for straying. However
the cusp groups, by virtue of their location on promontories, can either be
approached straight on heading south from the interior or by laying up and
following up *northwards* along the edge of the rocky promontories which
lead to the cusps. The coastline route, not unlike the coast of Maine, has
surprises in store.

Let's return to the most prominent cusps of Maskit's slice corresponding
to 0/1, at the value $\mu = 2i$. As we have seen, the limit set of this group is
none other than the Apollonian gasket. A direct approach from the interior of Maskit's slice would be to sail due south down the imaginary axis,
that is, through groups with $\mu = ti$ with real numbers t decreasing to 2.
We have already shown several pictures of this kind of direct 'pinching',
where gaps are slowly closed to pinch points (for example, see Figure 7.4).
Nothing terribly exciting happens during these trips.

However the coastal sailor reaches the 0/1 cusp by quite a different
route, by following the cusp groups

$$1/2, \ 1/3, \ 1/4, \ \ldots, \ 1/n, \ \ldots$$

northwards as n increases without limit. This route is rather like the
path we followed when we drew the boundary in the first place,
but now we are heading towards 0/1 rather than away from it. You
can see a close-up of the route we have to follow in the lower
right frame on p. 290. Approaching through the cusps at the μ-values
$\mu(1/n)$ is a way of tracing the boundary all the way up toward the
cusp $\mu(0/1)$.

We say we are approaching, because that is what is happening to the
trace parameter μ which we use to that we use to define the groups; we
mean that $\mu(1/n)$ tends to $\mu(0/1)$ as n tends to infinity. How the limit
sets behave is a completely different question. Does the limit set for the

$1/n$ cusp just meekly conform to the Apollonian gasket as n increases? Far from it!

Figure 10.15 shows the limit sets for $\mu(1/30)$ and $\mu(1/100)$. In the first, the centres of the two most prominent spirals are very close although still visibly distinct, and it is still quite easy to pick out the 1/30 circle chain. In the second, the value is so close to that of $\mu(0/1)$ that the largest spirals seem to be centred about the same point, and the 1/100 chain has virtually collapsed into a system of tangent loops of tangent circles.

 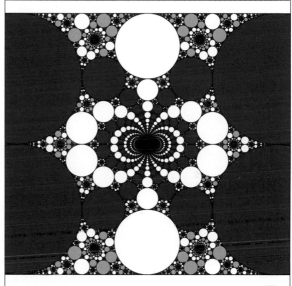

Figure 10.15. Limiting limit sets. The left frame is the limit set for the Maskit group corresponding to the 1/30 cusp, normalised so that $b(z) = z + 2$. The right frame is that for $\mu(1/100)$. In both pictures, the disks which are the remnants of the ordinary set on the two sides of the limit curve have been coloured black and white for greater contrast. The core circle chains have been coloured red.

The limit set of the 1/100 cusp appears to be approaching a 'limiting' limit set, but this limiting limit set is *not* the limit set of the 0/1 cusp the Apollonian gasket! In Figure 10.16, we have redrawn the 1/100 cusp group using Grandma's recipe for the generators. The centre point 0 is exactly the centre of the white spiral in Figure 10.15 (which was drawn using Maskit's recipe). You have to look rather carefully now to see the same pattern of loops of tangent circles. In the previous picture they were coloured alternately black and white; now the disks are all white. The easiest way to find the loops is to start at the centre point 0 from which they emerge and follow them round.

Although this 'limiting' limit set is obviously not the Apollonian gasket, you can nevertheless see the gasket emerging as a kind of hidden skeleton. In our coloured version in the lower frame, we show the same limit

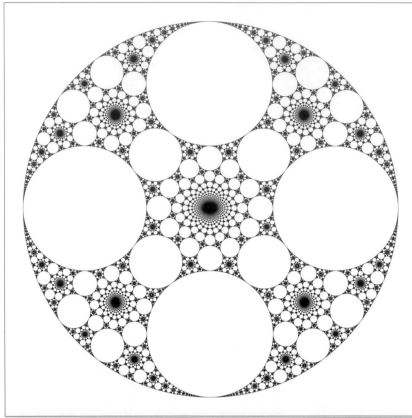

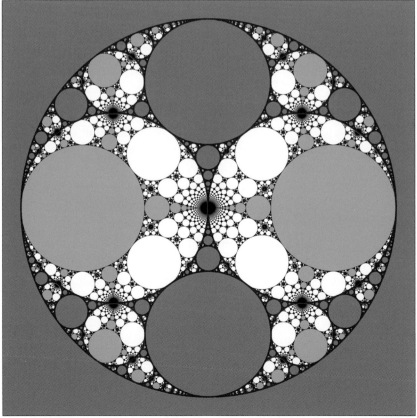

Figure 10.16. Differing limits. The upper frame is the limit set of the 1/100 cusp. It is imperceptibly different from the limiting shape as we move through the 1/n cusps as n tends to infinity. The lower frame has the limit set of the 0/1 cusp superimposed in red, with one side coloured pink. Another Apollonian gasket is coloured blue; it's a rotation of the pink one by 90°.

set with the gasket superimposed in red. See the extra fourfold rotational symmetry in the upper picture!

Since we are working in Grandma's normalization, let us write t_a in place of the Maskit parameter μ, thus $t_a = i\mu$. The $1/n$ cusp group, which we shall call G_n, is made from Grandma's recipe by setting $t_b = 2$ and $t_a = t_a(1/n) = -i\mu(1/n)$. Because the values of the parameters $t_a(1/n)$ tend to the value $t_a(0/1)$ as n tends to infinity, we say that the $0/1$ cusp group is the **algebraic limit** of the groups G_n. Indeed the generators of the $1/n$ cusp group G_n tend to the generators of the $0/1$ cusp exactly as you would expect. However, the limit sets of groups G_n do *not* approximate the limit set of the gasket group; as we have just seen, they tend to a more complicated geometry that seems to have an extra symmetry. This suggests that there is another kind of limiting group, which we might call a **geometric limit**, of groups G_n. This geometric limit group will be *different* from the algebraic limit, although, as we have seen, it actually *contains* the gasket group as a subgroup.

We can find lots of new symmetries of the limiting limit set in the following way. With Grandma's recipe the $0/1$ cusp group has generators

$$ a = \begin{pmatrix} 1 & 0 \\ -2i & 1 \end{pmatrix} \qquad b = \begin{pmatrix} 1-i & 1 \\ 1 & 1+i \end{pmatrix}. $$

Now the limiting limit set is symmetrical under rotation by $90°$, that is, the transformation $r(z) = iz$ (with inverse $R(z) = -iz$). Thus the conjugates of a and b by r should also be symmetries of the limiting limit set. Carrying this out, we find:

$$ raR = \begin{pmatrix} 1 & 0 \\ -2 & 1 \end{pmatrix} \qquad rbR = \begin{pmatrix} 1-i & i \\ -i & 1+i \end{pmatrix}. $$

These two new transformations generate a gasket group which corresponds to the blue circles in our figure. There is precisely one disk that is common to both families of pink and blue disks; it is the outer one which includes ∞. This disk is coloured pink in the figure, but it is also equivalent to the blue ones under the second gasket group.

What do these new symmetries have to do with the approximating groups G_n? Using Grandma's recipe, the $1/n$ cusp group G_n is generated by a pair of matrices a and b which depend on the trace parameters t_a and t_b. In all the groups, the generator b is parabolic, so $t_b = 2$. Thus we can write the generating matrices as $a(t_a)$ and $b(t_a)$ to show they only depend on t_a. The word $a^n B$ is also parabolic, so the value $t_a(1/n)$ for the group G_n is chosen to be a solution to the trace equation

$$ \text{Trace}(\, a^n B \,) = 2. $$

In terms of the μ-parameter, $t_a(1/n) = i\mu(1/n)$.

As n tends to infinity, the solution $t_a(1/n)$ of this equation tends toward the limit $t_a(0/1) = 2$. The matrices $a(t_a(1/n))$ and $b(t_a(1/n))$ also approach limiting matrices, namely the generators of the gasket group we wrote down above. If that were all there were to it, then the limiting limit set should be the same as the Apollonian gasket.

The truth is that anything in mathematics that involves limits of limits can behave in wild ways. For example, for any fixed positive integer m, the sequence $(1 - 1/m)^n$ tends to 0 as n tends to infinity. However if we fix n and let m tend to infinity, then $(1 - 1/m)^n$ tends to 1. Moreover, it is a remarkable fact that the sequence $(1 - 1/n)^n$ tends to $1/e$.

What we are seeing in these groups is rather similar. The trouble starts when we look at the limit of the elements $a^n B$. In each group G_n considered by itself, the elements $a^m B$ for increasing integers m are matrices whose entries rapidly increase to infinity. However, it turns out that the particular matrix $a^n B$ is not so far along in this behaviour as you might think. Remarkably, these matrices, whose entries ought by rights to be becoming larger and larger without bounds, actually have a sensible limit:

$$(a^n B) \text{ in the } 1/n \text{ cusp group } G_n \quad \longrightarrow \quad \begin{pmatrix} -1-i & 1 \\ -3-4i & 3+i \end{pmatrix}.$$

What is more, this limit matrix is a completely new transformation outside the gasket group! The presence of this extra transformation (and many others) accounts for the great difference between the algebraic and geometric limits of the groups G_n.

What has all this to do with the fourfold symmetry in Figure 10.16? Multiplying the matrices $a^n B$ on the right by b, we see that the limit of a^n as n approaches infinity is $\begin{pmatrix} -1 & 0 \\ -4 & -1 \end{pmatrix}$, which is very nearly the transformation raR we found above (actually it is $-ra^2 R$). This limit explains why the extra symmetry appears in the picture. (It also shows why the limit of $a^n B$ can't be in the gasket group, because certainly this new matrix isn't, so neither is its product with B.)

The transformations a and raR have a close relationship: they are both parabolic transformations which fix the point 0. That gives them at least one special property: they commute with one another. If we conjugate the group so that 0 is moved to ∞, we will see a striking pattern in the limit set, shown in Figure 10.17. Remarkably, we have returned to a genuine Euclidean square tiling just like the ones we encountered in our very first chapter. True, the tiles are fractals, but the

Figure 10.17. A doubly-periodic limit set. This group comes from the limit of the $1/n$ cusp groups in Maskit's slice. It is related to a three-dimensional manifold called the **Whitehead link complement** (see Chapter 12 and McMullen's paper in the ICM Proceedings, Berlin, 1998). This picture is actually a very close approximation to the limit, using the $1/100$ cusp group, with the generators a and b of Figure 10.16 conjugated by $z \mapsto -1/z$. The blue fundamental tile consists of all limit points beginning with B, while the yellow part comprises all those beginning with b. The red and green parts correspond as usual to infinite words beginning with a and A, respectively. What about the black parts? New nations have emerged!

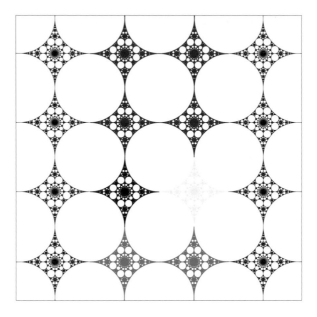

periodicity in the horizontal and vertical directions is the same. These extra symmetries give the limit set a noticeably thick density at the limit point 0.

Completely different limiting groups may be found by following a different sequence of cusps approaching the very same cusp $t_a(0/1)$. As a second and last example, we will follow the cusps at the fractions $2/(2n + 1)$ as n tends to infinity. These limit sets converge to the pattern shown in Figure 10.18.

If we conjugate by $z \mapsto -1/z$, the fixed point of a is sent to ∞ and the underlying lattice in this geometric limit is now generated by $z \mapsto z + 2i$, corresponding to the transformation a, and $z \mapsto z + 2 + \sqrt{3} + i$, corresponding to the limit of a^n in the $2/(2n + 1)$ cusp group as n approaches infinity. See the pattern of squares and equilateral triangles in the picture.

These few examples are the barest of beginnings of investigating the possibilities for limits of limit sets. These are still the most time-consuming of all the drawings in this book, and there remain many questions to be asked and answered.[1] It is exactly these sort of investigations that made it so difficult to stop somewhere and write this book!

[1] For a complete mathematical description of the possibilities, see the penultimate section of Minsky's paper cited on p. 291.

Slicing it to ribbons

In the mathematical world, one occasionally encounters some rare individuals who seem to have the ability to hold an image in their minds of

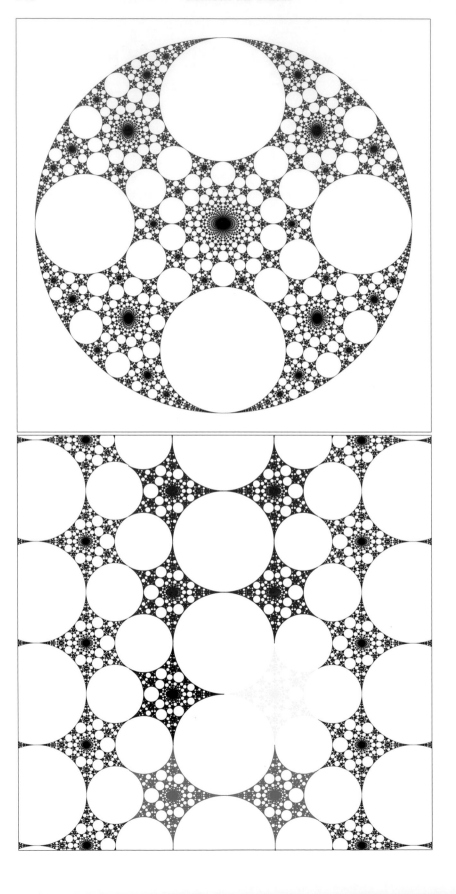

Figure 10.18. Another geometric limit. This is an approximation to the geometric limit of the $2/(2n + 1)$-cusp groups. The lower frame shows the limit set after conjugation by $z \mapsto -1/z$.

a four or higher-dimensional object. Even the ability to visualise three-dimensional geometry is a reasonably uncommon gift, and most of us have to get by with two (and sometimes much less) dimensional images in our heads.

The space of quasifuchsian once-punctured torus groups is described by two complex or four real parameters (the traces), which means it is one of these high-dimensional objects. Our approach to studying it has been to look at two-dimensional 'slices', meaning that we specify one of the complex trace parameters and then plot those values of the remaining parameter which correspond to quasifuchsian groups (or single cusps, in the case of the Maskit slice).

The samples we have given, Maskit's slice and the trace 3 slice, offer an impression of a object with a somewhat pointy boundary, which is however not terribly complicated otherwise. In recent years, as more detailed plots have emerged, this simple picture has begun to change. Exactly how the slices all fit together is a puzzle of very current interest.

Here we offer just two 'Maskit-type' slices in which the trace of b and $abAB$ are fixed and the trace of the first generator a is left to vary. The original Maskit slice corresponds to taking $\operatorname{Tr} b = 2$ and $\operatorname{Tr} abAB = -2$. It consists of all values of $\operatorname{Tr} a$ for which the group generated by a and b is free and discrete and represents the right sort of surface (a once-punctured torus and a triply-punctured sphere). The picture we made of this slice was periodic because for these exact values of $\operatorname{Tr} b$ and $\operatorname{Tr} abAB$, it turns out that $\operatorname{Tr} ab = \operatorname{Tr} a \pm 2i$. This gave the Maskit slice a translational symmetry because the group generated by ab and b is the same as the group generated by a and b, so that if $\operatorname{Tr} a$ corresponds to a free, discrete group, then so does $\operatorname{Tr} a \pm 2i$.

Figure 10.19. A cave of discrete groups. For $\operatorname{Tr} b = 1.91 + 0.05i$ and $\operatorname{Tr} abAB = -2$, our boundary-tracing algorithm produces this image of a slice of quasifuchsian space.

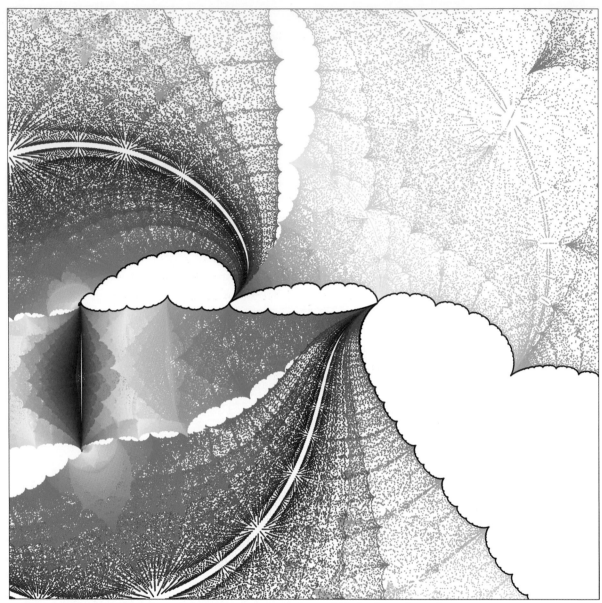

If we make a slight change in the constant value of Tr b, there is a dramatic change in the geometry of the slice. The reason is that Tr ab will not simply be a translation of Tr a, but a number obtained from Tr a by solving a more complicated quadratic equation. In Figure 10.19, we see a portion of such a slice in which we have fixed Tr $b = 1.91 + 0.05i$. The boundary has folded over on itself. It still has a 'cuspy' feel, and the repetitive effect arising from mapping Tr a to Tr ab is still evident, but this non-linear symmetry now causes the boundary to follow a double spiral pattern out to infinity.

The one defect with our boundary-tracing algorithm is that it does

Figure 10.20. Parabolic dust. Instead of just trying to find the solutions of the trace equations on the boundary, we compute all the solutions we can find and plot them as clouds of dust billowing in the plane. The boundary-tracing algorithm produces just the black curve.

not check the nature of every single point in the slice. Like the early maps of the new worlds discovered by intrepid European explorers, these slice maps have proved to be incomplete. In recent years, much more intense computations have revealed that there is quite a bit more to these slices than the boundary-tracing pictures indicated. The first suggestion that quasifuchsian space is more complicated than we might have thought was in a paper by Curt McMullen.[1] Since then, based on more of Jørgensen's ideas, a Japanese group based in Kyoto, Osaka and Nara[2] have produced many more exotic slices of quasifuchsian space. Here we show a version of McMullen's picture in Figure 10.20. Within the clouds of parabolic dust, we see white islands of free, discrete groups. The colour of the dust varies with the fraction p/q. In other slices, the number and arrangement of these islands appears to become extraordinarily complex.

It is only just now that mathematical pioneers are exploring the detailed shape of quasifuchsian and Schottky space. The exact geometry is the answer to the question of when two Möbius transformations generate a free and discrete group. The question may seem simple at first; the answer appears to require a whole world of detail.

[1] *Complex earthquakes and Teichmüller theory*, J. Amer. Math. Soc., 1998.

[2] Mainly Yohei Komori, Makoto Sakuma, Toshiyuki Sugawa, Masaaki Wada, and Yasushi Yamashita, see vivaldi.ics.nara-wu.ac.jp/~yamasita/Slice/

Projects

10.1: Infinite continued fractions

The number represented by any infinite periodic continued fraction satisfies a quadratic equation. Show that the fraction

$$x = 1 + \cfrac{1}{1 + \cfrac{1}{1 + \cdots}} = 1 + \frac{1}{x}$$

satisfies the equation $x^2 = x + 1$ and then explain why it is equal to the golden mean. Now let

$$y = \cfrac{1}{10 + \cfrac{1}{1 + \cfrac{1}{10 + \cfrac{1}{1 + \cdots}}}}.$$

Show that y satisfies a quadratic equation with a solution $(\sqrt{35} - 5)/10$.

10.2: Curt's similarity factor

How did McMullen come up with that amazing self-similarity factor $\frac{5+\sqrt{21}}{2}$ on p. 318? It's best to work in terms of the three traces (t_a, t_b, t_{ab}). Show that the

L and R moves on the generators change these three traces like this:

$$(t_a, t_b, t_{ab}) \xrightarrow{\;L\;} (t_a, t_{ab}, t_a t_{ab} - t_b)$$

$$(t_a, t_b, t_{ab}) \xrightarrow{\;R\;} (t_{ab}, t_b, t_b t_{ab} - t_a).$$

Thus we can easily (!) work out the effect of any composition of these maps on the traces, such as RL or RL^{10}. In particular, RL is the mapping

$$\begin{pmatrix} t_a \\ t_b \\ t_{ab} \end{pmatrix} \xrightarrow{\;RL\;} \begin{pmatrix} t_a t_{ab} - t_b \\ t_{ab} \\ t_{ab}^2 t_a - t_{ab} t_b - t_a \end{pmatrix}.$$

We are trying to work out a scaling factor in the parameter space in the vicinity of fixed points of this map. Scale factors involve derivatives, and there is a general principle underlying how to calculate scaling factors like this based on multivariable calculus and linear algebra. You can think of R and L as maps from complex three-dimensional space to itself. The scaling is given by a very nice recipe which finds what are called the *eigenvalues* of the *Jacobian* of RL. In our case, it has three eigenvalues: 1 and $\frac{5 \pm \sqrt{21}}{2}$. The one of absolute value greater than 1 is the answer.

10.3: Left and right turns

We defined 'left' and 'right' turns among pairs of generators as

$$L : (x, y) \mapsto (x, xy)$$

$$R : (x, y) \mapsto (xy, y).$$

Here is a variation which might also be considered:

$$L' : (x, y) \mapsto (x, yx)$$

$$R' : (x, y) \mapsto (yx, y).$$

These are turns with 'flips'.

- Show that for any sequence of left and right turns, if we replace some of the L's by L''s and some of the R's by R''s, then the final pair or words is unchanged except by possibly cyclically permuting the letters in the words.
- Show that every cyclic permutation of the final generator words may be achieved by replacing some turns by their flips.

10.4: Drawing the lightning

Here is a recipe for finding the infinite word described by Figure 10.8. Consider the points $(0, 0), (1, r), (2, 2r), \ldots$ on the line of irrational slope r and use a variable h which will be the fractional part of their y-coordinates, $kr - \text{floor}(kr)$ (where 'floor' (r) is the greatest integer less than or equal to some real number r). We find the successive values of h by starting at $h = 0$, and adding r until $h > 1$, in which case we subtract $(1 - r)$. Then keep subtracting unless this would bring us to a negative value for h, in which case we revert to adding. We associate a to adding r and B to subtracting $1 - r$. Because r is irrational, this process will go on forever.

Another way to think about this algorithm is that we are taking a limit of the circle chain patterns like the one described on p. 272. Starting from the initial circle C_0, we went through a cycle of adding q's and subtracting p's. It makes no difference in doing this if we scale everything so that we imagine the $p + q$ circles placed at $0, 1/(p+q), 2/(p+q), \dots, (p+q-1)/(p+q)$ along a line of length 1. Now take a limit! Instead of circles (which have now shrunk to points) think of points along the interval, so C_h just means the point whose real number coordinate is h along the lightning curve. The above recipe tells us to go through the same pattern, but adding r in place of p/q.

This gives a routine for drawing the yellow lightning streaks in Figures 10.4 and 10.6. To start drawing from the *right* end of the path (because $C_0 = 1$ is at the righthand end of the chain), we should reverse the order in which we go through the path, so we should take 1 as the initial seed and subtract r corresponding to applying A and add $1 - r$ corresponding to b. To implement this algorithm in practice, take a very large p/q approximating the irrational value you want. Start at 1, which we label z_0. Now generate z_j for $j = 0, 1, \dots, p + q$ by the rules

$$z_{j+p} = A(z_j) \text{ if } j \le q, \qquad z_{j-q} = b(z_j) \text{ if } j > q.$$

When you have calculated all $p + q$ points, plot them in order $z_0, z_1, z_2, \dots$. They will trace out exactly the curve you want.

You could have started at -1 and gone through the path in the other direction. To get good pictures, it turns out it is best to run through the plot twice, once in each direction.

10.5: Limit sets with mirror image insides and outsides

We want to explore quasifuchsian groups generated by a and b such that the transformation $T(z) = i/\bar{z}$ maps the limit set to itself and interchanges the inner and outer parts of the ordinary set. To make this happen, we will require that $T(a(z)) = b(T(z))$ and $T(b(z)) = A(T(z))$. (Note $T^2(z) = r(z) = -z$, see Project 8.9.) Using the fact that $\operatorname{Tr} abAB = -2$, show that all solutions are given by

$$a = \begin{pmatrix} t/2 & s_1 \\ s_2 & t/2 \end{pmatrix} \quad \text{and} \quad b = \begin{pmatrix} \bar{t}/2 & i\bar{s}_2 \\ -i\bar{s}_1 & \bar{t}/2 \end{pmatrix}$$

where $t^2 - 4s_1 s_2 = 4$, and $|s_1|^2 + |s_2|^2 = 2$. Now show that if a and b generate a quasifuchsian group, then T must map its limit set to itself. Therefore T maps ordinary points to ordinary points. To see finally that T must interchange the two parts of the ordinary set, use a famous classical result from topology, namely the Brouwer fixed point theorem, which states that a continuous map from a disk to itself must have a fixed point. (The inside part of ordinary set can be continuously deformed to a round disk.) Another way to see this is based on looking at the orientation of the generators, rather as in Note 9.6.

10.6: The extra symmetry

The group drawn in Figure 10.11 has a remarkable extra symmetry. The traces were found by solving the equations $t_{aaB} = t_a$ and $t_{aB} = t_B$. Notice this does *not* imply $aaB = a$ or $aB = B$, which would imply that a and B both equal I. Why is there a conjugating Möbius transformation c such that

$$caC = aaB \quad \text{and} \quad cBC = aB?$$

The conjugating map c is a new symmetry of the limit set which is *not* in the group generated by a and b! This conjugation gives a recursive definition of the special infinite word w_r – simply conjugate a over and over by c:

$$a \longrightarrow caC = aaB$$
$$\longrightarrow caaBC = aaB\ aaB\ aB$$
$$\longrightarrow caaBaaBaBC = aaB\ aaB\ aB\ aaB\ aaB\ aB\ aaB\ aB.$$

Check that, in contrast, conjugation by c leaves $abAB$ unchanged. Why does this imply that the c commutes with $abAB$? Compare with Project 6.1 to show that c must be parabolic with fixed point equal to the fixed point of $abAB$. This gives another reason why the limit points corresponding to w_r and $\overline{abAB}$ are the same.

   This process is an example of *rewriting rules*. We think of a word in a and B as a 'generation,' and we 'evolve' from one generation to the next by rewriting each a as aaB and each B as aB. The rules generate the sequence of Fibonacci words on p. 284.

10.7: Markov's conjecture

In Note 6.4, we discussed the Markov conjecture concerning the integer solutions of the equation

$$x^2 + y^2 + z^2 = xyz.$$

Suppose you have one integer solution, for example $(3, 3, 3)$. Show that our left and right moves, carried out on the level of the three traces (t_a, t_b, t_{ab}), lead to other solutions of this equation. If you want, try making a program to test out the truth of Markov's conjecture, explained on p. 191.

   Some of the more sophisticated algorithms for testing discreteness of a group are based on watching what happens when you run through the tree of traces obtained by making the L and R moves starting from any given initial set of traces (t_a, t_b, t_{ab}). There are a lot of interesting phenomena to explore here, too!

CHAPTER 11

Crossing boundaries

He also manifested hundreds of trillions of quadrillions of inconceivable numbers of subtle adornments, which could never be fully described even in a hundred trillion quadrillion inconceivable number of eons...

Avatamsaka Sutra[1]

Despite our many adventures, there remain certain boundaries we have not yet ventured to cross. Across the peaky Maskit boundary is indeed a sea of chaos; but it sparkles with islands of mystery. Here, for many experts, lie the only interesting groups. Another boundary is imposed by our rather artificial restriction to groups with only two generators a and b. Not having further eons at our disposal, all we can do in this short chapter is give a brief glimpse of these further vistas, taking, as Maskit has it, 'a trip to the zoo'.

Kleinian groups acquired their name from Poincaré. We shall tell more about this story in our epilogue. For our purposes, a **Kleinian group** will be any discrete group of Möbius transformations After seeing the plane-filling degenerate limit sets in the last chapter, you will appreciate the delicacy involved when we slip in that little word 'discrete'.

Closer relations between generators

We begin with Kleinian groups with only two generators. Taking a deep breath, let's venture out to some of those beckoning islands. Figure 11.1 shows what happens if we pick the values $t_a \doteq 1.924781 - 0.047529i$, $t_b = 2$ and $t_{abAB} = 0$ and use Grandma's four-alarm special recipe in Box 23. Is this another accident? In a manner of speaking, yes: we picked these values so as to get a double cusp group in which both b and the $1/10$ word $a^{10}B$ are parabolic, but in addition organised things so that the commutator $abAB$ has trace 0. From our work way back in Chapter 3, we

---

[1] From the *The Weaving of Mantra*, by Ryuîchi Abé, © Columbia University Press. Reproduced by permission of the publisher.

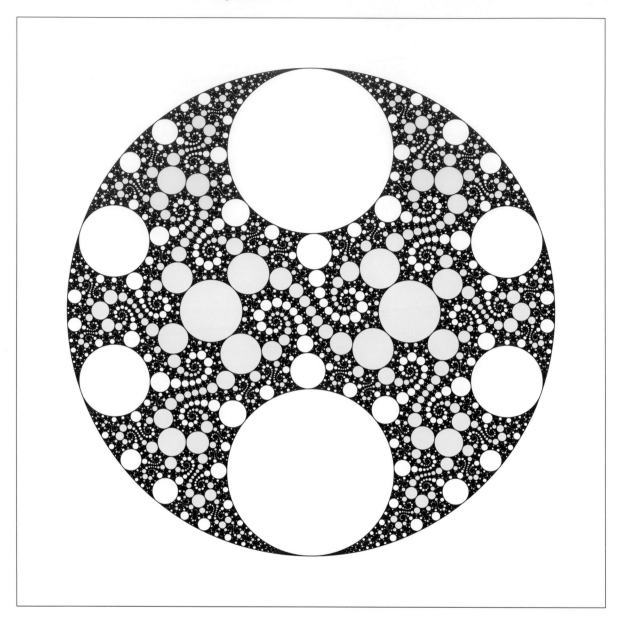

know that if a Möbius map T satisfies $T^2 = I, T \neq I$, then $\operatorname{Tr} T = 0$. Conversely, whenever $\operatorname{Tr} T = 0$, then $T^2 = I$. So our choice of traces means that $abAB$ is elliptic and, in fact, satisfies the relation $abABabAB = I$. In other words, the transformation represented by this particular product of generators is equal to the identity transformation I. There are subtle differences between this picture and the true double cusps in Chapter 9; for example, the yellow 'inner' circles never actually touch the 'outer' white ones.

Figure 11.1. Giving up freedom. This limit set corresponds to a two-generator group in which the generators a and b satisfy the relation $abABabAB = I$. For that reason, this group is not free. The two words b and $a^{10}B$ are parabolic.

[1] When we are thinking about equality of Möbius transformations, we write $(abAB)^2 = I$. However, since we are now mainly thinking about alphabet soup, it is more convenient to write $(abAB)^2 = 1$.

The really new ingredient here is the relation: the equation $(abAB)^2 = 1$.[1] This makes a dramatic difference. In our *a-b-A-B* language, when we excluded the taboo phrases *aA*, *Aa*, *Bb* and *bB*, we got what we called reduced words, one word describing each Möbius transformation in our group. Such true freedom of speech has been a foundation of all our efforts thus far. Of course in some specific group it might happen by chance that several reduced words all represent the same 2×2 matrix. Groups for which each reduced word actually represents a really *different* Möbius transformation are called free, in honour of this one-amendment bill of rights.

In our new group on the other hand, the relation *abABabAB = I* implies that, for example, the word *abABa* is actually *equal* to the shorter word *BAb*. This plays havoc with our entire plotting algorithm, since moving eight steps forward will be tantamount to standing still. Enumerating all the reduced words in the generators results in an enormous duplication of effort.

Just about the only non-free group we have met thus far was actually the very first group we studied – the symmetry group of the Iowa fields, which you can think of as generated by the two transformations $a : z \mapsto z + i$ and $b : z \mapsto z + 1$. These two transformations commute, which is to say that as maps, $ab = ba$. The symmetry group of the Iowa fields is not free!

We did meet one other example of a non-free group in Chapter 8, when a short plunge across the boundary led to the two-generator group with traces $t_a = 1, t_b = 1.90211303259032$. As we saw in Figure 8.10, the limit set of this group is not a random black scribble but a rather delicate lacework of circles. (The picture we produced isn't actually very good – it looks as if there are lot of gaps.) The value for the traces is as good an approximation as we can get to $2\cos(\pi/10)$, and it is at just such special numbers that freedom of word generation breaks down. In this group, both the generators *a* and *b* have the property that their tenth power is the identity transformation. That is, $a^{10} = aaaaaaaaaa = 1$, and similarly for *b*. If we ever see a word with ten *a*'s in a row, we can simply delete the string without changing the corresponding Möbius map at all. Worse yet, a row of seven *a*'s represents the same transformation as three *A*'s, because *aaaaaaaaaa = 1* is equivalent to *aaaaaaa = AAA*.

Not only does running our algorithm on a non-free group result in an enormous duplication of effort; in addition, the ordering of limit points spewed out by our algorithm will probably no longer be consistent with their location on the plane. Thus the program will not draw the limit set continuously as it works its way round the DFS labyrinth.

It does still work to the extent of drawing the limit set as a cloud of dust[1]; that is how we drew Figure 8.10. But for any accuracy at all, it is not a happy situation waiting in front of the monitor for the plot to terminate. Besides, the failure to progress down some of the branches has caused those annoying gaps.

In a group generated by a and b, all elements will still be represented by a word in a, b, A and B; the trouble lies in the fact that many words may represent the same transformation. Our existing algorithm already prevents the four forbidden two-letter strings (Aa, aA and so on) from ever appearing, and for free groups that is sufficient to ensure that we enumerate precisely one word for each transformation in the group. In non-free groups, we want to find a broader definition of forbidden words which will ensure that every matrix in the group is given by exactly one non-forbidden word. What we need is a pair of garden shears to prune our tree of words of all the forbidden branches. The 'shears' we have in mind is the theory of **automatic groups** introduced and developed in the book *Word Processing in Groups* (A.K. Peters, 1992) by D.B.A. Epstein, J.W. Cannon, D.F. Holt, S.V.F. Levy, M.S. Paterson and W.P. Thurston.[2]

In abstract group theory, groups are usually specified by a (hopefully short) list of generators and relations, that is, expressions like $aabab = 1$, $a^{10} = 1$ and so on. The idea is, that these relations engender all other possible equations between group elements, just as the equation $a^7 = A^3$ was engendered by the relation $a^{10} = 1$. The problem of deciding whether two different words represent the same group element is known as the **word problem**, and is a lot harder than you might think. For example, suppose we have two generators a and b for a group with relations

$$aaaaa = \text{identity element}$$
$$bbb = \text{identity element}$$
$$abab = \text{identity element.}$$

How many different elements are there in this group? In this book, we have become accustomed to dealing with *infinite* groups, but simple relations such as these may imply that the group actually contains only finitely many different elements. You are a master of organization if you can show that this particular group contains exactly 60 different ones. We met an example of a finite group with 8 elements in Project 2.6.

An **automatic group** is essentially a group which comes supplied with a computer program for listing precisely one word in the generators for each element of the group. It is an amazingly simple addition to our drawing algorithm, provided that we know the program (called an **automaton**) for the group we are working with.

The automaton includes a list of generators (a, b, A and B for our 2 generator groups; more generally we just number them g_1, g_2, and so on) and a finite list of 'states' which a word in the group may occupy. The term 'state' is used in the sense that a word (like $aBBBA$ or $g_1 g_5^2 g_2^{-1} g_1^{100}$) may be in a state of happiness (because it is acceptable) or a state of despair (because it has failed). We lump all the failures together as the **FAILURE** state, which we'll write as ☠. The other states are numbered, say from 1 to N.

The most important part of the automaton is a set of rules for building words, which says something like the following: if a given word is in the state i and we multiply that word on the right by the jth generator, then the new word will be in the state k. There should be one rule for each choice of state and generator. A convenient way to summarize these rules is in the form of a table in which the rows are labelled by the states, the columns are labelled by the generators, and the entries in the table give the new states.

We'll start with the simplest possible example with which we are already very familiar: the free group on a and b. For this group there is a very simple automaton with precisely six states. The identity word is the only word in the first state (sometimes called the 'start' state). The next four states consist of all words which end with a, b, A and B, respectively. The final state is the failure state. Here is the table which describes the automaton:

	a	b	A	B
Identity	a-state	b-state	A-state	B-state
a-state	a-state	b-state	☠	B-state
b-state	a-state	b-state	A-state	☠
A-state	☠	b-state	A-state	B-state
B-state	a-state	☠	A-state	B-state
☠	☠	☠	☠	☠

For example, suppose you are in the A-state, and you multiply on the right by b, then you get into the b-state. If you multiply by a on the other hand, you go to the ☠-state, which is the automaton's way of preventing the unwanted string aA. This automaton prevents all four strings aA, Aa, bB and Bb from appearing, since any of these sequences land us in the state of failure. The last line is sadly the same for all automata: once you're a failure, there is no hope of recovery, so usually we just eliminate the bottom line.

For our punctured torus groups, the last letter of a word is sufficient to tell us to which state the word belongs. In more complicated groups, this is not enough information, and we have to introduce some new states. In straightforward cases, this is done by subdividing the words which end in one generator, say a, into several different states, depending on a few letters which precede the last one.

Even with just the simple relation $(abAB)^2 = 1$, it is still a substantial task to work out a suitable automaton. It will have to list all the words in the group, spewing out precisely one word to represent each different transformation. For example, it will need to bar us from accepting the word *abABa*, because it is equal to the shorter word *BAb*. Table 11.1 shows the automaton we used. As you see, the set of all words ending in a have been divided into four states: those which end in the strings *ABa*, *Ba*, *ba*, and all the rest. In all there are nineteen states, as listed in the second column. We tell the state of a longer word by the longest possible string on its righthand end which is included in this list, so for example *aabaBA* is in the 12th state *aBA*, while *aabaBa* is in the 10th state *Ba*. This automaton was provided for us by the program `automata` written by D. Epstein, D. Holt and S. Rees at Warwick.[1]

Suppose for example we want to check out the word *abABa*. Proceeding generator by generator from the left, first, the generator a moves us from the start state 1 to state 2. Then looking at the line for state 2, we see that b moves us to state 6. Following through A, B and a, we move through states 12, 18 and finally ☠. The automaton does indeed bar us from moving to the node corresponding to the bad node *abABa*, which is more than halfway through the relation *abABabAB* = 1. In other words, *abABa* is a ☠.

Fortunately, although this automaton looks pretty complicated, it is actually very easy to build it onto our program. First, we number all the states $1, 2, 3$ and so on up to state N. Then we set up a two-dimensional array FSA (for *Finite State Automaton*) which gives the encoding of the automaton as a table. The entry `FSA[i,j]` gives the state that we move to from the current state `i` if we multiply by generator `j` on the right.

The automaton table will have N rows but the columns are still labelled by the generators, so in the simple case of the free group with generators a, b, A and B there are four. If we number the identity-state and the a, b, A and B-states 1 through 5 and label the failure state more mundanely by 0, then the array *FSA* would be

[1] There is now available a new more powerful version KBMAG by Holt.

$$\text{FSA} = \begin{bmatrix} 2 & 3 & 4 & 5 \\ 2 & 3 & 0 & 5 \\ 2 & 3 & 4 & 0 \\ 0 & 3 & 4 & 5 \\ 2 & 0 & 4 & 5 \end{bmatrix}.$$

Table 11.1. *The automaton for groups with commutator of order 2.*

State		a	b	A	B
1	I	a	b	A	B
2	a	a	ab	☠	B
3	b	ba	b	bA	☠
4	A	☠	b	A	AB
5	B	Ba	☠	BA	B
6	ab	ba	b	abA	☠
7	AB	ABa	☠	BA	B
8	ba	a	ab	☠	baB
9	bA	☠	b	A	bAB
10	Ba	a	Bab	☠	B
11	BA	☠	BAb	A	B
12	abA	☠	b	A	abAB
13	ABa	a	ABab	☠	B
14	baB	Ba	☠	☠	B
15	bAB	☠	☠	BA	B
16	Bab	ba	b	☠	☠
17	BAb	☠	b	bA	☠
18	abAB	☠	☠	BA	B
19	ABab	ba	b	☠	☠

Thus the number 0 in the 3rd row and 4th column tells us that following state 3 (which is *b*) by the 4th generator (which is *B*) gets us to the 0 state (which is failure).

In our original DFS algorithm, we made decisions about following branches in the word tree based solely on the rightmost generator in our current word. We kept track of this (see Chapter 5) by means of an array tags which recorded the number of the ith generator in the word from left to right as tags[i]. In these more complicated groups, this is not enough information, and we must keep track of more of the history of our travels. To do this we set up a new array state in which we store the state of the subword formed by the initial i letters as state[i] (so this subword automatically ends in the letter tags[i]). The subwords for *abABa* are

$$a, \ ab, \ abA, \ abAB, \ abABa,$$

in states 2, 6, 12, 18 and ☠ respectively.

Here's how we shall use the automaton in our enumeration of the tree of words. Suppose we have reached the word whose tags from left to right are:

`tags[1]`, `tags[2]`, ... , `tags[lev]`.

To go forward, we are proposing to multiply the current word on the right by `gens[j]`, which would mean setting `tags[lev+1]` = `j`. First, we evaluate

`FSA[ state[lev] , j ]`

where `state[lev]` is the state of the current word. The output will be the state of the new extended word. If this new state is 0, the extended word fails, and we do not allow this forward direction. In this case, we turn to the next generator `gens[j-1]` in the usual way and repeat the process. If the new state is acceptable, on the other hand, then we store it in `state[lev+1]`, increment `lev`, and then proceed forward to consider further extensions to the new word. This is all that is necessary to modify our drawing programs to run well for non-free groups.

By far the hardest part of all this is finding the automaton at the outset. Given an arbitrary set of generators and relations, it is not at all clear that one should be able, even theoretically, to do this. A famous result of Novikov in the 1950's demonstrated conclusively that there are groups for which the word problem is **unsolvable**, meaning that there is no algorithmic procedure (in a technical sense developed in mathematical logic) for deciding whether two words in the generators represent the same element of the group. The theorem also shows there are groups without automata, although Derek Holt has given a simpler, independent proof of this fact. The good news is that all Kleinian groups are automatic, and you can find the FSA in a flash using KBMAG.

Exploring the islands. Now we have gone to all this effort to set up this automaton, let's use it to explore some islands in the sea of chaos.

In earlier chapters, we have seen how to generate groups from an arrangement of four non-overlapping or tangent Schottky disks, and we have deformed these groups until their Schottky curves become non-circular and rub right up against each other, even in a fractally entangled manner. However, there is one thing which we have not done, which is to let the Schottky circles smash right through each other. Well, it's time to give up all propriety. In Figure 11.2, we made two 'Schottky crossing groups' starting from a chain of four thick black circles which are no longer tangent, but which intersect at an angle of 45°.

As usual, the generator *a* pairs the inside of the lower circle to the

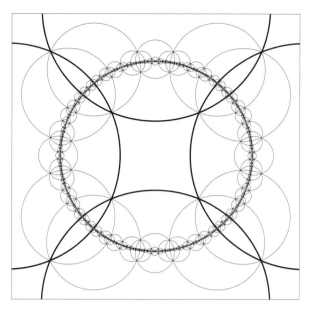
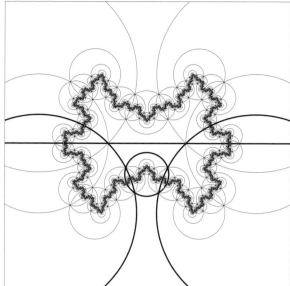

Figure 11.2. Intersecting Schottky circles. The four original Schottky circles are marked with thick black curves. The fact that they cross at an angle of 45° implies that there is a relation in the group, namely, $(abAB)^2 = 1$. The image circles nest down as before on the red limit sets.

outside of the upper one, and b maps the inside of the left circle onto the outside of the right one. The only difference from the arrangement in Chapters 4 and 6 is that this time, the original four circles cross. The thinner black circles in the picture are the images of the original four under all possible elements of the group. As usual, these image circles nest down to the limit set (shown in red).

As in our discussion of the 'necklace conditions' in Chapter 6, certain conditions must be met to ensure we get a nice picture. There, attention focussed on the tangency points of the four initial Schottky circles, this time, we need to look at where they cross. Look at the inner quadrangle in the two pictures, bounded by the four Schottky circles. Following through the matching conditions this time, the image circles when we repeatedly apply the generators along the 'forward' word $abAB$ also all pass through one of the four corners of this quadrangle, rotating round the corners as they go. Because we chose the angle as 45°, there are exactly four circles through each corner. Correspondingly, instead of requiring that the commutator $abAB$ be parabolic, now we have to insist that $abABabAB = I$. Notice that $abABabAB$ corresponds to eight forward steps in the tree of words.

The limit set of both these groups is still a Jordan curve. The two quadrangles, one inside and one outside the four Schottky circles, are still basic tiles whose images tessellate the regions inside and outside the limit set. Any crossing point of circles is a meeting point of eight quadrangles.

If we had started with a quadrangle whose angle was a different fraction of 90°, say $90°/n$, then the circles would match up after n cycles of the basic commutator, and we would be in a group with relation $(abAB)^n = I$. Now you can see that these groups aren't really islands at all but rather isolated sailors – if you change the angle (and hence the trace values) just ever so slightly, then *infinitely many* different circles would go through each of the four corners, cycling round and round forever and never landing exactly on top of a circle which had been placed down before.

As in our first example, it is quite easy to calculate matrices a and b which generate groups like this. The condition for the commutator to have order 2 is simply $\mathrm{Tr}\, abAB = 0$. We just have to choose $\mathrm{Tr}\, a$ and $\mathrm{Tr}\, b$, substitute $t_C = 0$ in Grandma's four-alarm special recipe and leave it to Grandma to cook up the rest. In the right frame of Figure 11.2 we took $\mathrm{Tr}\, a = \mathrm{Tr}\, b = 2.2$. The left frame comes from arranging the generators as in our original special recipe way back on p. 170 in Chapter 6, making allowance for the fact that the commutator is no longer parabolic. The traces for this group are

$$\mathrm{Tr}\, a = \mathrm{Tr}\, b = 2\sqrt{2} \cos \frac{\pi}{8} \doteq 2.6131259.$$

All we have to do is plug these values into our program-plus-automaton, and off we go.

All our work about cusps and boundaries in the last chapters can be carried out just as well with the assumption that the commutator has order 2. The upper picture in Figure 11.3 is the analogue of the Apollonian gasket, with both a and b parabolic.

The lower frame has traces $\mathrm{Tr}\, a = 1.9247306 - 0.0449408\, i$ and $\mathrm{Tr}\, b = 1.91 + 0.2\, i$; feeding on loxodromy as before, a degenerate dragon has roared to life! The parts of the limit set beginning with a, b, A and B are coloured red, blue, green and yellow respectively. It is very entertaining to compare these with similar pictures in earlier chapters and trace some of the differences in structure. The very first picture of this chapter was a group with order 2 commutator where b and the $1/10$ word $a^{10}B$ is parabolic.

Finally, let's return to the place where we dived off the edge of quasi-fuchsian space into the sea of non-discreteness. The group we mentioned earlier on, with $\mathrm{Tr}\, a = \mathrm{Tr}\, b = 2 \cos \frac{\pi}{10} \doteq 1.90211303$ and $\mathrm{Tr}\, abAB = -2$ is serendipitously discrete, because both generators a and b satisfy the relations $a^{10} = b^{10} = 1$. As we mentioned earlier, the plot of this group on p. 241 had some unsightly gaps. The key to a proper plot is the automaton. The outcome is shown in in Figure 11.4.

Figure 11.3. Two examples of order 2 commutator groups. Above *a* and *b* are both parabolic and below Tr *a* = 1.9247306 − 0.0449408 *i* and Tr *b* = 1.91 + 0.2 *i*.

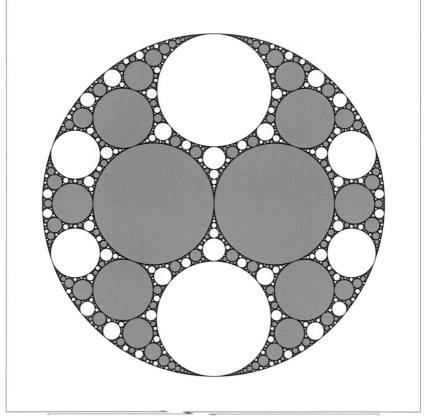

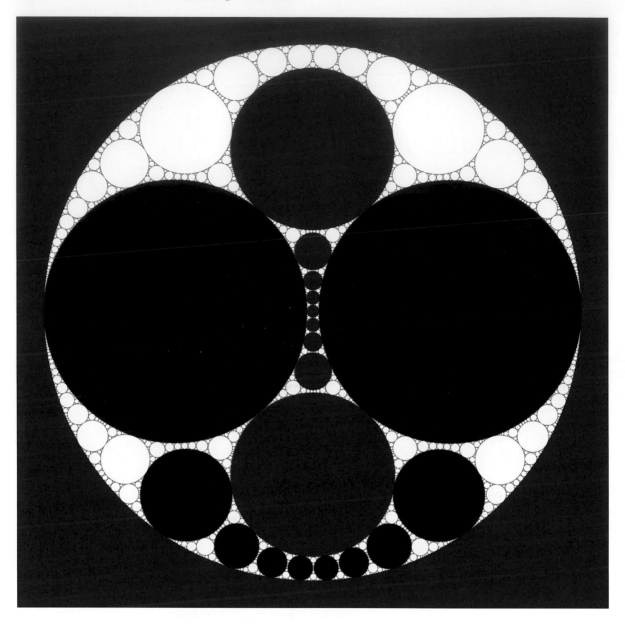

Non-free, discrete groups like this are very abundant outside quasi-fuchsian space; we chose this one just as an example. (They are also very interesting to topologists, as they are closely connected to knots!) Looking at Figure 9.11, you can pick out some of them sitting at points on the vertical line between $-2i$ and $2i$. They are at the centres of the small white holes where a star shape forms in the red part of the dust. Just how other discrete groups may be arranged here has not yet been fully explored.

Figure 11.4. A group generated by two elements a **and** b **of order 10.** The red disks are equivalent under powers of a, while the blue disks are equivalent under powers of b. If you start with any red disk and apply the powers a, a^2, a^3, etc., you will obtain all the red disks. There are only ten red disks because $a^{10} = I$.

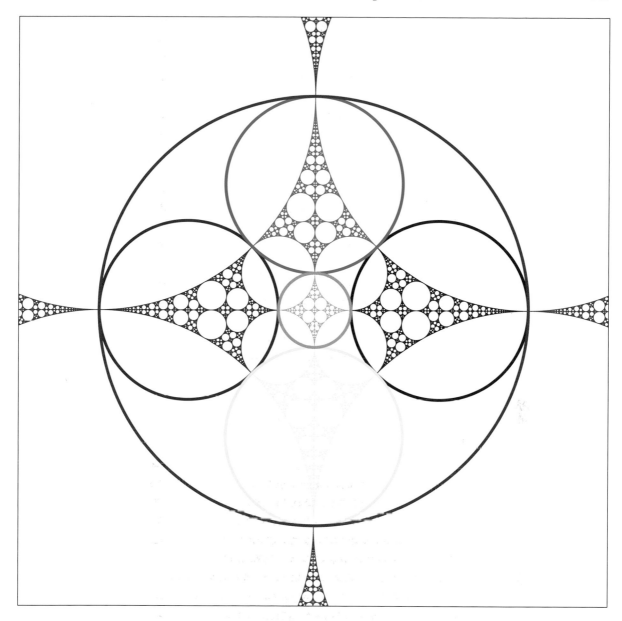

Figure 11.5. Three pairs of tangent circles. The generators *a*, *b* and *c* identify the pairs of circles which are not tangent in this diagram. The parts of the limit set which begin with *a*, *b*, *c*, *A*, *B* and *C* are coloured blue, green, lavender, red, yellow and purple, respectively.

Adding the letter 'c' to our language

One can grow a little tired of babbling $abba - bAABB - aBaa - Baba \cdots$. Certainly, language should get far more interesting if we allow *c*, and then *d*, and perhaps even *e*! What we mean is that we should extend our considerations to groups with more than two generators. The algorithm for enumerating the tree of words hardly changes: with *a*, *b* and *c*, we have a six-letter alphabet (because we have to include the inverses *A*, *B* and *C*).

In a free group, we must of course avoid cC and Cc; in non-free groups, one would use an automaton to enumerate only the acceptable words in the tree.

Formulas for the matrices for interesting groups with more than two generators are slightly harder to come by. It is true that we can perhaps conjugate the group to obtain formulas with some special symmetry, but in general we have to expect that groups with more than two generators will exhibit much greater variety than two-generator ones.

One way to find formulas is to return to our original framework of Schottky groups and try to find interesting ways to pair up circles in the plane. This is precisely what we did to make the glowing limit set picture on p. xviii; it comes from a group with three generators, arranged as in Figure 11.5. (In the version in the Introduction, we left off everything outside the outer purple circle, to make it fit well on the page.) The generator a maps the red circle onto the blue circle, b maps the yellow circle onto the green circle, and c maps the large violet circle onto the small central lavender one.[1] For each generator, we choose the simplest purely hyperbolic 'stretch' map which carries the inside of one circle onto the outside of the other. As you might expect, this gives us certain matching conditions, which imply that certain combinations of a, b and c must be parabolics, fixing various of the tangency points between the circles.

The version in the Introduction shows you all the image Schottky circles. In the version here, the parts of the limit set which correspond to infinite words beginning with a, b, c, A, B and C are coloured according to the same scheme as the initial circles. We should point out that the DFS algorithm does not trace this limit set as a curve: no matter how we order the generators, we have to plot the limit points as dust.

We can try to vary the way in which we put three pairs of circles touching each other. It happens that all arrangements of six circles which are tangent in the manner of Figure 11.5 are the same, after possibly applying a Möbius transformation. This is because all the pieces of the plane outside all six circles in Figure 11.5 are ideal triangles, and when they are glued together into surfaces the result is a collection of four triply-punctured spheres.[2] In Figure 11.6, we show another arrangement of tangent Schottky circles, which has a little more flexibility. Outside our new arrangement of disks is a 'hexagonally' shaped tile whose boundary consists of arcs of all six circles. If the paired edges of this hexagon are glued together in the correct pattern, we get a torus with two cusped punctures! Such a surface has greater flexibility than a torus with just one puncture, and we should be able to get quite a variety of different groups of this sort.

[1] Notice that this particular arrangement of circles has octahedral symmetry if you plot it on the Riemann sphere; to get an arrangement with icosahedral symmetry you would need six generators.

[2] For the experts: this is a special case of a Steiner pattern; from an even more sophisticated viewpoint, it is a special case of the Andreev–Thurston theorem on circle packings!

Figure 11.6. Six kissing Schottky disks.

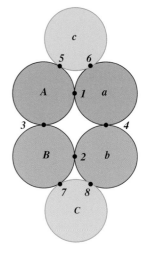

A method of actually finding groups which do this was developed by Maskit and Irwin Kra, who call them 'plumbing constructions', because they attach handles to holes on spheres in the manner of a plumber connecting pipes. We can find nice matrices for the generators a, b and c because our old friend the modular group is concealed in this picture – in several different disguises! The easy one to spot comes from the blue and red circles, which are paired by the generators a and b and situated in exactly the modular group pattern we met in Chapter 7. This group will be there provided we set up the matching conditions properly, so that, using the numbering in the picture, $a(1) = 1$, $b(2) = 2$, $a(3) = 4$ and $b(3) = 4$. In addition, of course, a, b and aB should all be parabolic. As long as we are willing to conjugate, our work in Chapter 7 allows us to take the first two generators to be

$$a = \begin{pmatrix} 1 & 2 \\ 0 & 1 \end{pmatrix} \qquad b = \begin{pmatrix} 1 & 0 \\ 2 & 1 \end{pmatrix}.$$

All that's left is to find a formula for the third generator c. The key is that the new tangency points should be mapped as follows:

$$c(7) = 5 \qquad c(8) = 6 \qquad a(5) = 6 \qquad b(7) = 8,$$

following the numbering in the picture. If we follow these transformations around, we find that $acBC(6) = 6$. To make sure the Schottky circles nest properly (see Box 18) we had better also assume this element is parabolic, enabling us to find out information about the matrix representing c. Leaving further hints to the Project 11.3, here we'll just give the answer:

$$c = \begin{pmatrix} 1 & \mu_1 \\ 0 & 1 \end{pmatrix} \begin{pmatrix} 1 & 0 \\ \mu_2 & 1 \end{pmatrix} = \begin{pmatrix} 1 + \mu_1\mu_2 & \mu_1 \\ \mu_2 & 1 \end{pmatrix}.$$

Any group with three generators like this is called a Maskit group for a twice-punctured torus. It depends on two complex parameters, μ_1 and μ_2. The lacework of such a limit set can be substantially more complicated than for the once-punctured torus groups for which there is only one complex parameter μ.

Let's see these formulas in action. All we have to do is choose some complex numbers for μ_1 and μ_2 and run our program (modified to handle three or more generators) to draw the limit set. With more complicated groups, the space of acceptable values for the μ's is also more complicated, and we are reduced to the dangerous practice of exploring without maps. Anyway, the principle of looking for accidentally parabolic elements is still a good guide. After some trial and error, we found that the values $\mu_1 = i$ and $\mu_2 = 4i$ causes both c and $cbCB$ to become parabolic. This group is something like the 'main cusp' $\mu = 2i$ in Maskit's slice, which gave the Apollonian gasket. The limit set is drawn in Figure 11.7. The parts corresponding to infinite words beginning with the six different generators are coloured yellow for a (mostly off to the right of the page), purple for A (mostly off left), blue for b, orange for B, red for c and green for C.

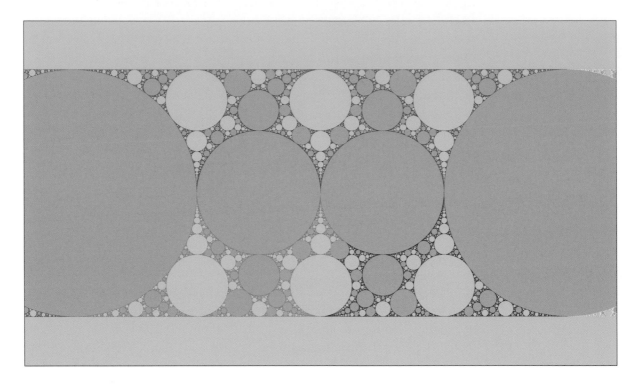

Once we found this main promontory to sail around, it was easier to probe for wilder groups. Slightly altering our parameters to

$$\mu_1 = -0.04 + 0.97\,i \qquad \mu_2 = -0.16 + 3.87\,i$$

we found the limit set in Figure 11.8. The delicate lacework of the main cusp has been ripped apart into flailing spiral arms. This group is not a cusp group, and we can see the connected pink region of the ordinary set on which live tiles which get glued up by the group into a twice-punctured torus.

If we gingerly vary the μ-values which determine Figure 11.8, we can tiptoe deeper and deeper into the recesses of twice-punctured torus space. The result is that the limit set becomes ever 'hairier', approaching what we called a state of degeneracy in the previous chapter. Figure 11.9 shows an extremely hairy limit set for

$$\mu_1 = -0.04 + 0.9651\,i \qquad \mu_2 = -0.16 + 3.861\,i.$$

Figure 11.7.
Twice-punctured torus gasket. A limit set for a cusp group on the boundary of the space of twice-punctured torus groups. Where are the Schottky circles?

Figure 11.8. A wilder twice-punctured torus. A limit set for a group very near the boundary of the space of twice-punctured torus groups.

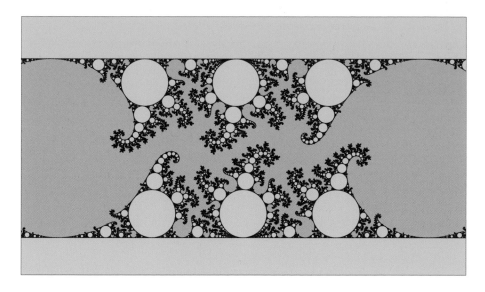

Figure 11.9. Degeneracy. As we probe ever deeper in the nooks and crannies of the boundary, our limit set becomes 'hairier.' This one is quite close to a degenerate group.

All we have done is slightly lower the imaginary parts of the parameters from their previous values. We coloured the upper and lower portions of the limit set differently to highlight the extremely tortuous channel which has developed in the remaining pink ordinary set. At an accurately drawn and truly degenerate boundary point, all the pink region would completely disappear.

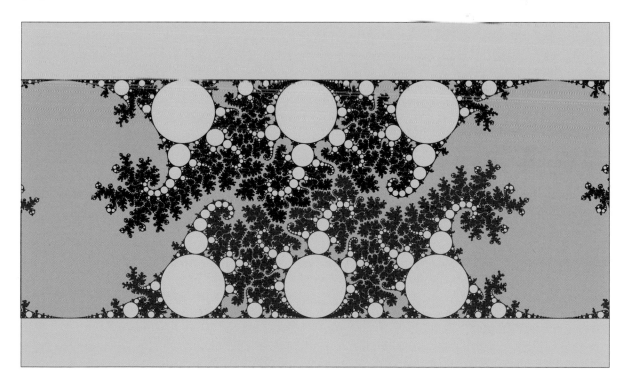

Of course everything we have just done can be repeated for more than three pairs of Schottky circles. Besides, there are many more ways to arrange for the Schottky curves to touch each other. Figure 11.10 shows an example made from four generators (*a*, *b*, *c* and *d*) arranged to produce a two-holed torus (surface of genus two) with one puncture. The generators have been further chosen to produce several accidentally parabolic words.

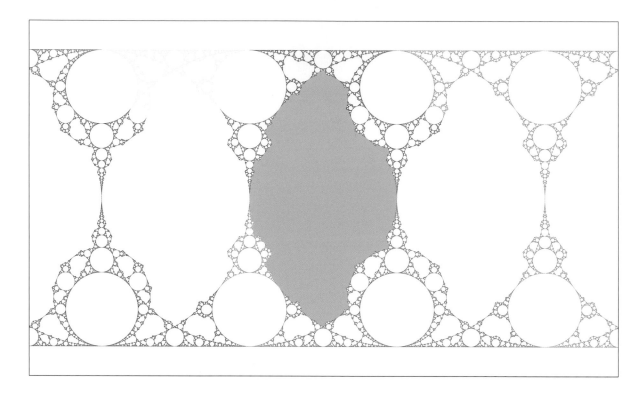

One phenomenon visible in these pictures that wasn't present for two-generator groups is that certain 'sides' of the limit set seem decidedly more rugged than others. A small part of the limit set in Figure 11.8 reminds us of the classic artistic metaphor for fractals, Hokusai's *Great Wave* (see Figure 11.11, with apologies to Mandelbrot for borrowing the metaphor from *Fractal Geometry of Nature*). Forgive us for trying to colour it accordingly.

Figure 11.10. More and more generators. This is the limit set of a group with four parabolic generators *a*, *b*, *c* and *d*. The eight basic parts of the limit set are coloured differently. The pink region is left invariant by a quasifuchsian subgroup, which represents a five-times punctured sphere.

Projects

11.1: Back to Chapter 1

After labouring so diligently to work with Möbius transformations, our final touch of adding the Finite State Automaton to the program at last allows us to include

Figure 11.11. The Great Wave. A portion of the limit set in Figure 11.8 showing a 'cresting wave'.

all the Euclidean symmetry patterns described in Chapter 1. This is because they are based on groups generated by two independent translations a and b with one key relation: $ab = ba$, equivalently $abAB = 1$. Here we encourage you to find an automaton that enumerates the words in this group, and then utilize it in the DFS program just as described in this chapter. Believe it or not, the hexagonal tilings in Chapter 1 were drawn by precisely this method.

11.2: Generators of finite order

The example we gave of a group with generators of order ten can be easily changed to allow different powers of the generator to be the identity transformation. Work out an automaton for the group generated by a and b with only the relation $a^4 = 1$, or more generally with $a^n = 1$. Then apply it to groups specified by Grandma's formula with traces $t_a = 2\cos(\pi/n)$ (guaranteeing $a^n = 1$), $t_b = 2.2$, and $t_{abAB} = -2$, for example.

11.3: Twice-punctured tori: the sequel

Here are a few more details of finding the third generator in the twice-punctured torus group. We already know that a and b are parabolic. That means cbC, as a conjugate of b, is also parabolic. Once again, a and cbC and the product $acBC$ are all parabolic. Use what you know about the modular group to prove there is a transformation v (with inverse V) such that

$$Vav = \begin{pmatrix} 1 & 2 \\ 0 & 1 \end{pmatrix} = a \qquad VcbCv = \begin{pmatrix} 1 & 0 \\ 2 & 1 \end{pmatrix} = b.$$

Use the first relation to show v is (up to a minus sign) of the form

$$v = \begin{pmatrix} 1 & \mu_1 \\ 0 & 1 \end{pmatrix}$$

for some complex number μ_1. Use the second relation in a similar way to show

$$Vc = \begin{pmatrix} 1 & 0 \\ \mu_2 & 1 \end{pmatrix}.$$

Now put it all together to get c.

Can you find a fundamental tile for this copy of the modular group in Figure 11.6?

11.4: Four-times punctured spheres

Figure 11.12 shows another arrangement of six kissing Schottky circles. The a and b disks are exactly the same as in the twice-punctured torus case, but the c disks are attached in a new way. This time, the b and c disks also follow the pattern of the modular group, so we have two conjugates of the modular group which share a common generator b. Work out a formula for the third generator c assuming that a and b are conjugated to their standard position. There will be only one complex parameter's worth of freedom.

The common exterior of these Schottky curves glues up to make a sphere with four punctures. This family of groups follows recipes of Kra and Maskit for parametrizing four-times-punctured spheres.

Sadly, despite our labour to add a third pair of circles, groups created in this pattern are not very different from the groups in the original Maskit slice. A precise relation was proved by Kra,[1] and the limit sets are also almost the same.

11.5: Draw the Schottky curves

In Figure 11.10, the parts of the limit set corresponding to infinite words beginning with a, b, c, d, A, B, C and D are coloured with different colours. Given that d is translation by 2 and that all four generators are parabolic, try to sketch in plausible Schottky curves to be matched by the generators.

Figure 11.12. Six more kissing Schottky disks.

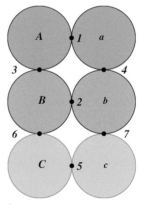

[1] Journal A.M.S., 1990; see also C. Matthews, Ph.D. Thesis, Oklahoma State University, Stillwater, Oklahoma, 1994.

CHAPTER 12

Epilogue

At that moment when I put my foot on the step the idea came to me, without anything in my former thoughts seeming to have paved the way for it, that the transformations I had used to define the Fuchsian functions were identical with those of non-Euclidean geometry.

L'Invention Mathématique, Henri Poincaré

In the course of the last few chapters, we have been making increasingly frequent reference to two as yet unexplained topics: **hyperbolic**, otherwise called **non-Euclidean geometry**, and **Teichmüller theory**. To conclude the development of our story we want to offer some brief explanations, as it is on these two mathematical pillars which all the deeper developments of our subject rests.

Before we come to that though, let's return briefly to the work of Felix Klein. In the preceding chapters, we have done no more than scratch the surface of Klein and Fricke's epic books. As we have already mentioned, the work which led to the full understanding of Kleinian groups was the topic of an intense rivalry between Klein and Poincaré.

The background to Klein and Fricke's volumes is a subject called at that time *Funktionentheorie*, the study of differentiable functions of a complex variable. We first meet the familiar trigonometric functions sine and cosine as functions of a real variable, that is, if x is a real number, then so are $\sin x$ and $\cos x$. The great significance of these functions is their *periodicity*, thus $\sin x = \sin(2\pi + x)$, $\cos x = \cos(2\pi + x)$ and so on. As anyone acquainted with Fourier series will appreciate, this periodicity makes the trigonometric functions very important for solving all sorts of physical problems.

Now the variables x and y can just as well be thought of as complex, and when extended in this way these functions take on a whole new life. They become complex-valued functions of a complex variable, and as such the periodicity is reflected in the fact that they are invariant under the translation $T : z \mapsto z+2\pi$. This can be neatly expressed in our language of group theory. The group generated by T is the set of all powers of T, that is T^2, T^3, T^{-1} and so on. Saying that $\sin z$ is periodic under T is exactly the same as saying that sin has the same value on all points in the same T orbit, in other words, $\sin T^n z = \sin z$ for all n and for all complex numbers z. The trigonometric functions are invariant under the group generated by T. This group has a basic tile which is a vertical strip in the plane of width 2π. If you know the value of a trigonometric function on this strip, then you know it everywhere. The function perfectly reflects the symmetries of the group.

A major development in the nineteenth century was the theory of **elliptic functions** which are invariant under a *pair* of translations in different directions. The basic tile for a group generated by two such translations is a parallelogram whose sides are the two translation vectors. Copies of this basic parallelogram fill up the whole plane. In the simplest case, the generators are just translation by one unit in horizontal and vertical directions. The basic parallelogram is a square and the tiling is our picture of Iowa fields in Chapter 1. An elliptic function is a function periodic or invariant with respect to the group generated by the two translations, so it has the same value at a point and at its image point in any other tile. From another point of view, emphasized by Klein, the elliptic function really lives on the torus you get by gluing up the basic tile.

Both Klein and Poincaré were interested in looking for analogous functions for a Kleinian group, so called **automorphic functions**. These would be functions which lived on tilings identical to those we have been studying. Their periodicity would be with respect to the tiling, in other words, just like elliptic functions, they would have the same value at a point and at its image point in any other tile. Klein already had examples in the shape of the **modular function** discovered by Gauss. The modular function comes up in number theory – it is a complex-valued function invariant under the modular group we met in Chapter 7.[1]

To make Figure 12.1, we took the kissing Schottky group illustrated in Figure 8.4 and used a recipe of Poincaré to calculate an automorphic function, that is, a complex-valued function invariant under this particular group. Then we coloured the plane according to the argument of its values – as the argument cycles from 0 to 360°, so the colours cycle according to a preset chart. The special points around which the colours spin through a complete cycle are the places where the function takes either the value zero or infinity. The function we used is the ratio of two infinite sums:

$$\left(\sum \frac{1}{(az+b)(cz+d)^3}\right) \Big/ \left(\sum \frac{1}{(cz+d)^4}\right)$$

[1] Klein and Fricke's first two volumes entitled *Vorlesungen ueber die Theorie der elliptischen Modulfunctionen* were published in 1890–2. The modular group is connected to elliptic functions by the fact that changes of generators for the doubly-periodic tiling groups are implemented by the Möbius transformations in the modular group.

Figure 12.1. A picture Klein and Poincaré would have envied. The group is the quasifuchsian group determined by the traces Tr a = Tr b = 2.2, whose limit set is shown in Figure 8.4. The plane has been coloured according to the argument of the value of an 'automorphic' or group-invariant function, which Poincaré would have called a 'fonction kleinéenne'.

where in each sum there is one term for each transformation $z \mapsto (az + b)/(cz + d)$ in the group. Just as in plotting a limit set, we did not in fact add up all the infinitely many terms, just a very large number. This accounts for inaccuracies in the picture near the limit set. Because the function is invariant under the group, the colouring reveals a natural tiling of the ordinary set; the thin black lines show the Schottky circles, superimposed from Figure 8.4.

As we described on p. 179, Poincaré's interest in this question arose out

Jules Henri Poincaré, 1854–1912

Henri Poincaré was a mathematical giant. Often described as the last universalist in mathematics, he can be said to have originated the theories of algebraic topology, functions of several complex variables, modern dynamics and chaos, not to mention three-dimensional hyperbolic geometry with its connection to our own story. He made major contributions to number theory and algebraic geometry, and left his mark on numerous branches of mathematical physics, including celestial mechanics, fluid mechanics, optics, electricity, telegraphy, capillarity, thermodynamics, potential theory and quantum theory. He is acknowledged as a co-discoverer, with Einstein and Lorentz, of the special theory of relativity. After achieving prominence as a mathematician, he turned his superb literary gifts to the challenge of discussing the meaning and importance of mathematics and science in such well known works as *Science and Hypothesis*.

Poincaré came from a distinguished family. One of his cousins Raymond Poincaré was prime minister of France several times and was president during the First World War. Born in Nancy where his father was Professor of Medicine, as a schoolboy Poincaré excelled in every topic he studied, having been partially taught as a young and rather sickly child by his gifted mother. His school in Nancy is now named the *Lycée Henri Poincaré* in his honour. Described by his teacher as a 'monster of mathematics', he won first prizes in the national *concours général*.

Going on to study at the prestigious École Polytechnique in Paris, Poincaré graduated in 1875 well ahead of his class. He continued his studies at the École des Mines and then worked briefly as a mining engineer, simultaneously finishing his doctorate under Hermite. He taught briefly at Caen, moving back to Paris in 1881. Appointed in 1886 to a chair at the Sorbonne and then also at the École Polytechnique, his work expanded taking ever new areas into its scope. From 1887 he was a member of the French Académie des Sciences, and in Klein's words: '...from then on grew ever greater in stature, so that soon he became the chief representative of French mathematics, ever more known and admired, the glory of his fatherland.' He died suddenly following an operation at the early age of 58.

of his study of certain differential equations following the work of Lazarus Fuchs. You can think of these equations as the analogue of the simple equation $d^2y/dx^2 = -y$ satisfied by the sine and cosine functions. Fuchs already realised that the 'periodicity' of solutions to his equations had a certain invariance under linear fractional transformations. Poincaré initially restricted his study to what he called 'groupes fuchsiens', that is, groups whose tilings filled up the inside of the unit disk. When Klein wrote to him in May 1880, pointing out the existence of other types of groups of linear fractional transformations (among them the Schottky groups we met in Chapter 4), there ensued an intense and, at least on Klein's side, highly competitive correspondence. With astonishing rapidity[1], Poincaré expanded his ideas and, in a masterful combination of geometry, algebra and analysis, laid down the foundations of the modern theory of what he now called 'groupes kleinéens'. Klein was not at all pleased with Poincaré's designation. Here are his words on the subject:

> Poincaré, who when he began was very scantily acquainted with the situation in Germany, had called the groups with a limit circle 'groupes fuchsiens', despite the fact that Fuchs' contribution did not merit this. When I made him aware of the general functions, he called them 'fonctions kleinéennes'. Thus prevailed great historical confusion.

A blow by blow account of all these developments is to be found in *Linear Differential Equations and Group Theory from Ricmann to Poincaré* by Jeremy Gray.[2]

To understand more about Poincaré's great insights, let us step back for a brief look at what the subject of non-Euclidean geometry is all about.

[1] In 1881 Poincaré published no less than 13 papers on this subject in what was described by Klein as *'eine stürmische Publikationsserie'*.

[2] Birkhauser, 1986.

Non-Euclidean geometry

From antiquity, the geometry laid down by Euclid had been regarded as self-evident. Starting from some simple and indisputable assumptions or **axioms**, all true facts about geometry can be deduced. The Euclidean view of the universe so dominated thinking that to doubt it would have been tantamount to doubting the solidity of the world itself.

In the early years of the nineteenth century, a shocking discovery, made more or less simultaneously by the great Gauss, the Russian mathematician Nikolai Lobachevsky and the Hungarian János Bolyai, revolutionized our mathematical (and later our physical and philosophical) world view. They discovered that it was possible to construct a new system of geometry in which one of Euclid's fundamental assumptions was just not true. The revolution this implied for mathematics was extraordinary. A whole new version of geometry existed, completely internally consistent and plausible, which simply did not conform to Euclid's rules.

On the scale of your own home neighbourhood, this brave new world looks and feels exactly like the Euclidean one. It has points and lines, circles and triangles, all measured in the usual way using distances and angles. The main difference is that if you move any substantial distance away from 'home', you find this new world is much *bigger* than the one you are accustomed to. Fortunately, we can aid our imagination by finding curved surfaces in three-dimensional Euclidean space which behave exactly as if they were chunks of a non-Euclidean plane. Such a surface is shown in Figure 12.2. If you were living on this surface, it would seem flat. The shortest path across the surface between two points would be a non-Euclidean line, just as arcs of great circles are the shortest distance paths or 'lines' on the surface of a sphere. The dark curve in the first picture is a circle. In the second picture, this non-Euclidean circle has been cut out and folded flat. See how much longer the rim of the circle is than a normal Euclidean one. The circumference of a circle of radius r turns out to be $2\pi \sinh r$ which, unless r is quite small, is roughly πe^r, considerably more than $2\pi r$![1] In fact a small strip around the rim contains most of the area of the circle. As Thurston puts it, if you build a camp fire in non-Euclidean space, you had better get very close to it, because the warmth dissipates very fast if you move just a little bit away.

A more traditional way to model non-Euclidean geometry is a method analogous to using stereographic projection to represent points on the

[1] Notice however that if r is very small then $\sinh r$ roughly equals r, so for very small circles the Euclidean and non-Euclidean circumferences are almost the same.

Note 12.1: **The parallel postulate**

To understand what Gauss, Lobachevsky and Bolyai had discovered, suppose you have an infinitely long straight line L – one side of an infinitely extended railroad track if you will. Pick a point P at distance, say, 2 metres from L. Common sense says you should be able to lay down another straight rail through P parallel to L, in such a way that, however long you continue the second rail, it never meets L. Furthermore, if you alter the direction of the second rail even slightly, then the rails won't be exactly parallel and eventually the second rail will bump into the first. Euclid's famous **parallel postulate** is the assertion that this is indeed the way things are: his geometry assumes that for any line L and any point P not on L, there is *just one* line through P which never meets L. The trouble with the postulate is that

you can't really check it because it involves idealised *infinite* straight lines. For over two millennia, mathematicians tried, unsuccessfully, to deduce this proposition from other, more self-evident, assumptions. These three men discovered that the reason for this lack of success was that the parallel postulate *did not necessarily have to be true*. They showed that there was a world in which you could do perfectly good and consistent geometry, in which all the rest of Euclid's assumptions were valid, but in which the parallel postulate was simply not true. The world in which this geometry existed, came to be known variously as non-Euclidean or hyperbolic. In the non-Euclidean plane, there can be not one but *many* lines through our point P which, no matter how far they are extended, miss the given line L.

Figure 12.2. Measurements on this surface in three-dimensional space are identical with those on the non-Euclidean or hyperbolic plane. The second picture is a non-Euclidean circle. These pictures were made by Jeff Weeks, and originally appeared in *The mathematics of three-dimensional manifolds*, Scientific American, July 1984. We thank Jeff Weeks and Scientific American for permission to reproduce them.

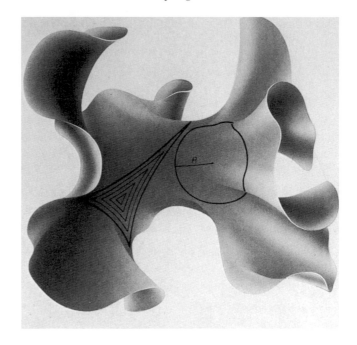

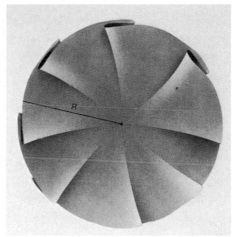

[1] Two of Beltrami's papers on this subject have been translated, with commentary, in the excellent little book *Sources of hyperbolic geometry* by John Stillwell, published in the American and London Mathematical Societies joint History of Mathematics Series, 1996. The same volume contains translations of Klein's 1871 paper on non-Euclidean geometry and the most important of Poincaré's memoirs of 1880–1.

sphere by points on the plane. Since hyperbolic space expands so rapidly, we have to squeeze its infinite outer reaches very radically to achieve this. A way to do this was first discovered by Eugenio Beltrami in 1868.[1] In Beltrami's model, the whole hyperbolic plane gets compressed into the (Euclidean) unit disk. Distances are rescaled so that when we draw something small near the boundary of the disk, its true non-Euclidean size is really pretty large. Actually we have already seen pictures of Beltrami's model, for example Dr. Stickler walking in a Fuchsian tiling on p. 163. The image Dr. Sticklers look as if they shrink radically near the limit set (that is, the boundary circle), but measured with a non-Euclidean ruler *they are all the same size.*

The non-Euclidean 'straight lines' in Beltrami's model appear as arcs of circles which meet the circumference of the disk at right angles. It is quite easy to see that with this interpretation the parallel postulate is just not true. However, angles are correct, so for example perpendicular non-Euclidean lines are represented by arcs of circles which meet perpendicularly. By one of the usual unfair twists of history, Beltrami's disk model is nowadays almost always called the **Poincaré disk**.

A good way to understand this new geometry is to look at some Fuchsian tilings – for example, the beautiful pictures displayed in Figures 12.3 and 12.4. There are lots of non-Euclidean 'lines' in our picture – all the boundaries of all the tiles! Measured with a non-Euclidean ruler, *all the tiles are exactly the same size.* In this new strange world, various facts which we always took for granted are just not true. For example, everyone knows that the angles in a triangle add up to 180°. In the non-Euclidean world, not so! There, the angles in a triangle can *never* add up to 180°. The angle sum can be anything, as long as it is *less* than 180°.[1] You can check this in Figure 12.3 – the angle sum in radians of each of its triangles is

$$\frac{\pi}{2} + \frac{\pi}{8} + \frac{\pi}{8} = \frac{3}{4}\pi.$$

On various levels, controversy about the true status of non-Euclidean geometry raged through the nineteenth century. Klein's first important work played a definitive role. Inspired by work of Cayley, and having learnt about the new geometry for the first time in 1869 from Stolz in Berlin, Klein became convinced that the distance and angle measurements necessary for non-Euclidean geometry could be extracted as special cases of general procedures in projective geometry, the geometry of perspective. His 1871 paper[2] showed that if Euclidean geometry was consistent, then so was non-Euclidean. Klein found himself caught up in disputes with philosophers as well as mathematicians and it took some time for the question as to whether or not non-Euclidean geometry 'really existed' to finally be laid to rest.

Looking back, it is hard to understand what all these controversies were about. To the modern mind, Beltrami's work already demonstrates the consistency. What Klein did was catalyse the 'paradigm shift' required to put thinking about all geometries on a equal footing. To Klein's sorrow, Cayley always maintained that his arguments were circular and never accepted his ideas.

[1] On a sphere, the angle sum of a triangle is greater than 180° and the parallel postulate also fails – there are *no* lines through a point *P* which never meet a line *L* not containing *P*. 'Non-Euclidean geometry' always refers to the case in which the angle sum is less than 180° and there are many parallels to *L* through *P*.

[2] *Ueber die sogenannte Nicht-Euklidische Geometrie*, Math. Annalen 4, 1871.

Figure 12.3. A tiling of the hyperbolic plane modelled by the circular disk. From the hyperbolic point of view, all the 'triangles' are exactly the same shape. The (orientation-preserving) symmetries are Möbius maps which preserve this tiling and map any coloured triangle onto any other. This and the following figure were drawn by our DFS tiling program with the aid of an automaton provided by the program KBMAG by Derek Holt. The pattern of the colouring relates to the automaton.

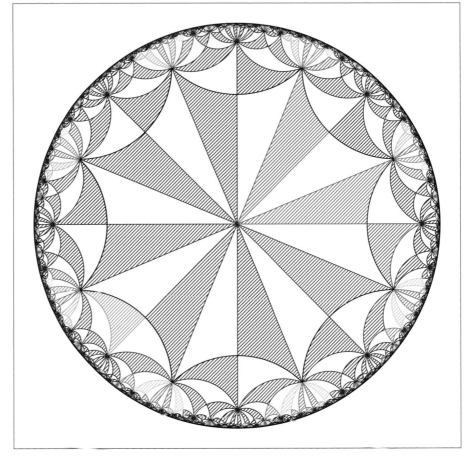

In 1881, Poincaré, taking part in a geology course organised by the École des Mines, was mounting a bus to go on a field trip. A thought flashed into his mind – retold in the famous quote at the beginning of this chapter. The Möbius transformations he had been working with were exactly the symmetries of non-Euclidean geometry![1] Looking at Figure 12.4, and knowing about Beltrami's interpretation of non-Euclidean geometry, it is not hard to see what this means. The Möbius symmetries of the picture are just the manifestation in the model of symmetries of the non-Euclidean plane.

The rivalry between Klein and Poincaré centred on the proof of the so-called **uniformization theorem**, which in one of its many guises says roughly speaking that you can find a way to measure distance on any two-dimensional surface so that the surface 'rolls out' to match exactly, distances and all, either a piece of ordinary Euclidean plane, a piece of sphere, or a piece of the hyperbolic plane. 'Rolling out' is the opposite procedure to cutting out and gluing up; for example, a torus rolls out

[1] Poincaré's account, originally delivered as a lecture on *L'Invention Mathématique* to the Société de Psychologie in Paris, of the circumstances leading to this insight is probably the most famous account ever written of the process of mathematical discovery by a great creative genius. Extracts appear in many sources and the 1913 translation by G. Halstead is reprinted in full in Volume 4 of *The World of Mathematics* by James Newman.

onto a tiling like the Iowa fields. They proved that any surface more complicated than a torus rolls out onto the hyperbolic plane. We can understand this quite easily in terms of our pictures. To get two-dimensional surfaces you only have to look at Fuchsian groups; for example take the Dr. Stickler picture on p. 163 in Chapter 6. The four-sided inner tile is a non-Euclidean square (with 0° angles at the corners!) which glues up to make a surface like a torus with a puncture.

Actually, there are surfaces hidden among the symmetries of the two

Figure 12.4. The 2–3–7 tiling. Our version of a classic illustration from the works of Klein and Fricke. The name comes from the angles at the vertices of each of the triangles, which, in radians, are $\frac{\pi}{2}$, $\frac{\pi}{3}$, and $\frac{\pi}{7}$. Thus the angle sum of each triangle is $\frac{41}{42}\pi$.

tilings we have displayed. In Figure 12.3, we have highlighted a curvy octagon in thick black lines. A subgroup of the symmetries of the tiling glues the curved edges of the octagon together to form a two-holed doughnut, or surface of genus two. The 2–3–7 tiling in Figure 12.4 hides a very symmetric surface of genus three, or three-holed doughnut. It is a challenge to find a patch of triangles that is sewn up into this surface; we give one hint in that such a patch must contain $2 \cdot 168 = 2^4 \cdot 3 \cdot 7$ triangles, half coloured and half white.

Riemann and Teichmüller: spaces of surfaces

The second strand in our story goes back to foundations laid by the great mathematician Bernhard Riemann (1826–1866). The problem is to describe in some sense just 'how many' different surfaces there are. We are interested in the glued-up surfaces, the ones with no edges, which means surfaces like spheres, tori and pretzels perhaps with several holes or 'handles'. The number of handles is called the **genus** of the surface, usually denoted by the letter g. A sphere has no handles, so its genus g is 0, a torus has one handle so $g = 1$, while a two-holed doughnut or pretzel has $g = 2$. Riemann's idea was to look for maps between two surfaces of the same genus which, like Möbius maps, preserved angles. These are called **conformal maps**. When such a map existed, Riemann thought of the two surfaces as close cousins, something like two triangles in Euclid's world which are similar. (This idea has many applications, for example in studying the distribution of electric charge on a surface.) People nowadays say that two such surfaces, though different in size and shape, are examples of the same **Riemann surface**.

Riemann's discovery was that counted in this way, there weren't that many Riemann surfaces: a surface of genus g can be described using exactly $3g - 3$ complex numbers or parameters he called its **moduli**. Klein and Fricke found another way to count moduli using the theory of groups of Möbius maps. The simplest example is this. Consider Schottky groups with g generators, associated to $2g$ disjoint disks which are paired by the maps. Just as in Chapter 4, the points outside all the disks form a swiss cheese-like tile and, glued up, this makes a surface of genus g. Actually, we can get all possible surfaces – up to conformal maps – in this way. Note that each Möbius map depends on exactly four complex numbers. Making the determinant 1 reduces the parameter count to 3 for each map, or $3g$ in all. And a conjugation of the whole set produces a conformal map between the glued up surfaces, so this reduces the parameter count by 3!

(For $g = 1$, not all conjugations change the one Möbius map which makes the torus (see p. 85), and one parameter remains.)

In the years immediately before the Second World War, a young German mathematician Oswald Teichmüller introduced an entirely new idea into this subject by looking, for any two surfaces of genus g, for maps between them which were as close to conformal as possible. These he called **extremal quasiconformal** maps, see Note 12.2. They proved a powerful tool in developing the theory, giving a completely new way of parameterizing Riemann surfaces. His results give the same dimension count of $3g - 3$ but in addition show, remarkably, that the space of Riemann surfaces has many properties in common with hyperbolic space itself.

Teichmüller's work brings us to two great mathematicians whose work has dominated many developments in the modern era, Lars Ahlfors and Lipman Bers. The trail that led Bers to his study of quasiconformal maps and Teichmüller theory began during two years spent at the Institute for Advanced Study in Princeton: in trying to prove a result about fluid flow he came across work of Ahlfors which showed him the power of quasiconformal maps. Through the 1950's Bers embarked on the study of Teichmüller's papers and then from the early 1960's, Bers and Ahlfors rapidly began establishing new links between this theory and Kleinian groups.

One famous result, called the **Bers' simultaneous uniformization theorem**, states that given any two Riemann surfaces of the same genus, there exists exactly one quasifuchsian group such that the surfaces associated to the two halves of the ordinary set (the part inside and the

---

Note 12.2: **Quasiconformal maps**

One of the basic properties of Möbius maps is that they preserve angles. In other words, if two curves meet at a certain angle, then the images of the curves under the map meet at the same angle. We have used this property many times in our work. A map of a part of the plane can have this property without being Möbius, in which case it is called **conformal**. (Readers who have studied complex analysis will recall that conformal maps are exactly those which are holomorphic or complex differentiable.) A

quasiconformal map by definition has to map the whole Riemann sphere to itself in a one-to-one and continuous way, but it is allowed to distort angles by a bounded amount. If the distortion factor were 2 for example, then two curves which met at an angle of θ in the original picture could meet at any angle between $\theta/2$ and 2θ in the new picture, but no more, and no less. A quasiconformal map is called **extremal** if it distorts as little as possible given the pair of surfaces it is trying to match.

Oswald Teichmüller, 1913–1943

Oswald Teichmüller arrived in Göttingen as a seventeen-year old student in 1931. He displayed unusual mathematical talent, but at the same time, the lonely young man found comradeship by becoming a zealous and fervent Nazi. Disliked by almost all his contemporaries, he exhibited shocking rudeness to his professors, and by 1932 had joined the infamous *Sturmabteilung* or 'brownshirts' and become a ringleader of Nazi students in Göttingen. By the time he was finishing his studies in 1935, most of the senior professors had been dismissed or had fled. After his graduation, Teichmüller moved to Berlin to work with Bieberbach (also a Nazi) where he gained his 'Habilitation' in 1937. In 1939 he served in the invasion of Norway and then began to work for the OKW or army high command. Through all this he worked intensely hard on mathematics, producing no less than 34 papers before returning to active service 'for the liberation of his Fatherland' in 1943. He was reported missing presumed dead following heavy fighting along the Dneiper in autumn of that year. Many details of Teichmüller's life and work can be found in an article by William Abikoff, Mathematical Intelligencer, Volume 8, No. 3, 1986.

part outside the limit curve) are the surfaces you chose. Subsequently, together with his former student Maskit, Bers began exploring groups on the boundary of quasifuchsian space as described in Chapters 9 and 10.

The three-dimensional approach

In 1881, Poincaré originated a completely different and very important geometrical approach to Kleinian groups. This approach, which we have completely ignored in this book, is based on looking at the Riemann sphere as the boundary of a solid ball inside which we model three-dimensional hyperbolic geometry, in just the same way as Beltrami modelled two-dimensional hyperbolic geometry inside the unit disk. In this model, the non-Euclidean planes look like hemispheres or bubbles sitting over circles in the Riemann sphere. The beautiful fact which Poincaré realised is that using inversions you can extend a Möbius map from being a map of the sphere to being a map of the whole ball, and if you do, the map you get in the ball preserves non-Euclidean distance! In other words, the messy distortions we get when Möbius maps act in the plane disappear

Lars Ahlfors, 1907–1996

Lars Ahlfors was born in Finland and died in Massachusetts. He received his doctorate in 1930 in Helsingfors (now Helsinki) University and taught at Harvard in 1935–8. In 1936 he became one of the two first winners of the Fields medal. In 1938, he was offered a chair at Helsinki. After difficult and sometimes dangerous times during the war, he managed to get to Zürich in 1944 (pawning his Fields Medal for the trip – times were hard) and then to Harvard in 1946 where he remained until his retirement in 1977. His textbook *Complex Analysis* has remained the student's bible on the subject for fifty years. Ahlfors made major contributions to every area of complex analysis, his early work being described by Hermann Weyl as 'a vineyard planted by that grand gardener from the North'.

He enters our story with his work on quasiconformal maps, starting with a geometrical definition which allowed all of their basic properties to be established with much more ease than with the former cumbersome analytical definitions. He was able to give the first complete proof of Teichmüller's important theorems, making Teichmüller's work accessible to the mathematical public. The famous *Ahlfors conjecture*, made in 1965, is that the limit set of any finitely generated Kleinian group either has area zero or is the whole sphere. Contemplating some our pictures in Chapter 10 you will see that this statement is not at all obvious, although the last twenty years of work have confirmed it in very many cases.

Bers and Ahlfors continued their work long into retirement, integrating their approach with Thurston's new ideas. An NSF grant application by Ahlfors famously contained only one sentence: 'I will continue to study the work of Thurston'. Those were days when bureaucracy was not so pernicious.

and they metamorphose into maps like rotations and translations of non-Euclidean three-dimensional space, preserving size and shape.

This is a very powerful simplification. From many viewpoints, it makes Kleinian groups much easier to analyze. Just as in two dimensions a discrete group of symmetries has a two-dimensional tile, so a discrete group of symmetries of hyperbolic 3-space has a basic three-dimensional tile – perhaps better thought of as a solid crystal. Turned around, we can say that the group of symmetries of a three-dimensional tessellation of hyperbolic 3-space is a discrete group of Möbius maps, exactly what we have called a Kleinian group.

To see the connection with our two-dimensional tiles on the Riemann sphere, think of the Fuchsian Schottky tiling on p. 180. The region *inside* the disk, the universe of two-dimensional hyperbolic geometry, is tiled by an open ended 'half' of our swiss cheese tile, cut off from its outer half by the unit disk which now represents points at infinite hyperbolic distance from the origin. The two-dimensional tile has four open ends where it meets this 'circle at infinity'. In just the same way, the groups in this book, seen three-dimensionally, have three-dimensional tiles or crystals which stretch out to the 'sphere at infinity'. The two-dimensional tiles we have been looking at are just the places where these open-ended three-dimensional crystals meet the Riemann sphere. We saw an example in the very first picture of this book! The three-dimensional effect in the frontispiece is actually the manifestation of the three-dimensional hyperbolic geometry which goes with the limit set we drew in Figure 11.5. To understand this, just as we used the Cayley transform to map between the plane and the disk, think of the sphere at infinity as the complex plane and the hyperbolic three-ball as that half of three-space which sits above it. The basic three-dimensional Schottky crystal is the region above the plane and outside the red hemispherical bubbles you see in the picture. With a little imagination, you can see that the crystal is open-ended and meets the plane in the black regions which are just the tiles we have been studying in the two-dimensional plane.

Many Kleinian groups are not quite like the ones we have focussed on in this book, because their crystals are entirely contained inside the ball (possibly with some isolated 'ideal vertices' meeting the sphere at infinity). The limit set of a group like this will be the whole Riemann sphere. Figure 12.5, a still from the prizewinning video *Not Knot!*[1] shows part of a tessellation of hyperbolic 3-space by dodecahedra (regular solids with 12 pentagonal faces). The dodecahedra have been so designed that the angle between the two faces which meet along each edge is 90°, a feat manifestly impossible in Euclidean space. The picture has been carefully calculated so you are viewing it from the proper perspective, as if you were an inhabitant of this hyperbolic three-dimensional universe. See how exactly 8 copies of the dodecahedron fit together in a 'cubical' fashion at each corner. The symmetry group of this pattern of crystals is generated by the 180° rotations about each of the red, blue and green lines. When you use the symmetries to glue up points on the various faces of one dodecahedral crystal you get the three-dimensional analogue of a surface, called a **3-manifold**.

Now anyone can imagine a surface, but it is not so easy to imagine a three-manifold. One of the simplest examples is ordinary three-dimensional space to which one extra point 'at infinity' has been added. This wraps up three-dimensional space into a manifold called the **3-sphere**, in much the same way as wrapping up the plane and adding the point ∞ converts the complex plane into the Riemann 2-sphere.

[1] Stills from www.geom.uiuc.edu/video/NotKnot/; DVD available from A K Peters/CRC Press.

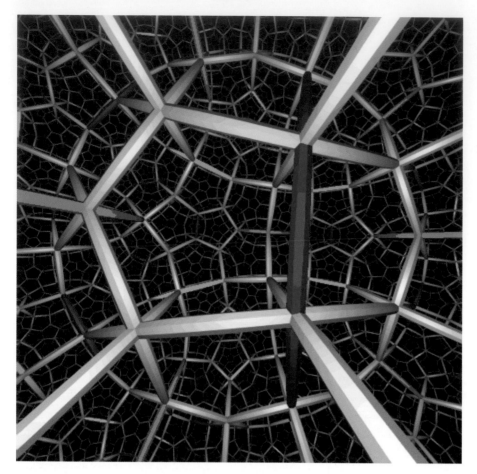

Figure 12.5. A tiling of hyperbolic 3-space. This is a view of a tiling of three-dimensional hyperbolic space by dodecahedra, *as seen from inside the hyperbolic world.* We are grateful to Charlie Gunn and the creators of the film *Not Knot!* for permission to include this still.

You can make lots of other examples of 3-manifolds by twisting up a piece of string into a complicated knot, joining the ends smoothly to make a continuous loop, and then declaring the manifold to be all of the 3-sphere except for the knot you just made. Manifolds like this are called **knot complements** and can be played with by gluing and more complex operations known as surgery.

About 1970, Jørgensen and Riley began independently, and from different viewpoints, exploring some very interesting examples of 3-manifolds whose symmetries were Kleinian groups. We have met their main example already – the doubly-degenerate group in Chapter 10. Riley showed that this group (or rather, the group generated by *a*, *b* and the extra symmetry we called *c*) has a basic crystal which in hyperbolic 3-space glues up to be topologically the 3-sphere minus the simple hitch knot illustrated in the margin, known to mathematicians as a **figure of eight**. Just as on a two-dimensional hyperbolic surface you have to travel an infinite non-Euclidean distance to get out to the puncture, so in three-dimensions you

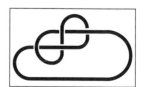

have to travel an infinite distance to get to the hole left where you took out the knot. Thus the name by which this group usually goes is the **figure of eight knot complement**, whose space filling limit set we drew on p. 335. Part of the reason this group is so complicated, is that its basic three-dimensional crystal *has infinitely many sides*. At about the same time, Albert Marden, a former student of Ahlfors, was laying the foundations for the study of Kleinian groups from the viewpoint of three-dimensional topology.[1]

[1] *The geometry of finitely generated kleinian groups*, Ann. of Math., 1974.

The young American mathematician William Thurston stunned the mathematical world in 1977 with the announcement of his **hyperbolization theorem**. This theorem is really the three-dimensional analogue of the uniformization theorem, which as we have seen says that all two-dimensional surfaces can be endowed with a concept of distance which allows one to 'unroll' them onto either the sphere, the Euclidean plane or the hyperbolic plane. Up until that time, no one had dreamed that a similar result might be true for 3-manifolds, but this is exactly what Thurston showed. His work demonstrated that the groups being studied by Jørgensen and Riley were not just isolated examples; on the contrary, a very large class of 3-manifolds are associated to discrete groups of Möbius transformations in a similar way. Thus not only surfaces, but also 3-manifolds, can be understood in the beautiful setting of the symmetry groups introduced by Klein and Poincaré. An inspiring exposition is to be found in Thurston's book *Three-dimensional Geometry and Topology*, Princeton 1997.

Mathematics has barely yet digested the influx of Thurston's revolutionary new ideas, for which he won the Fields medal in 1982. Indeed many of his results remain only partially published, having been disseminated by his students and by word of mouth. Nevertheless, his ideas have engendered a huge output of work over the last 25 years whose conclusion is yet far from reached. Based on his hyperbolization theorem, Thurston went even further proposing his **geometrization conjecture** that *all* 3-manifolds can be cut up into natural pieces each of which has one of eight basic geometrical types, of which by far the most common is hyperbolic. Much progress has been made towards proving this, however its difficulty is such that it includes the three-dimensional Poincaré Conjecture[2] as a special case.

[2] That any three-dimensional manifold on which all paths can be shrunk to a point must be a sphere. A proof of the Poincaré conjecture will win a recently announced million dollar prize, see www.claymath.org. Thurston's approach to this has been described by a fellow mathematician as God's method: make a list of *all* 3-manifolds and go through them one by one.

Let us end by showing how some of Thurston's ideas have influenced our own study. Thurston introduced a key three-dimensional object closely related to the circular Fuchsian structures we have been noticing in our limit sets. Thinking of the sphere at infinity as the complex plane, take a quasifuchsian group and find all the disks contained in one side of the ordinary set which meet the limit set in at least three points. Over each of these, erect a hemispherical 'soap bubble' in hyperbolic 3-space (the Euclidean half space sitting above the plane). As you see in Figure 12.6, the meeting of all these soap bubbles creates a dome which roofs over the inside of the ordinary set. The edge of the dome reaches down to meet the horizontal plane in the limit set of the group. There is another dome which covers the outside region of the ordinary set. The two domes together form what is called the **convex hull boundary** of the group, because the region between them is the smallest convex[1] set in hyperbolic 3-space which meets infinity in the limit set.

[1] A set is convex if it contains all the straight line segments between any two of its points.

Recently, the convex hull boundary has become an important tool. It closely reflects the shape of the ordinary set, but it carries the extra feature of easily visible hyperbolic geometry. It is also symmetrical under the group, and it too carries a symmetrical tile. When you glue up this tile, you create a 'pleated' surface made up of pieces of hyperbolic planes which meet at an angle along the 'bending lines' you can see where the domed spheres meet and which on the glued-up surface become **geodesics** or shortest paths on the surface.

It turns out that the bending geodesic has to be either a loop corresponding to one of our special p/q words, or, in in-between cases, a 'geodesic lamination' represented by an irrational 'infinite word' of the kind we met in Chapter 10. The bending lines belong to the p/q word exactly when the limit set is made up of overlapping circles in the p/q circle chain pattern. This theory has been developed by the middle author of this book and her collaborator Linda Keen[2] and is the key to our unexplained colouring of the Maskit slice in Chapter 10.

[2] *Pleating Coordinates for the Maskit embedding of the Teichmüller space of punctured tori*, Topology, 32, 1993.

Each of the roughly vertical 'rays' in Figure 12.7 traces through all the groups in the Maskit slice whose limit sets contain a chain of circles arranged in the p/q pattern – in other words, whose convex hull boundary is 'pleated' along the loop which belongs to the p/q word. The ray for the p/q word ends in the p/q cusp on the boundary where the curve has shrunk to zero length. As we move upwards along a ray, the p/q curve gets longer and longer, tending to ∞ as we reach the 'top' of the picture which is, of course, way off our page.

Now we can explain the approximately horizontal coloured 'latitudes' which we used to colour the Maskit slice in Chapter 10. They are just the contours of all points for which the pleating curve – that is, the p/q curve belonging to ray we are on at the time – have equal length. Well, almost. The problem is, if you try to move sideways across the picture, then of course the traces (which are closely related to the lengths) get enormous as you approach any 'irrational' ray through fractions p/q whose denominator q gets equally enormous. The way to resolve this problem is to 'normalize' by dividing the p/q 'length' by q. The horizontal colouring

**Figure 12.6. The convex
hull boundary.** You can
analyze the limit set from a
three-dimensional view by
finding all circles whose
interior is in the ordinary set
and which go through at
least three points on the
limit set and then drawing
the hemispherical bubbles
which sit above them. This
creates a 'roof' over the
region inside the limit set,
which in recent years has
become an important tool.
These pictures contributed
by Yair Minsky show that
this complicated roof is in
fact a wonderful dome.

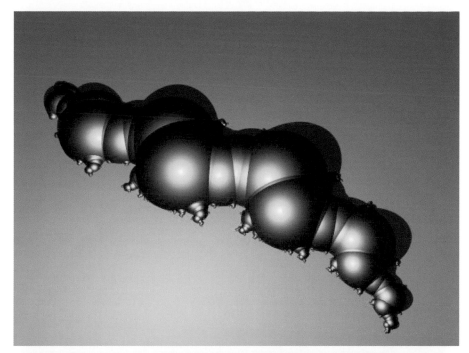

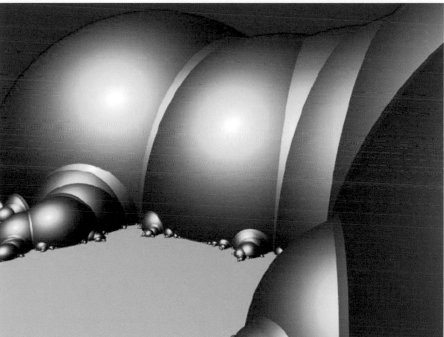

traces the contours along which the pleating curve has constant normalized
length.

The rays and the contours give a set of 'coordinates' which are sufficient to locate the precise position of any group in the Maskit slice. Using

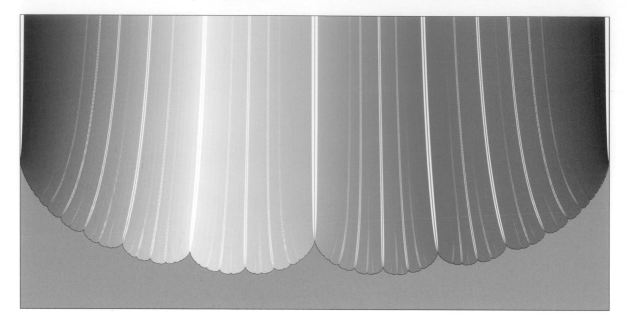

them, Keen and Series were able to demonstrate that the points calculated by our Newton algorithm which we believed to be on the boundary of the Maskit's slice really are the cusp groups we think. There are more details of the formulas for drawing the rays and contours in Note 12.3.

Figure 12.7. End of the rainbow. This is the Maskit slice with the pleating rays thickly drawn and coloured by their corresponding fraction p/q.

Note 12.3: **Colouring the Maskit slice**

The principle behind finding the vertical 'rays' in Figure 12.7 is that if the p/q word is the bending line, then its trace turns out to be a real number which gives a good measure of its length. (Actually the length is the logarithm of the multiplier we met in Chapter 3.) The solutions to the equations $T_{p/q}(\mu) = t$ for real numbers $t \geq 2$ trace out the vertical 'ray' that begins at the p/q cusp (where $t=2$) and ascends in the vertical direction, eventually coming closer and closer to the vertical line $\mathrm{Re}(\mu) = 2p/q$. To draw these rays, use the newton solver to find the p/q cusp on the boundary, and then in small steps in t starting at 2, use Newton's method to solve $T_{p/q}(\mu) = t$, with the previous point plotted as the initial guess. (Note: if you just try solving

this equation without starting at the cusp you may well find yourself on the wrong 'branch' of the curve $T_{p/q}(\mu) = t$ and the curves you trace will just be a mess.)

To draw the 'lines of latitude' represented by our colouring, we have to normalize the length by dividing by q, which allows for the fact that if q is large, the curve wraps many times around the surface. To draw them, first pick a number $s > 1$, then carry out the boundary tracing algorithm just as described in Chapter 9, except that the trace equation we solve is

$$T_{p/q}(\mu) = s^q + s^{-q}$$

for the next result $\mu(p/q)$.

INDEX

Road Map of Two–Generator Groups

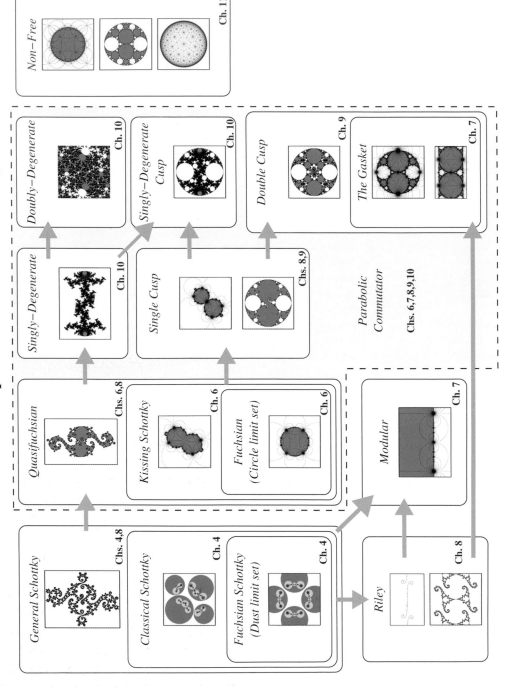

Non–Free — Ch. 11

Doubly–Degenerate — Ch. 10

Singly–Degenerate Cusp — Ch. 10

Double Cusp — Ch. 9

The Gasket — Ch. 7

Singly–Degenerate — Ch. 10

Single Cusp — Chs. 8,9

Parabolic Commutator — Chs. 6,7,8,9,10

Quasifuchsian — Chs. 6,8

Kissing Schottky — Ch. 6

Fuchsian (Circle limit set) — Ch. 6

Modular — Ch. 7

General Schottky — Chs. 4,8

Classical Schottky — Ch. 4

Fuchsian Schottky (Dust limit set) — Ch. 4

Riley — Ch. 8